ABRAMS

A History of
the American Main Battle Tank

Volume 2

by
R.P. Hunnicutt

Line Drawings
by
D.P. Dyer and Dan Graves

Color Drawing
by
Uwe Feist

Foreword
by
Brigadier General Philip L. Bolté USA-Retired

E P B M
ECHO POINT BOOKS & MEDIA, LLC

Published by Echo Point Books & Media
Brattleboro, Vermont
www.EchoPointBooks.com

Copyright © 1990, 2015 R. P. Hunnicutt
ISBN: 978-1-62654-166-5

Cover image by Uwe Feist

Cover design by Adrienne Núñez,
Echo Point Books & Media

Editorial and proofreading assistance by Christine Schultz,
Echo Point Books & Media

Printed and bound in the United States of America

CONTENTS

ACKNOWLEDGEMENTS

Contributions to the research for this armored vehicle history came from a wide variety of sources. Among these was Brigadier General Philip L. Bolté, U.S. Army, Retired, who also kindly agreed to write the foreword for the book. During his military service, he was familiar with many of the projects described as both a user and a developer of armored vehicles.

Leon Burg of the Technical Library at the Tank Automotive Command (TACOM) was instrumental in tracking down key sources of information on the early experimental programs. Also at TACOM, Clifford Bradley, Oscar Danielian, Major General Oscar Decker, Dr. Herbert Dobbs, Colonel Thomas Huber, Dan Smith, and Joe Williams, all now retired, were extremely helpful. Roland Asoklis and J. B. Gilvydis provided much of the background leading up to the M1 development program. Eugene Trapp was a prime source of information on the MBT70/XM803 project. Arthur Volpe of the TACOM Public Affairs Office located photographs of the ROBAT and various M1 modifications.

At General Dynamics Land Systems Divsion, formerly Chrysler, Dr. Philip W. Lett, Joseph Yeats, and Briggs Jones were extremely helpful in finding both data and photographs of the various M1 modifications. Also at GDLSD, David H. Bartle located information on the armament of several vehicles and Steve Payok tracked down the dates of various design changes.

John Purdy, Steven Maxham, and David Holt made the resources of the Patton Museum available to the author during several trips to Fort Knox. Larry Bolls, Mark Falkovitch, David Holliday, and James Montgomery of the TEXCOM Armor Engineer Board were a great help in finding rare photographs. Mr. James Schroeder of BMY, Division of Harsco Corporation, provided photographs and data on BMY's counter obstacle vehicle and the program to develop the M88A1E1 recovery vehicle. As usual, Phil Dyer and Uwe Feist did the line drawings and the dust jacket painting respectively and Dan Graves contributed the four view drawing of the combat engineer vehicle XM745. Michael Green, Russell P. Vaughn, and Greg Stewart provided numerous photographs for the research program.

At Aberdeen Proving Ground, Major General Andrew H. Anderson made available the resources of the Test and Evaluation Command (TECOM) and Robert Lessels was extremely helpful in locating data and photographs.

Other contributors included Joseph Avesian, Carl Bachle, Lieutenant General Robert Baer, Timothy Balliett, George Bradford, William D. Bunnell, Major Fred Crismon, Paul Denn, Gregory Flynn, Jr., Robert W. Forsyth, Donald F. Hays, Joseph Hayes, Robert Hesko, William Highlander, Edward R. Jackovich, Marvin Kabakoff, Glenn W. Kroge, Charles R. Lemons, Michael A. Leu, Jacques Littlefield, Lieutenant Colonel James Loop, Donald J. Loughlin, Dr. Richard E. McClelland, Jim Mesko, Colonel Samual L. Myers, Jr., Ernst Niederbuehl, Richard Ogorkiewicz, Colonel Jimmy Pigg, Larry Roederer, Michael E. Rogers, R. Paul Ryan, Richard D. Scibetta, Don Selby, Roger Smith, Walter J. Spielberger, Bruce Stainitis, Lyle Walcott, and Steven Zaloga. Also, special thanks go to Alan Millar and HMS Typography, Inc. for help with the final preparation of the book.

FOREWORD

by

Brigadier General Philip L. Bolté, U. S. Army (Retired)

It began in World War II: the feeling that perhaps the shaped charge fired in Panzerfaust and Bazooka would spell the end of the tank as a survivable weapon system. Guided missiles added to the belief that the days of the tank were numbered. Yet a series of wars in the Middle East reminded the world that there is no ultimate weapon. Tanks could survive and, in consort with other weapon systems, could carry the day in spite of concentrated efforts to drive them from the battlefield.

Meanwhile, disregarding popular opinion, a group of professionals—uniformed, civil service, and from industry—had set about after World War II to meet the tank requirements of the coming years as they perceived them. In his earlier books, Dick Hunnicutt traced the details of their efforts in meticulous detail: "PATTON" recounted the development of the entire Patton series of tanks, while "FIREPOWER" told the story of the Army's post-war venture into the realm of heavy tanks. Now he has reviewed the story of tank development in the fifties and sixties, and carried it into the seventies and eighties as the Army and its contractors sought the proper replacement for the Pattons.

There was no lack of imagination on the part of those who, in the 1950s, set about to capitalize on post-war technological advances in designing a replacement for the M48 series of tanks. The public, and even most of the Army, had no understanding of the starts and stops, and of the excursions experienced by the designers and developers as they proceeded with the work that finally led to fielding the M60 series of tanks. In the end, it was no great

leap forward, but, in fact, a relatively modest modernization step.

Under the close observation, and even guidance, of the Office of the Secretary of Defense, the Army then spent most of the sixties pursuing a joint effort with the Federal Republic of Germany as the two countries tried to develop a common tank that would satisfy their somewhat different requirements. Ultimately, the countries went their separate ways. The U. S. follow-on effort, the XM803, was terminated by Congress because of anticipated procurement cost.

In the early seventies, the Army initiated the XM1 tank program, a program that finally led to fielding the Abrams, a proper successor for the M60s, by the end of the decade. Progress has continued, with the fielding of the M1A1 and planned fielding of further improvements in the 1990s.

In this book, Dick Hunnicutt has pursued meticulously the step-by-step progress from the M48 to the Abrams. He has recounted almost forty years of effort by the Army through times when external support was sometimes there and sometimes not. While he has concentrated on the technical aspects of tank development, underlying those efforts is a story of dedication and technical competence.

While this book is a walk down memory lane for those of us who were participants in the efforts recounted, its real value is to offer those who come later the opportunity to learn from history. Mistakes were made, methodologies improved, and lessons were learned. The results of the years of effort should be more than hardware: engineers and managers at every level should find much food for thought in these pages.

INTRODUCTION

As any new technology approaches production, other more advanced concepts already are being studied as possible future replacements. Thus research projects were developing the requirements for future tanks as the design of the M48 was being completed in the early 1950s. The objective of this book is to trace the various research projects and tank development programs that were aimed at replacing the M48 Patton tank and its product improved version, the M60 series. The results of many of these projects also were applied as product improvements to the production tanks.

In addition to the numerous research and concept studies, three major tank development programs were initiated to provide a new main battle tank. The first of these was the T95 program which drew on the earlier work with the T42 to produce a lighter weight medium gun tank. Originally, a parallel program was initiated for a similar heavy gun tank designated as the T96. Later, both projects were combined using the T95 chassis. Official designations were assigned for 13 variations in the series ranging from T95 through T95E12 and the former T96 became the T95E4. However, only the T95 through the T95E3 versions were completed. The others were partially constructed or existed only as mock-ups. The T95 program included many advanced concepts such as smooth bore rigid mount guns and new fire control equipment. Many of these new components still required extensive development resulting in program delays. These delays combined with changing user requirements eventually resulted in the cancellation of the T95 program. It was replaced by the product improved Patton redesignated as the 105mm gun tank M60. At that time, the M60 was considered to be an interim tank pending the development of a new main battle tank incorporating the new technology available.

After the demise of the T95, work continued on the development of various components and their application to new main battle tank concepts. Also at this time, interest in MBT main armament shifted from the high velocity gun with a kinetic energy armor piercing round to a gun launched guided missile using a shaped charge warhead. The laser range finder and other new fire control components also began to appear on the scene. They soon were to be put to use in a major tank development program. On 1 August 1963, an agreement was signed between the United States and the Federal Republic of Germany to jointly develop a new main battle tank for production and use by both nations. Designated as the MBT70, this tank evolved into a highly complex and extremely expensive fighting vehicle. These facts, along with differences between the United States and Germany over certain components and design details eventually resulted in the joint project being terminated leaving each nation to pursue its own line of development. In the United States, the design was simplified to reduce costs and to improve the reliability. Retaining the MBT70s missile armament, the modified version was designated as the 152mm gun-launcher tank XM803. However by this time, preference in the armored force was shifting away from the guided missile and back to the high velocity gun. Also despite its simplification, doubts were being expressed about some of the complex features of the XM803. These circumstances, in addition to its high cost, resulted in the cancellation of the XM803 by Congress in December 1971.

Although the XM803 program was terminated, Congress recognized the Army's need for a new tank and funds were allocated for its development. This third try was successful. The Army organized a task force to specify the requirements for the new vehicle and contracts were awarded to the General Motors Corporation and to the Chrysler Corporation to design and build prototype tanks. These prototypes were evaluated competitively at Aberdeen Proving Ground in early 1976. Subsequently, the Chrysler tank was selected for full scale engineering development followed by production beginning in February 1980. Designated as the 105mm gun tank M1, it was named the Abrams in honor of General Creighton Abrams, the former Army Chief of Staff. The M1 was first fielded at Fort Hood, Texas later in 1980.

After the M1 was in production, work continued to upgrade the new tank. New components including a 120mm smooth bore gun, better armor, and an improved fire control system were introduced resulting in a nomenclature change to the 120mm gun tank M1A1. The first production M1A1s were completed at Detroit in August 1985 and they were fielded to units in Europe replacing the basic M1s already there. The development of the Abrams continues with a long list of improvements scheduled for future versions of the tank. It is rapidly replacing the M60A3 in the Army and will do the same in the Marine Corps. The Abrams, with its many modifications, will remain in first line service for the foreseeable future.

Two difficulties arise in writing the history of a tank that is still on active duty. In the first place, the story can not be completed and, in the second, security restrictions limit the data that can be presented. Thus the description and the data sheets covering the late model tanks are incomplete compared to the coverage of the earlier vehicles.

R. P. Hunnicutt
Belmont, California
December 1989

PART I

ADVANCING THE STATE OF THE ART

The concept model of the tank with the 100 inch diameter turret ring is shown in these photographs dated 17 August 1951. In the upper view, grill armor is installed on the front hull in an effort to defeat shaped charge rounds.

NEW DESIGN CONCEPTS

Parallel with the development of the T48 tank during the Summer of 1951, other studies were being conducted to consider various improved designs as its possible successor. Many of these concepts incorporated new features, some of which still required considerable development. Others involved the modification of conventional designs and could have been available in a relatively short period of time.

The T48 featured a turret ring with an inside diameter of 85 inches and one study investigated the advantages of a still larger ring with a diameter of 100 inches. Such a large ring permitted a greater slope on the turret armor, allowed the mounting of a heavier weapon, and increased the space available for the turret crew. However, the large ring limited the width to which the tank could be reduced for shipment and caused problems in locating the driver. Photographs dated August 1951 show a concept model with a 100 inch turret ring. In this case, the problem of the driver's position was solved by relocating him inside the turret ring. With this arrangement, his normal entrance and exit from the vehicle was through the turret hatch. For most positions of the cannon, this design permitted only indirect vision for the driver through periscopes fitted in the hull forward of the turret ring. With the gun to the rear in the travel lock, direct vision was possible through a port in the turret wall underneath the bustle.

The M-1 proposal is illustrated above in the sectional view at the left and the artist's concept at the right. Note the driver's location inside the turret ring with vision through the periscope in the front armor.

To review many of the new design concepts, a conference was sponsored by Detroit Arsenal in March 1952. Dubbed Operation Questionmark, this was the first of what became a series of conferences intended to stimulate greater interaction and exchange of ideas between the designer and the user of armored fighting vehicles. At this meeting, seven new medium tank concept studies were presented and compared with the T48. The first of these, designated M-1, was similar to the concept model previously described with a turret ring 100 inches in diameter. The driver was located in the center hull just inside the ring and only had direct vision when the gun was locked in the travel position. The commander, gunner, and loader occupied conventional positions in the turret. Protection on the front hull was

provided by four inches of armor at 60 degrees from the vertical and the tank was armed, like the T48, with the 90mm gun T139. Powered by the Continental AOS-895 engine with the XT-500 transmission, the estimated vehicle weight was 40 tons. As indicated by its designation, the six-cylinder, air-cooled, opposed, supercharged engine had a displacement of approximately 895 cubic inches. This was the same power plant installed in the medium tank T42 and the light tank T41 series. It developed a maximum of 500 gross horsepower at 2800 revolutions per minute.

The second and third medium tank concepts, designated M-2 and M-3, had estimated weights of 46 and 43 tons respectively. Both were armed with the 105mm gun T140 and, like the M-1, the front armor was four inches

The M-2 concept can be seen above and at the right. The driver's direct vision port is shown underneath the turret bustle. Below, the M-3 concept locates the driver in the turret.

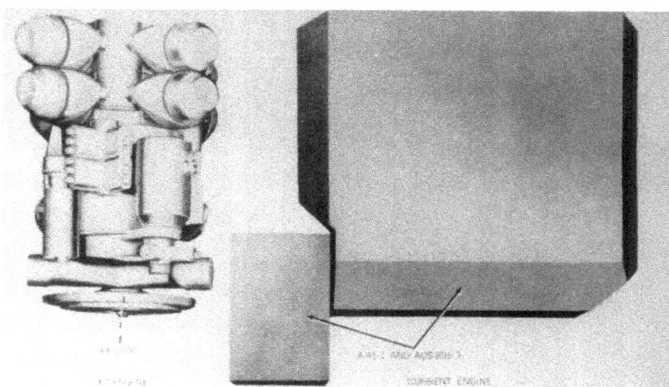

Above, a model of the AX-1100 engine is at the left and, at the right, the space requirements of this power plant are compared with those for the AOS-895 engine and the A41-1 auxiliary engine. The two cylinders intended for use as an auxiliary engine can be seen on the bottom of the AX-1100.

thick at an angle of 60 degrees from the vertical. On both vehicles, the turret ring diameter was increased to 108 inches. The most obvious difference between the two proposed tanks was the location of the driver. In the M-2, he was seated in the hull inside the turret ring, similar to the arrangement in the M-1. In the M-3, he was relocated to the left front of the turret. The 108 inch diameter turret ring provided ample room with all four of the crew members in the turret. The height of the M-3 was reduced to 92 inches compared to 101 inches for the M-2. Linkage caused the driver's seat to counterrotate so that he always faced forward regardless of the turret position.

The M-2 and M-3 were powered by different engines, but both used the XT-500 transmission. The M-2 was driven by the Continental AVS-1195. This was a supercharged eight-cylinder version of the AV-1790 engine used in the T48. At that time, it was estimated to develop 685 gross horsepower. The proposed engine for the M-3 was the AX-1100. This was an experimental air-cooled, two stroke cycle diesel designed by General Motors Corporation with ten cylinders in three radial banks. Two banks contained four cylinders each in an X configuration with the remaining two cylinders in a third bank. Originally, the latter were expected to be able to operate independently when the remaining eight cylinders were shut down, thus serving as an auxiliary engine to drive a generator. At the time of these proposals, the AX-1100 was expected to

develop 717 gross horsepower. Its extremely compact design permitted the length of the M-3 chassis to be reduced to 236 inches compared to 262 inches for the M-2. The shorter and lower M-3 resulted in the three ton lower estimated weight compared to the M-2. Although it initially appeared promising, the program for the AX-1100 was dropped before development was complete.

Medium tank concepts M-4 and M-5 were the same except for the power plant. The M-4 used the ill-fated AX-1100 and the M-5 was equipped with the AOS-895. Both vehicles had the engine mounted with the XT-500 transmission in the left front hull alongside the driver. The turret with its T139 90mm gun was installed on the rear of the chassis with the gunner, tank commander, and loader in their usual positions. Front hull armor was four inches at 60 degrees from the vertical on the upper glacis. With the front mounted power plant, the final drives and sprockets were at the front of each track. The weights for the M-4 and M-5 were estimated as 38 and 38.5 tons respectively. The turret ring was 85 inches in diameter for both vehicles, but the overall width was 127 inches for the M-4 and 135 inches for the M-5. The length without the gun was 217 inches for the M-4 and 212 inches for the M-5. The rear mounted turret reduced the gun overhang in the front decreasing the overall vehicle length. However, the driver's vision was reduced and access for maintenance of the power train was restricted by the front armor.

The M-4 medium tank concept is shown below. With the rear mounted turret, the overhang of the cannon in the forward position is greatly reduced.

Concept M-6 with the pod mounted remote control cannon can be seen above. The vision problem for the crew located low in the hull is obvious.

Medium tank proposals M-6 and M-7 presented the most radical of the new design concepts. Utilizing a chassis similar to the M-4 with the front mounted AX-1100 engine and XT-500 transmission, both vehicles were armed with an automatic cannon in an overhead oscillating turret. All three crew members were located in the hull with the driver at the left front alongside the engine. The gunner and the tank commander were seated side by side to the rear of the driver in a rotating basket. The M-6 was armed with the 105mm gun T140 and the M-7 was fitted with the 90mm gun T139. The 105mm gun increased the estimated weight of the M-6 to 37 tons compared to 32 tons for the M-7. The ring diameter for mounting the automatic turret was 73 inches. The chassis for the M-6 was lengthened to 241 inches compared to 217 inches on the M-7, but the width was 127 inches for both vehicles. The heights for the two tanks differed by only two inches with 106 inches for the M-6 and 104 inches for the M-7. Recoil energy was used to operate the automatic loading mechanism. This arrangement isolated the crew from the danger of flashback or fumes, but it greatly restricted the vision of the gunner and the tank commander. The complexity of the automatic loading system also would have presented serious problems during any development program.

Below is a concept study dated 2 February 1953 of an unmanned turret armed with a 105mm gun and fitted with an automatic loader.

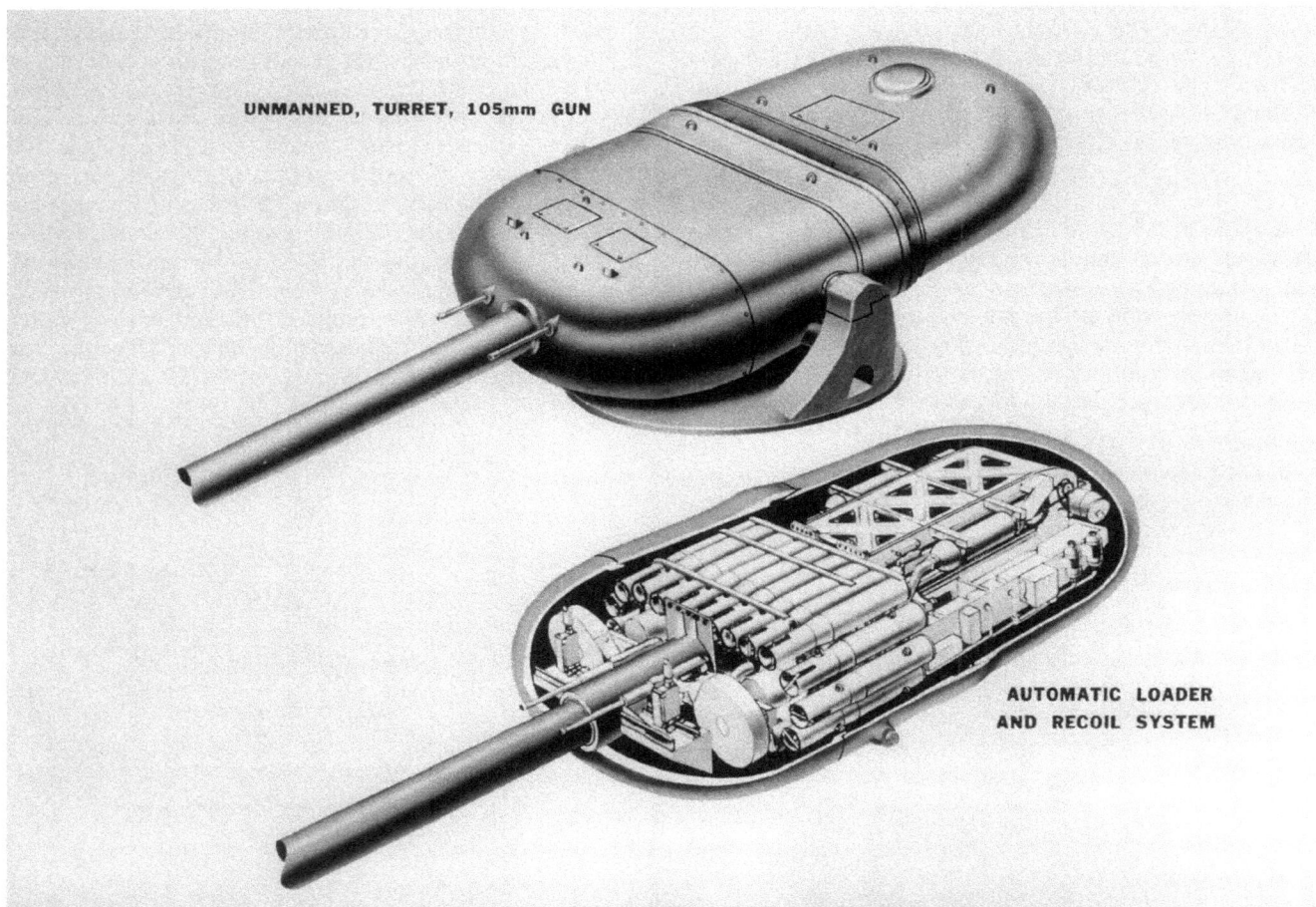

UNMANNED, TURRET, 105mm GUN

AUTOMATIC LOADER
AND RECOIL SYSTEM

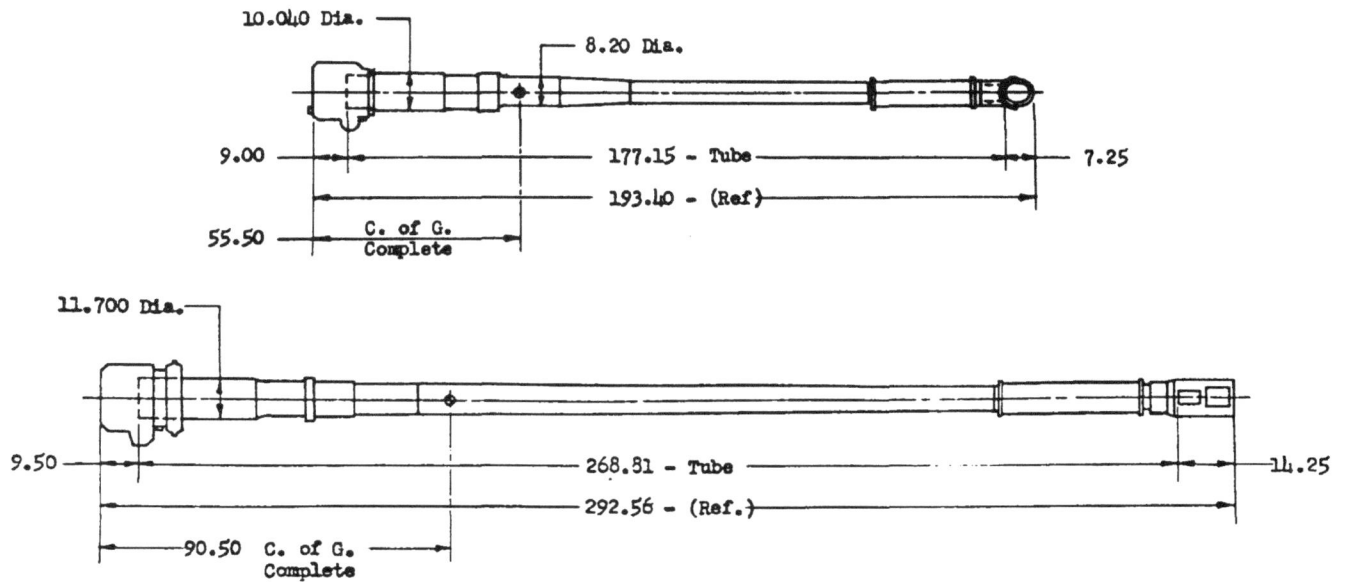

The dimensions of the 90mm gun T139 (upper) and the 105mm gun T140E3 (lower) are shown in the sketches above.

Although none of the Questionmark proposals were selected for development, the conference served a very useful purpose in exchanging ideas. The user obtained a view of the many new design concepts being considered and the designer was acquainted with the many practical requirements necessary for troop use.

To obtain alternate design concepts from industry as well as from Detroit Arsenal, a research and development program was initiated in April 1952. Under this program, contracts were awarded to the H. L. Yoh Company, Inc. of Philadelphia, Pennsylvania, Chrysler Corporation of Detroit, Michigan, and Associated Engineers, Inc. of Springfield, Massachusetts.

The H. L. Yoh Company, Inc. presented a report dated 1 June 1953 outlining seven preliminary design concepts for a new medium tank and several special feature proposals. The tank concepts ranged from a conventional design to several unusual arrangements and all were armed with the 105mm gun T140. The maximum front armor was five inches thick at an angle of 60 degrees from the vertical. The power plant proposed for all seven tanks was the AOSI-1195-5 with an XT-500-1 transmission except for the seventh concept which required a special transmission. At this time, the eight-cylinder AOSI-1195-5 was expected to deliver 675 gross horsepower.

The first design proposed by H. L. Yoh Company, Inc., designated as the M-I-Y, located the crew in their usual positions with three men in the turret and the driver in the front hull. The turret was mounted on a ring 89 inches in diameter and the estimated weight of the vehicle was 46 tons. The concept included several new features such as a machine gun in the turret side wall for use by the

Below, an artist's concept of the M-I-Y tank proposed by H. L. Yoh appears at the left and a model of the vehicle is at the right. Note the machine gun mounted in the turret side wall. The track within a track can be seen in the artist's concept drawing.

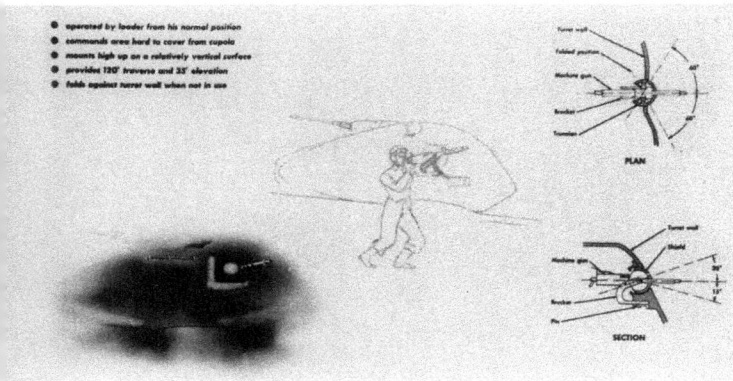

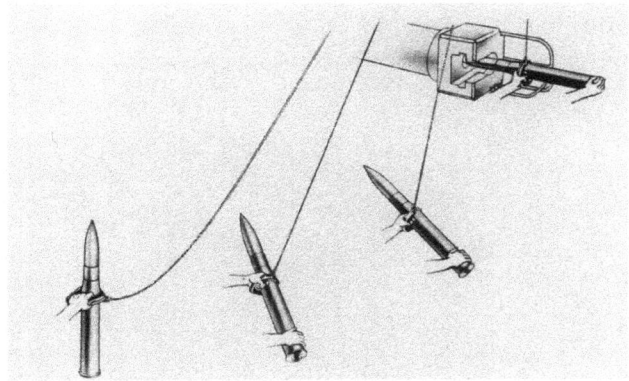

The special features of the H. L. Yoh proposals can be seen in these views. Above, the loader's machine gun mounted in the turret side wall is at the left and the hoist arrangement for handling ammunition is at the right. Below, the horizontal shock absorbers are shown at the left and the track within a track concept is illustrated at the right.

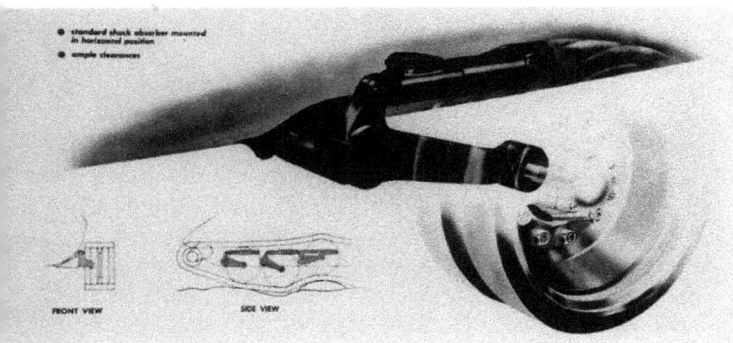

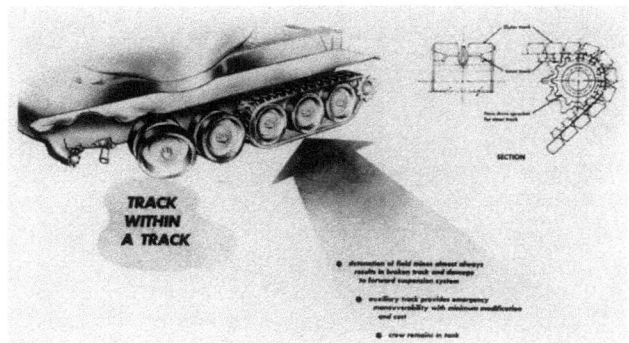

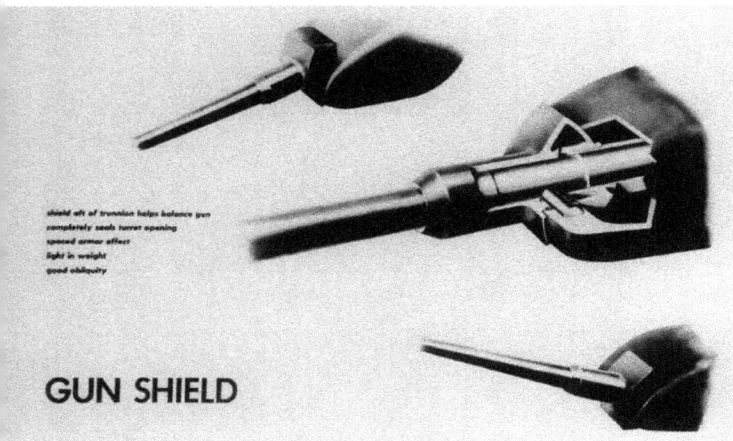

GUN SHIELD

At the left is the hollow gun shield design proposed by H. L. Yoh.

loader, a cable type ammunition hoist to assist the loader, a suspension with horizontally mounted shock absorbers, and a track within a track arrangement for emergency use. The latter consisted of a narrow track mounted inside the main track enclosing the drive sprocket and only the three rear road wheels. In the event of mine damage to the main track and front road wheels, this would permit the tank to maneuver out of danger under its own power. A new

hollow gun shield also was proposed with most of its weight concentrated behind the trunnions to help balance the cannon. The hollow shield also was expected to improve protection by acting as spaced armor.

An unusual crew arrangement was proposed in the second design, designated as the M-II-Y. In this vehicle, only the tank commander and the loader were located in the turret. The gunner was positioned in the right front hull alongside the driver. This arrangement required remote sighting equipment for the gunner, but it permitted a narrow front on the turret with improved protection. This concept also included the new shield design, horizontal shock absorbers, emergency inner track, ammunition hoist, and an auxiliary bow mount machine gun for the gunner or driver. The estimated weight was 45 tons.

Proposals M-II-Y and M-III-Y appear above and below respectively. Note that the M-II-Y is fitted with a bow machine gun for use by the gunner, but the machine gun mount in the turret side wall has been eliminated.

Design concept M-III-Y also proposed a turret with a small frontal area for good ballistic protection, but the gunner was retained in the turret along with the tank commander and the loader. This vehicle had the same special features as M-II-Y and also was estimated to weigh 45 tons.

Medium tank proposal M-IV-Y installed the main armament in an oscillating turret mounted inside a shield ring. The four crew members were located in their conventional positions and the estimated weight of the tank was 47 tons. The shield ring prevented installation of the loader's auxiliary machine gun, but the vehicle was fitted with the horizontal shock absorbers, emergency inner track, and cable ammunition hoist.

The M-V-Y concept featured a small oscillating turret supported by trunnions on a yoke mount rotating on a 53 inch diameter ring. Only the tank commander and the gunner rode in the turret. The loader was located in the hull behind the driver and below the turret ring. From this position, he transferred ammunition stowed in the hull up to the semiautomatic loader in the turret. The tank was armed

Below is the M-IV-Y concept by H. L. Yoh. The oscillating turret design prevented the installation of a machine gun in the turret side wall.

15

Two oscillating turret proposals appear in these views with the M-V-Y above and the M-VI-Y below. Note the two hull mounted machine guns on the M-V-Y.

with two coaxial machine guns in addition to the tank commander's .50 caliber weapon. Two additional auxiliary machine guns were installed, one in each side of the front hull for use by the driver and the loader. The suspension used the horizontal shock absorbers and the emergency inner track was specified. This design was intended to completely balance the gun and turret and permit an extremely small frontal area on the latter. The estimated weight of the vehicle was 40 tons.

The proposed medium tank M-VI-Y also used an oscillating turret similar to the previous concept. However, the ring on which the yoke rotated was enlarged to a diameter of 63 inches and the loader was relocated into the turret with the tank commander and gunner. A semiautomatic loader was provided with 16 ready rounds in the magazine. As before, the design permitted a narrow turret front with good ballistic protection. The estimated vehicle weight was 46 tons.

The most radical concept proposed by the H. L. Yoh Company, Inc. was designated as the medium tank M-VII-Y. This design located the AOSI-1195-5 engine in the turret bustle transmitting the power through a ring gear and a special transmission to the rear mounted drive sprockets. The three man turret crew occupied their usual positions and the driver was located in the right front of the hull. The turret ring diameter was 89 inches and the estimated weight of the vehicle was 43 tons. A large fuel tank was located to the left of the engine in the turret bustle. Advantages claimed for this configuration included the balancing of the cannon by the engine weight and a shorter lighter hull. However, it is highly doubtful if any such advantages could offset the problems with such a complex drive system. The center of gravity also would have been quite high in such a vehicle. However, it certainly indicated that the concept studies were completely unhampered by conventional practice.

The M-VII-Y concept by H. L. Yoh can be seen in these three illustrations. Note the turret mounted engine in the cutaway drawing.

Above is the Chrysler concept of a four track tank. The band type track proposed for this vehicle can be seen in this view.

Chrysler Corporation received two contracts under the research and development program. One was to produce a concept of a new medium tank with a four track suspension system and the other considered a new medium tank with the usual two track suspension. Preliminary studies of a four track tank had been in progress for some time at Detroit Arsenal. Sketches of such an arrangement appeared in a presentation on tank and automotive development to the Research and Development Board on 16 May 1951 and a preliminary study of a four track tank with a hydrostatic drive was carried out at the Arsenal by Joseph Williams and Clifford Bradley. One of the most attractive features of the four track system was its reduced vulnerability to mine damage. Even if the two front tracks were disabled, the two remaining rear tracks would permit the tank to maneuver under its own power. The four track arrangement also permitted a better armor configuration when using very large diameter turret rings.

On 26 March 1953, Chrysler presented reports to Detroit Arsenal covering their concept studies for a four track medium tank and a new medium tank with a two track suspension. The initial reports were followed by additional data in May and July, however, both studies were preliminary concepts and did not contain detailed design information.

The same turret design was used on both of Chrysler's proposals. The outside diameter of the turret base was 136 inches and it was mounted on a ring with an inside diameter of 93 inches. Later, the proposal was modified to reduce the outside diameter to 125 inches for shipping purposes, but the original inside ring diameter of 93 inches was retained. The entire four man crew was carried in the turret with the gunner at the right front and the tank commander at the right rear. The loader worked in his usual position at the left rear. The driver was located at the left front and his station automatically rotated so that he always faced forward regardless of the turret position. The 105mm gun was installed in a rotor-shield combination mount with a .30 caliber coaxial machine gun on the left and an articulated telescopic sight for the gunner on the right. A radar range finder was specified, but no details were provided. The combination gun mount with integral trunnions was installed as one assembly. It was to be doweled and bolted in place from outside of the turret to reduce assembly costs. A .50 caliber machine gun was mounted in the commander's cupola. Provision was made for the installation of a .30 caliber machine gun with a parabolic deflector in a turntable on top of the turret. The armor protection on the hull and turret of both vehicles was equivalent to that on the T48 tank.

Below are two early sketches of the four track configuration. Note the large turret ring diameter possible with this design.

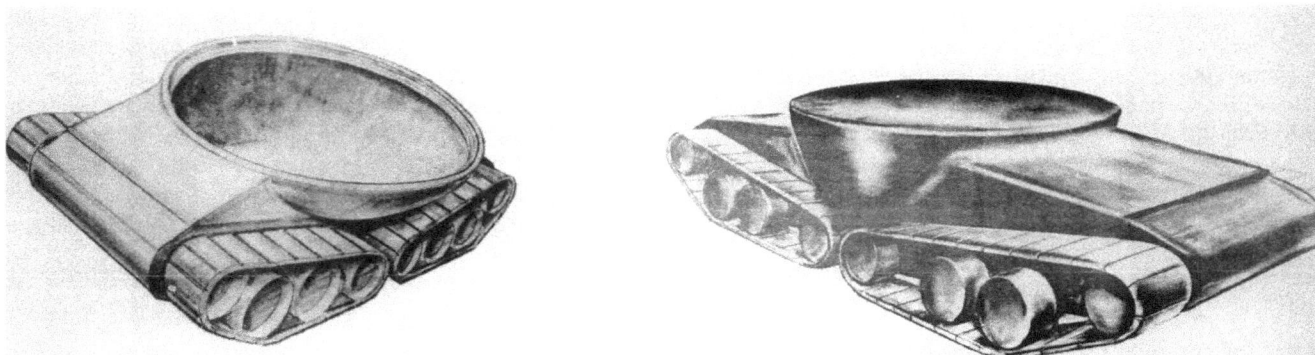

Chrysler's two track tank design proposal is shown above. Like the four track vehicle, this tank features the driver-in-turret arrangement.

The four track medium tank differed from its more conventional mate in the power train as well as the suspension. A gas turbine rated at 500–600 horsepower was proposed for the four track vehicle. The turbine was mounted transversely in the rear hull and coupled to an electric generator through a reduction gear box. Four electric traction motors, two in front and two in the rear, transmitted power to the tracks through spur gear final drives. Three amplidyne exciters, one for exciting the generator and one for exciting each pair of traction motors, were driven by an auxiliary drive from the turbine reduction gear box. A resistor for absorbing energy during braking was attached to the deck grill in the rear of the hull.

Each of the four separate suspension units consisted of a drive drum (replacing the usual sprocket), three wheel assemblies, one track support roller, and one track compensator. A new sectional band type track was proposed for this vehicle. This track, of steel and cable construction, was 28 inches wide. The center guides were an integral part of the track and in addition to their guiding function, they transmitted power to the track by engaging the drive pins in the center of the drive drum. Two of the wheel assemblies utilized two dual road wheels 11½ inches in diameter attached to a suspension arm. The third wheel assembly consisted of one 11½ inch diameter dual road wheel and one 17 inch diameter dual wheel also attached to a suspension arm. The wheel pairs could rotate up to three inches about their common attachment point by compressing rubber pads on the suspension arm. The movement of the suspension arm itself was resisted by the torsion bar. Friction snubbers were installed on the suspension arms and movement was limited to five inches by bump stops. The track compensator consisted of two 10 inch diameter wheels at each end of a short beam pivoting about a center axle. It was intended to equalize track tension by pivoting about its center.

Steering with the electric drive was by the use of controllers which varied the speed of the traction motors. A dual drive system was provided so that either the driver or the tank commander could drive the vehicle.

An air-cooled eight cylinder, horizontally opposed engine developing about 700 horsepower was specified with the XT-500 transmission for Chrysler's concept of a new two track medium tank. Although the proposal did not refer to any particular engine, it was indicated that a diesel power plant was preferred. The suspension on each side of the proposed vehicle consisted of seven 24 inch diameter dual road wheels with torsion bar springs and three track support rollers. A dual 24 inch diameter fixed idler was at the front of each track with a drive drum at the rear. The 28 inch wide sectional band type track utilized cast steel and cable construction with integral center guides. The latter also propelled the vehicle by engaging the drive pins in the center of the drive drum. Friction snubbers were installed on the first three and last two road wheel arms on each side.

With the driver located in the turret, an electrical system was required for steering, braking, and transmission control. Manual brake controls also were provided for emergency use. The estimated weight for both Chrysler medium tank concepts was about 45 tons.

The contract with Associated Engineers, Inc. also investigated the concept of a four track medium tank and their work produced a far more detailed analysis than any of the other concept studies. Their initial work considered a four track electric drive vehicle with four separate traction motors similar to the Chrysler proposal, although the design was developed in much greater detail.

The 105mm gun T140 was installed in a large turret mounted on a ring with an inside diameter of 110½ inches. A coaxial .30 caliber machine gun was fitted to the left of the cannon with a telescopic sight for the gunner to the

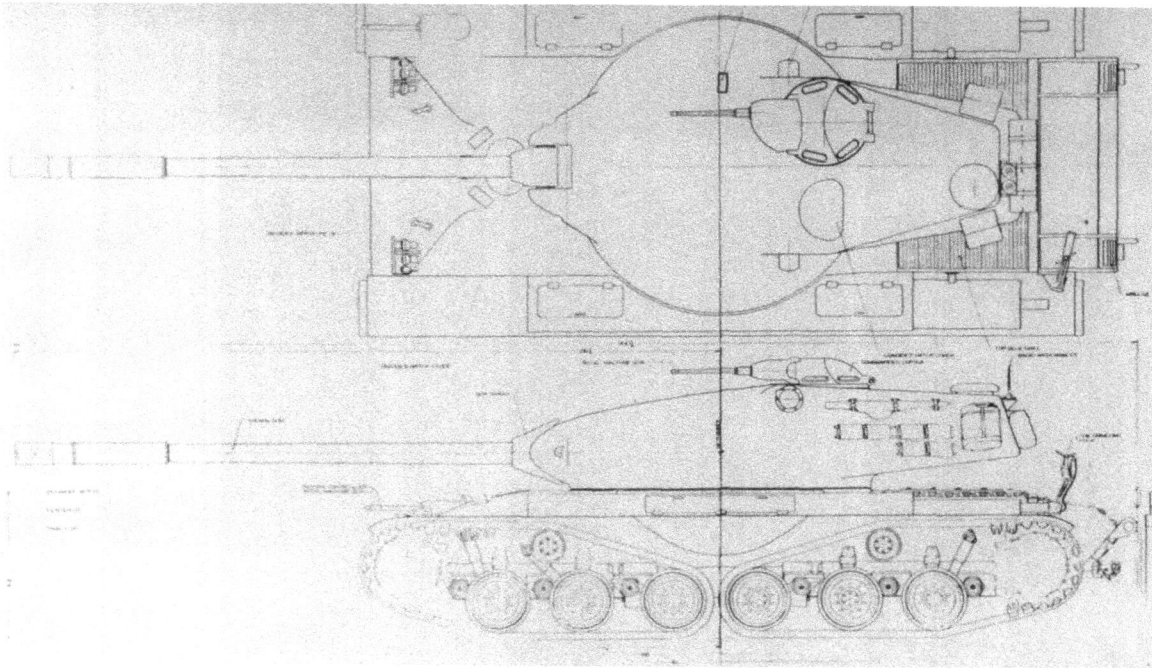

The phase I design of a four track tank by Associated Engineers, Inc. is illustrated in the drawings above and at the bottom of the page. Note the fender mounted machine gun featured in this design.

right. The crew members were located in their normal positions with the tank commander, gunner, and loader in the turret and the driver in the front center of the hull. The commander's cupola with its .50 caliber machine gun was installed at the right rear of the turret roof, but still within the turret ring. A commander operated optical range finder was located just in front of the cupola. A remote controlled .30 caliber auxiliary machine gun was mounted on the right front fender for use by the driver. The estimated combat weight of the tank was 54 tons.

As mentioned previously, the sprocket in each of the four suspension units was driven by a separate direct current electric motor providing 25 per cent of the total tractive effort available. In addition to the sprocket, each suspension unit consisted of three 26 inch diameter road wheels and one track support roller. Ordnance standard torsion bars were used to support the sprung weight of the vehicle. External shock absorbers were fitted on the two front road wheel arms of the forward suspension units and on the rear road wheel arms of the aft suspension units. The tank was powered by the experimental eight cylinder Continental AOSI-1195-5 air-cooled engine, then rated at 710 gross horsepower. The engine drove a large direct current generator with an alternator to supply power to the four traction motors. Further analysis revealed some major

deficiencies with this arrangement. Since each traction motor was capable of applying only 25 per cent of the maximum power from the generator, the loss of one track resulted in the utilization of only 75 per cent of the available output. Also, it could be shown that if one track was lost, it was necessary to remove power from its mate on the opposite side of the tank to permit adequate steering. Thus the loss of a single track resulted in a 50 per cent reduction in available power. It was not possible to switch full power to the remaining two tracks since size and weight restrictions made it impractical to install four traction motors, each with sufficient capacity to utilize 50 per cent of the generator output.

Another disadvantage of the individually powered tracks appeared when ascending a steep grade. Here, the greater portion of the vehicle's weight was supported by the rear suspensions increasing the power requirements at the rear sprockets. Also, when climbing an obstacle, the front suspension units might lose traction completely and full power would be needed at the rear. This could not be done with the four traction motor design.

As a result of the design analysis, the contract with Associated Engineers, Inc. was extended to investigate a phase II vehicle to overcome some of the problems encountered in phase I. The four traction motors in the earlier

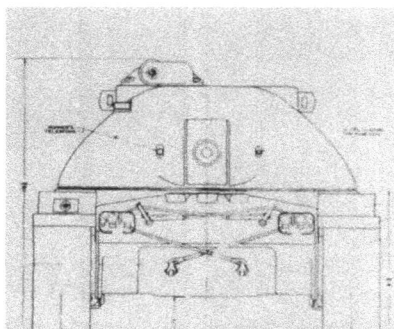

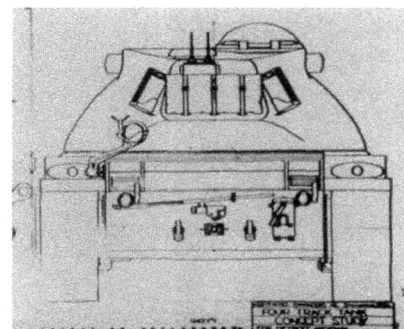

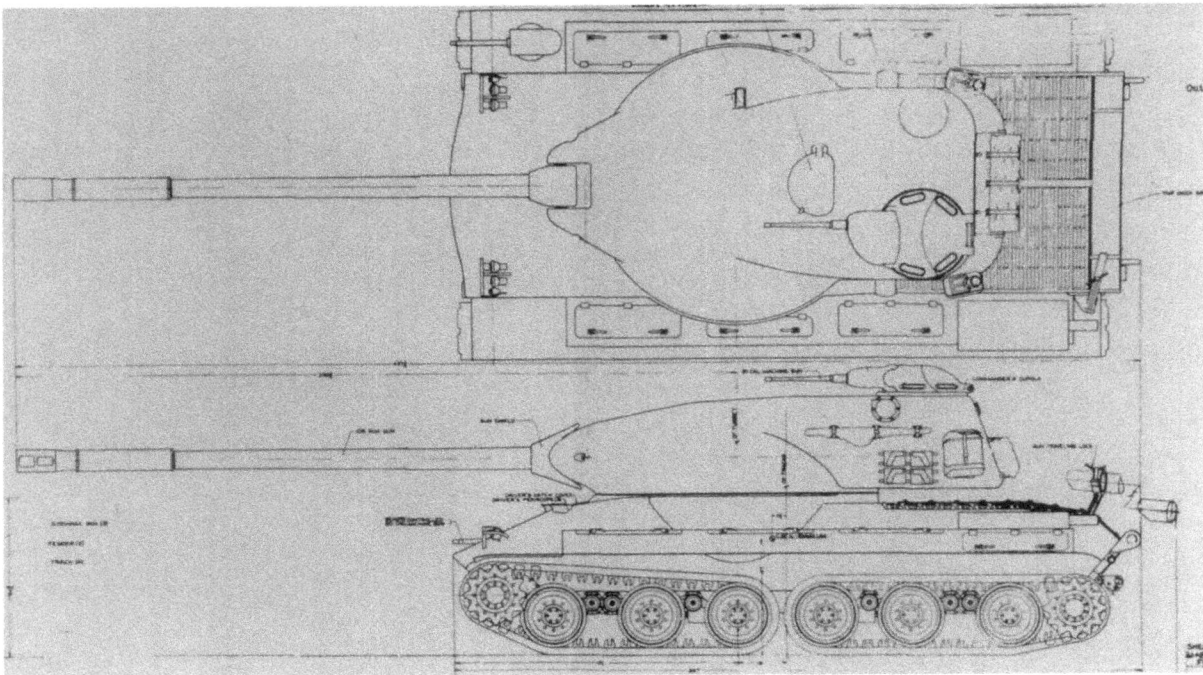

The phase II configuration of the four track tank proposed by Associated Engineers, Inc. is shown in the drawings above and at the bottom of the page. Note the change in the suspension design compared to phase I.

design were replaced by two larger electric motors, each intended to power the pair of tracks on one side of the tank. These motors were installed, one on each side of the generator, in the rear hull and connected to the drive sprockets by gears and shafts. If one track was damaged, it could be uncoupled by the driver and full power applied to the remaining track on that side. Also, it was unnecessary to uncouple the mating track on the opposite side. Thus the tank retained the full use of all available power. Full tractive effort also was retained when climbing a steep grade.

Because of the greater weight inherent in the electric drive system, the phase II study was extended to include the use of a hydraulic torque converter transmission. This arrangement replaced the generator and electric traction motors with an XT-500-1 transmission. The system of shafts and gears used to transmit the power to the four final drives was modified to accommodate the new transmission. As with the electric drive, the driver could uncouple a damaged track and transfer full power to the remaining track on that side. Early in the phase II program, a hydrostatic transmission was considered for use in the new tank. However, further study indicated that such an application would not be practical with the hydrostatic components available at that time.

The reduction in numbers of traction motors from four to two and their relocation permitted a redesign of the suspension on the phase II vehicle. Torsion bars still were used to spring the 26 inch diameter road wheels, but the sprockets were reduced in size and the track support rollers were eliminated. The shock absorbers on the first two wheels of the forward suspension unit and the last wheel of the rear suspension unit were relocated inside the hull. With less space required to house the two traction motors, the hull on the phase II tank was reduced in length. It still consisted of a large homogeneous armor casting with welded sections of rolled armor plate. The turret ring inside diameter was reduced to 100 inches and the cast turret was redesigned. As on the phase I vehicle, the armor protection was equal to that on the 90mm gun tank M48. On the phase II tank, the tank commander was relocated to the left rear in the turret bustle, but the gunner and loader remained in their usual positions. The new turret also featured an escape hatch in the rear wall of the turret bustle. The armament remained the same as on the phase I vehicle.

The phase II tank had an estimated combat weight of almost 51 tons for the electric drive version and just under 49 tons for the model with the hydraulic torque converter transmission. Both vehicles made far more effective use of the available power than the phase I tank. However, the study showed that neither was as efficient as a two track tank. Also, the gap between the two tracks on each side resulted in increased ground pressure and the mobility remaining after the loss of a track was impossible to predict. It would have depended upon the terrain involved and on the rolling resistance of the damaged suspension unit. Whatever it was, it was not considered worth the added complexity and reduced efficiency of the four track system. Further development of the four track tank was discontinued.

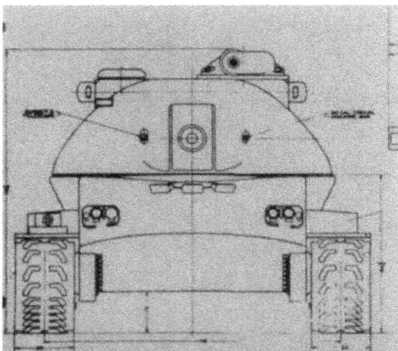

This photograph of a lightweight medium tank concept model was dated 7 March 1953.

Despite the wide range of design concepts considered, major interest both with the user and at Detroit Arsenal remained in a conventional tank that was lighter in weight and had better performance than the M48. After the medium tank T42 was rejected as being underpowered, work had continued to provide a lighter weight medium tank to meet the user's requirements. Several such concept studies were presented by Detroit Arsenal during 1953. In November, one such presentation compared six concepts of future medium gun tanks with the M48. Four of these concepts were considered as relatively short term projects making use of components already well along in the development cycle. The remaining two were long range programs depending upon the successful development of various key components.

The first of the short term proposals, designated as the lightweight medium tank, was armed with the 90mm gun T139 mounted in a turret with a 77 inch inside diameter ring. Powered by the AOI-1195 engine with the XT-500 transmission, this vehicle had an estimated combat weight of 39 tons. With a chassis length of 236 inches, the suspension consisted of only five road wheels and four track support rollers per side. Overall length with the 90mm gun

forward was only 317 inches and the width was 131 inches compared to 143 inches for the M48. This concept was a true successor to the 90mm gun tank T42. With the AOI-1195 engine rated at 560 gross horsepower, the power to weight ratio was improved over the earlier T42, although it did not equal the 16.4 horsepower per ton of the M48.

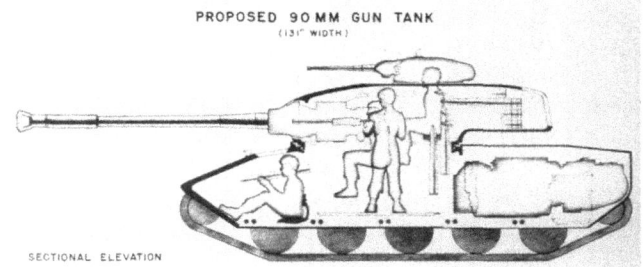

At the lower left is another model of a lightweight medium tank. The drawings above and below, dated 17 September 1953, show the short term proposal for the 131 inch wide lightweight medium tank armed with the 90mm gun T139 in a 77 inch diameter turret ring.

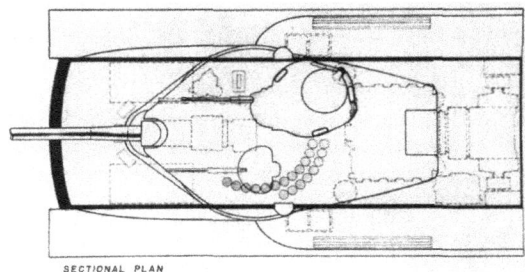

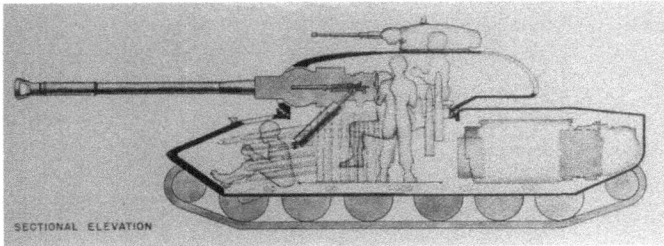

Above is the short term proposal, dated 5 November 1953, for a new medium tank armed with the 90mm gun T139 in a turret with an 85 inch diameter ring. The space requirement for the AOI-1490 engine with the XT-500 transmission can be seen in the sectional drawing.

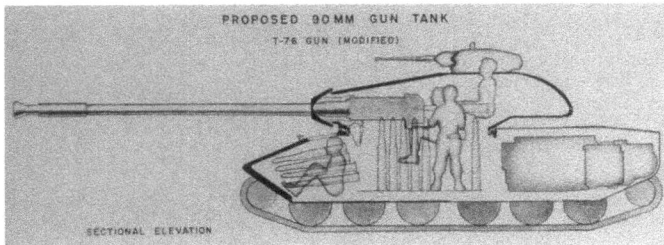

The short term proposal for a new medium tank armed with the 90mm gun T76 (modified) appears above. Note the large cartridge case required by this more powerful weapon.

The remaining short term proposal concepts utilized essentially the same chassis powered by the ten cylinder AOI-1490 air-cooled engine with the XT-500 transmission. The AOI-1490 was a new experimental engine expected at that time to develop 675 gross horsepower. Cooling air intakes were located in the rear sponsons of these vehicles eliminating the top rear deck grills. The cooling air, mixed with the engine exhaust, was ejected through a grill in the rear of the hull. This arrangement tended to cool the exhaust gases and reduce the infrared signature. Another new feature was a flat track torsion bar suspension with six road wheels per side. This type of suspension eliminated the track support rollers and allowed the track to ride on the top of the road wheels.

The three tanks differed in their turrets, all of which were installed on 85 inch inside diameter rings. This was considered to be the smallest practical size to mount a 105mm gun and all three vehicles were designed to be capable of mounting this weapon. As proposed, the three tanks were armed with the 90mm gun T139, the 90mm gun T76 modified, and the 105mm gun T140E3 which resulted in respective estimated vehicle combat weights of 41, 43.5, and 45 tons. Needless to say, the estimated performance of the three tanks declined with the increasing weight. The 90mm gun T139 was, of course, the same weapon as in the M48 tank. The 90mm gun T76 modified was a high powered weapon with a muzzle velocity of 3500 feet per second compared to 3100 feet per second for the T139 gun

The short term proposal for a new medium tank armed with the 105mm gun T140E3 can be seen in the model photograph above and the sectional drawing below. Like the other proposed tanks on this page, it had a turret ring diameter of 85 inches.

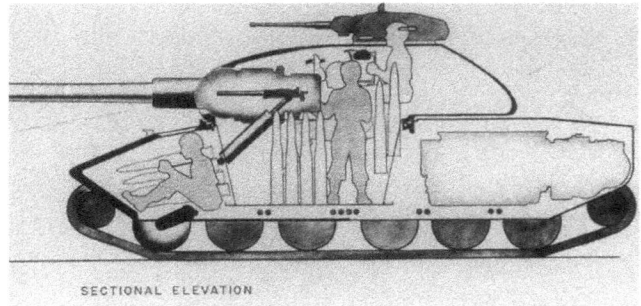

MEDIUM TANK GUN COMPARISON:

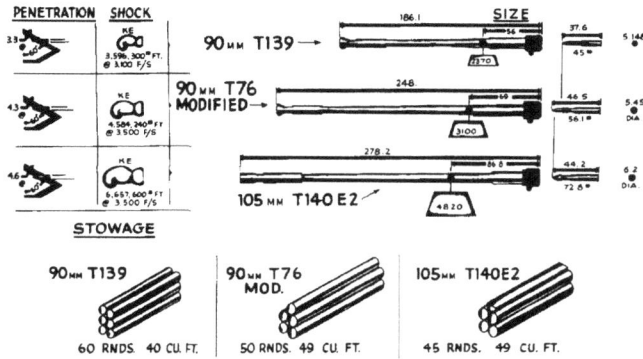

SHOCK FROM STRIKES WILL DISSIPATE IN TURRET BODY
TO ELIMINATE DISTORTION OR DEFLECTION AT FIRE CONTROL INSTALLATION

The high velocity guns proposed as armament for the new tanks are compared above. The arrangement for isolating the fire control equipment from the turret wall is illustrated in the sketch at the top right.

when firing early armor piercing (AP) shot. The T76 modified was considerably heavier than the T139, but it also could be balanced in the 85 inch turret ring and did not require an equilibrator. As mentioned earlier, the 105mm gun T140E3 was considered to be the largest weapon suitable for mounting in the 85 inch turret ring, but it could not be balanced and required an equilibrator.

All of the short term proposal concepts were of conventional configuration with the driver in the center front hull and the other three crew members in the turret. One new feature was an escape hatch in the rear wall of the turret bustle. This also was expected to be useful in stowing ammunition and for removing the cannon or other equipment for maintenance. Another innovation was a separate internal mount for the fire control instruments to isolate them from the shock of impacts on the turret.

The two long range designs proposed in the presentation at Detroit were both armed with the 105mm gun in an equilibrated mount and this mount was rigid without a recoil system. This feature, combined with a turret ring diameter of 90 inches, greatly increased the usable space inside the turret. Both vehicles were powered by an unspecified version of the new X type engine operating with the XT-500 transmission. It was expected that the use of the X engine would reduce the vehicle weight by about a ton. The two concepts differed primarily in the location of the driver. One version placed him in the usual position in the center front of the hull. In the other, he was relocated to the left front of the turret. The latter arrangement shortened the hull and lowered the vehicle height from 88 to 81.5 inches. This reduced the estimated vehicle weight from 44 to 42.5 tons. Other features considered for long range application included automatic loading equipment and a radar range finder. The proposals in this presentation were representative of the large number of new design concepts under study at Detroit Arsenal during this period.

SUSPENSIONS

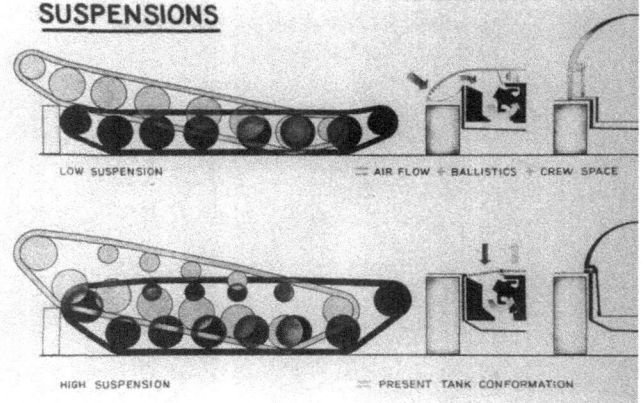

Above, the low flat track suspension is compared with the design using track support rollers. Note that the low suspension height allows the engine air intake to be located in the sponson instead of in the top deck.

The two long term proposals presented at the November 1953 meeting in Detroit are shown here. In the conventional arrangement above, the driver is seated in the hull. Below, all four crew members are located in the turret.

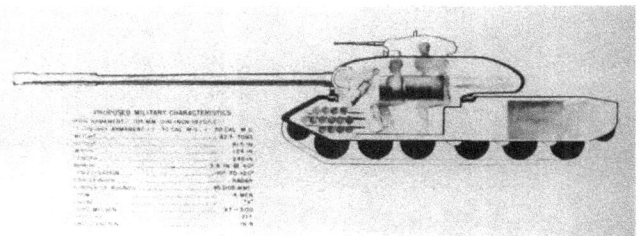

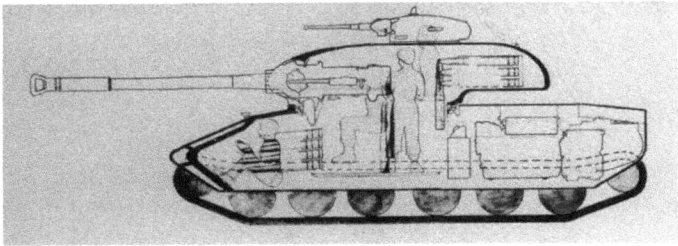

The TS-1 proposal at the Questionmark III conference is illustrated by the model photograph and the sectional drawing above. The siliceous cored armor can be seen in the hull front.

In June 1954, the third Questionmark conference was convened at Detroit Arsenal to review the progress made since Questionmark I in April 1952. Questionmark II had been held in September 1952, but it was restricted to the consideration of self-propelled artillery and antiaircraft weapons.

During the Questionmark III meeting, nine design concepts were presented as possible replacements for the 90mm gun tank M48. A number of special features also were proposed which were suitable for application to a wide variety of vehicles. All of these proposals were designated according to a special nomenclature system. Under this system, tracked vehicles were identified by a T followed by an S or an L indicating that either a short or a long development period was required. The S referred to items which might be available within two years. Those with an L in their designation required a development time of up to five years.

At that time, a major concern in the U.S. Army was the extensive logistic support required for armored and mechanized units. A large portion of this support was the supply of the necessary fuel. Thus an important consideration in any new tank design was the reduction of the fuel consumption. All of the concepts proposed to replace the M48 tank reflected this concern by reducing the vehicle weight and adopting more efficient propulsion systems. It was estimated that the short time range designs would

reduce the fuel consumption of an armored division by as much as 50 per cent. The long time development programs were expected to provide a total reduction of 65 per cent. All of the tank concepts were designed for easier maintenance and with reduced widths to ease the transportation problem. The flat track suspension was used, eliminating the track support rollers. The tanks also featured sponson intake grills and a rear exhaust with an insulated rear deck to suppress infrared radiation. A conventional crew arrangement was used with a three man turret crew and the driver in the center front hull. Main armament varied, but each concept was fitted with a single .30 caliber coaxial machine gun and a .50 caliber machine gun in the commander's cupola.

The flat track torsion bar suspension used six road wheels per side on the first three short development time proposals designated as the TS-1, TS-3, and TS-4. All three vehicles were powered by the AOI-1195 engine through the XT-500 transmission. The 90mm gun T139 armed all three tanks and, in the case of the TS-3, it was installed in a rigid non-recoiling mount. A new feature was the Optar (optical tracking, acquisition, and ranging) pulsed light range finder. This device measured the range by timing the reflection of a light pulse from the target. The major difference between the TS-1, TS-3, and TS-4 was in the armor protection which resulted in respective estimated weights of 43, 40, and 35 tons. The front armor on the TS-1 provided

Above are models of the XT-500 transmission (left) and the AOI-1195-1 engine (right). Below, the model and the sectional drawing depict the TS-3 tank proposal with homogeneous steel front hull armor.

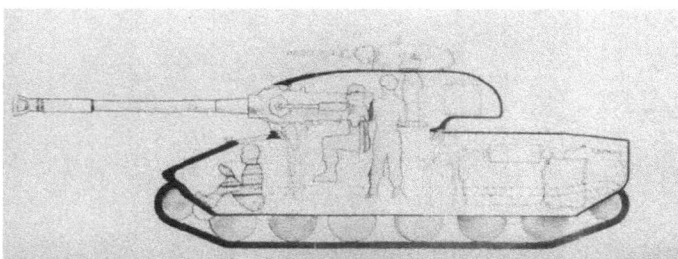

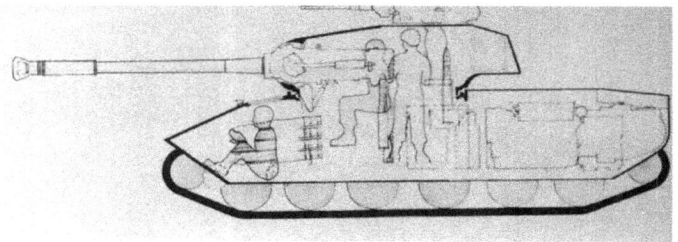

Above is the lightweight TS-4 tank proposal with the thin homogeneous steel armor. At the right below is a sectional drawing of the TS-7 tank with siliceous cored armor and the smooth bore 90mm gun T208.

protection against kinetic energy armor piercing projectiles (AP) equivalent to that on the M48 tank. However, against shaped charge rounds, it was far superior. This was due to the use of siliceous cored armor. The latter consisted of 3.8 inches of homogeneous steel cast around a fused silica core and sloped at 60 degrees from the vertical. On an equal weight basis, this combination was far more effective against shaped charge rounds than solid homogeneous steel.

The TS-3 and TS-4 were protected by solid homogeneous steel front armor 3.8 inches and 1.5 inches thick respectively. Both were sloped at 60 degrees from the vertical. The side armor on both tanks was reduced compared to the M48.

Design concept TS-7 was similar to the TS-1 utilizing the siliceous cored front armor, although the vehicle was somewhat lighter in weight. However, the main armament was changed to the 90mm gun T208. This was a high powered smooth bore cannon and it was installed in a rigid non-recoiling mount. Firing fin stabilized armor piercing projectiles, this weapon had extremely high penetration performance.

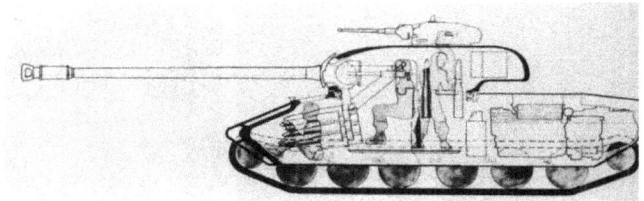

Proposal TS-9, like the first three short term concepts, was armed with the 90mm gun T139. However, the size and weight of the vehicle was drastically reduced. The front armor was three inches of solid homogeneous steel at 65 degrees from the vertical, but the side armor was vertical and only one inch thick. The tank was shortened so that the flat track torsion bar suspension had only four road wheels per side. This was possible because of the compact new power train. The power plant was the AOI-628-1 air-cooled engine driving through an XT-270 transmission. The engine developed 340 horsepower at 3200 revolutions per minute and it was mounted transversely in the tank hull. The result was an estimated vehicle weight of only 25 tons.

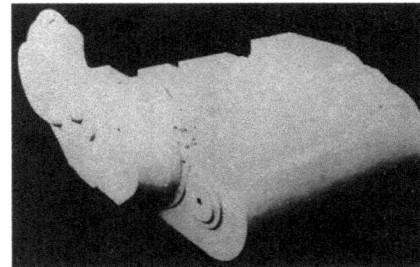

Above are models of the AOI-628 engine (left) and the XT-270 transmission (right). Below, the concept model of the TS-9 proposal is at the left and a sectional drawing is at the right. Note the extremely short vehicle length resulting from the use of the AOI-628-1 engine and the XT-270 transmission.

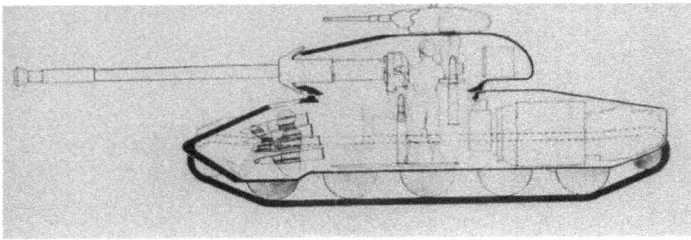

The TL-1 and TL-2 long term concepts appear above and below respectively. The TL-1 was the design that eventually developed into the 90mm gun tank T95.

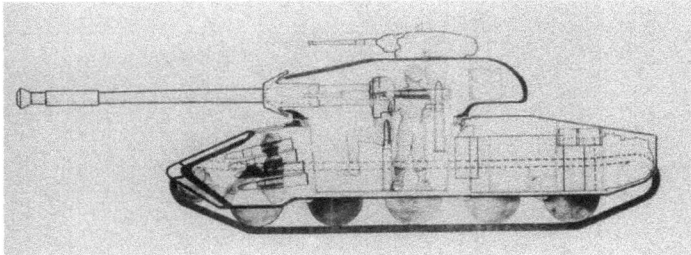

The four long range development proposals, identified as TL-1, TL-2, TL-5, and TL-7, were all armed with the 90mm T208 smooth bore cannon. However, in some cases it was proposed to modify the weapon by shortening the barrel by as much as three feet. Also, both rigid and recoiling type mounts were considered. All four concepts used the flat track torsion bar suspension and the first three were fitted with five road wheels per side. On the shorter TL-7, only four road wheels per side were required. The TL-1 and TL-5 were powered respectively by twelve and eight cylinder versions of the proposed General Motors X type diesel engine. Designated as the AX-1320 and the AX-880, they were expected to develop 600 and 400 gross horsepower respectively at 2500 revolutions per minute.

Further study had indicated that the integral two cylinder auxiliary engine, as on the AX-1100, was unnecessary because of the low fuel consumption of the diesel. At idle, the full diesel engine could drive a generator while using less fuel than a gasoline powered auxiliary generator set. Thus the later concepts of the X type engine only used multiples of four cylinders.

Proposal TL-2 made use of another new engine concept. This was the six cylinder, air-cooled H type engine designated as the AHDS-1140. This was a two stroke cycle diesel engine with opposed pistons. It was estimated to develop 600 horsepower at 2500 revolutions per minute. Like the power packages in the TL-1 and the TL-5, the engine in the TL-2 was coupled to the XT-500 transmission.

Above are models of the AX-880 engine (left) and the AHDS-760 engine (right). The latter was a four cylinder version of the six cylinder engine proposed for the TL-2. Below are the model photograph and the sectional drawing of the proposed TL-5. Note the very light armor.

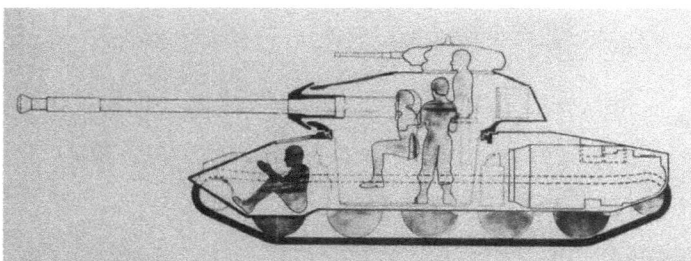

The long term concept TL-7 can be seen above. As on the TS-9, the new power train permitted a much shorter vehicle greatly reducing the weight.

The TL-7 was fitted with the same transversely mounted power package as the TS-9 consisting of the AOI-628 engine and the XT-270 transmission. As before, this resulted in a much shorter chassis helping to control the weight of the vehicle. Also, the armor on the TL-7 was the same as on the TS-9 with three inches of homogeneous steel at 65 degrees in front with one inch of vertical armor on the sides. This can be compared with the front of the TL-1 which was protected by 3.8 inches of homogenous armor at 65 degrees from the vertical. The TL-2 had the same 3.8 inches of steel, but it was siliceous cored armor for greater protection against shaped charges. The TL-5 only had armor equivalent to a light tank with 1.5 inches of homogeneous steel on the front at 60 degrees from the vertical. The differences in armor protection and other design features resulted in estimated weights for the TL-1, TL-2, TL-5, and TL-7 of 41, 43, 36, and 26 tons respectively.

Following the Questionmark III conference, the TL-1 concept was selected as the basis of a new medium gun tank. Ordnance Committee action on 27 January 1955 designated the proposed vehicle as the 90mm gun tank T95 and it was the major tank development program in the U.S. Army for the next five years. Details of this program are discussed in a following section.

Although not selected for full scale development, considerable interest remained in the TL-7 concept for a lightweight medium gun tank. Further study resulted in three versions, designated as the TL-7A, TL-7B, and TL-7C. Details of these three concepts, which differed mainly in the armor protection, were presented during a conference at Detroit Arsenal on 7 April 1955. The TL-7A carried armor equivalent to a light tank with only one inch of homogeneous steel on the front hull at 60 degrees from the vertical. On the TL-7B and TL-7C, the front was increased to 2.8 inches and 3.8 inches respectively, still at 60 degrees from the vertical. Because of its greater weight, the track width on the TL-7C was increased to 21 inches compared to 20 inches for the two lighter vehicles. The power package on the TL-7A and TL-7B consisted of the transversely mounted AOI-628 engine now coupled to the XTG-400M transmission. On the TL-7C, this engine was rebored to 5¼ inches increasing the displacement to 695 cubic inches. This change raised the estimated gross horsepower to 400 compared to 340 for the original AOI-628. All three versions of the tank retained the 90mm T208 smooth bore gun rigidly mounted in a turret with a 78 inch inside diameter ring.

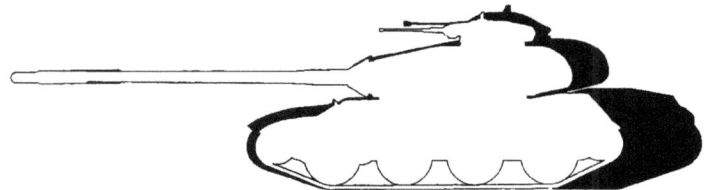

The short length of the TL-7 is obvious in the silhouette comparison at the right with the 90mm gun tank T95. At the bottom right is a three view layout of a typical TL-7 design. A later model of the TL-7 concept appears below. Compare the latter with the model at the top of the page.

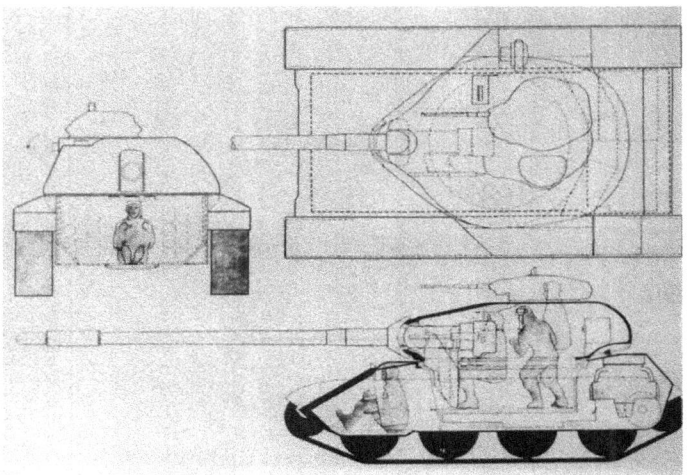

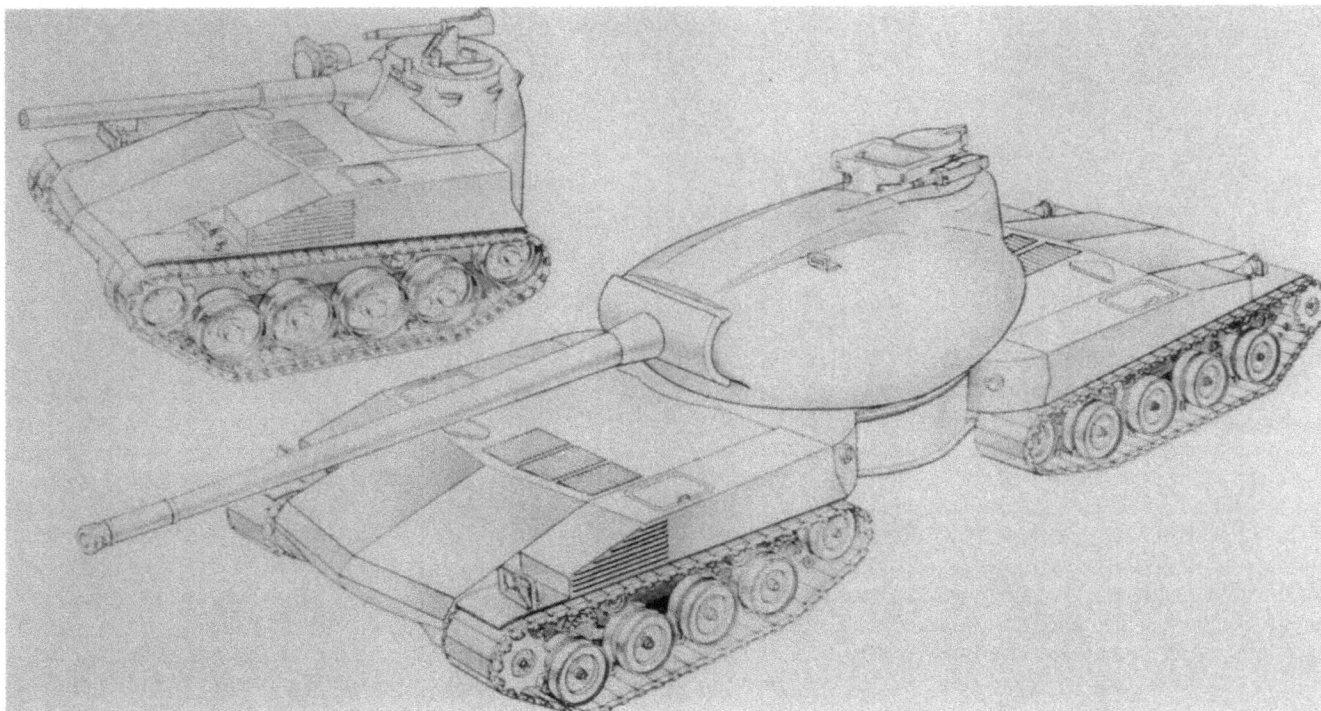

The two vehicles comprising Continental Motors Hen and Chick concept for the Astron Project are illustrated above.

A memorandum from the Army Chief of Staff, dated 10 January 1953, noted that policy required that research and development of new equipment follow two parallel paths. The first was the continuous improvement of existing materiel by a process of evolution. The second path was the long range development of new solutions to military problems utilizing entirely new types of equipment. In the latter case, it was considered that the usual system of submitting a list of military characteristics for a required item tended to inhibit the full exploitation of any new ideas that might be forthcoming. Therefore, it was directed that contracts be awarded to firms with recognized research and engineering staffs for the development of an X-weapon to perform the role of the medium tank. Such an X-weapon was to be available for production not later than 1958. This date was subsequently changed to 1961. These instructions were recorded in the Ordnance Technical Committee minutes by item 34753 dated 24 April 1953. OTCM 34753 assigned the name ASTRON to identify the project and indicated that at least two of the contracts should be with major automotive companies experienced in tank design and production. The contractors were to be responsible for studying the X-weapon as a complete unit without restrictions being placed on any components. Seventeen proposals and bids were received from industry in response to the request. These proposals were reviewed at Detroit Arsenal and at the Pentagon during May and June and Ordnance was directed to place contracts with the General

Motors Technical Center and Chrysler Corporation. However, on 7 December 1953, Chrysler withdrew from the competition because of other engineering commitments. Since Continental Motors Corporation had presented a very interesting proposal, they were awarded the second contract replacing Chrysler. In addition to the two contracts with industry, the engineering staff of the Ordnance Tank Automotive Command (OTAC) at Detroit Arsenal also participated in the project. The concepts developed by all three were presented at a meeting in the Pentagon on 17-18 May 1955.

The Continental Motors Corporation solution to the problem posed by the ASTRON project was a weapon system consisting of a pair of combat vehicles operating together. The primary vehicle, referred to as the Hen, was armed with a rigidly mounted 90mm liquid propellant gun and was powered by two separate tracked propulsion units. The small secondary vehicle, called Chick, used a single propulsion unit identical to those of the Hen and it was armed with a 105mm boosted rocket gun, also in a rigid mount.

The estimated weight of the Hen was 45 tons with the two propulsion units accounting for about 15 tons each. Thus the primary vehicle could be separated into three loads of about 15 tons each. Each propulsion unit was powered by an eight cylinder, air-cooled diesel engine proposed by Continental. Designated as the AVDS-450, it was expected to develop a maximum of 310 gross horsepower at 2800

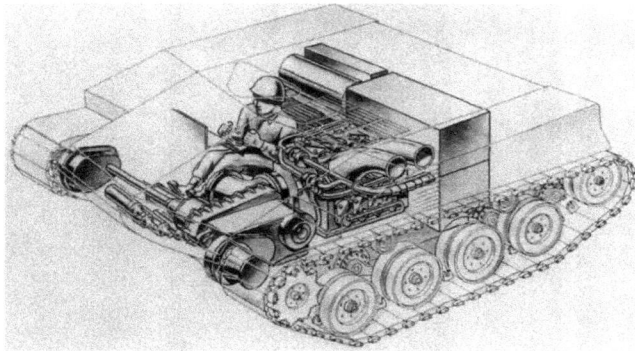

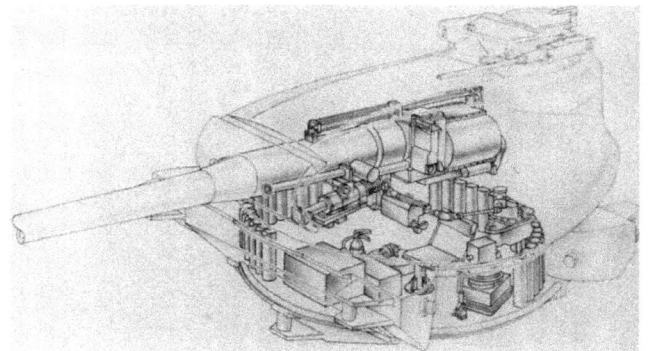

The cutaway drawings above show the interior arrangement of the Hen propulsion unit (left) and the fighting compartment (right). The dimensions of the complete vehicle are given in the sketch at the right.

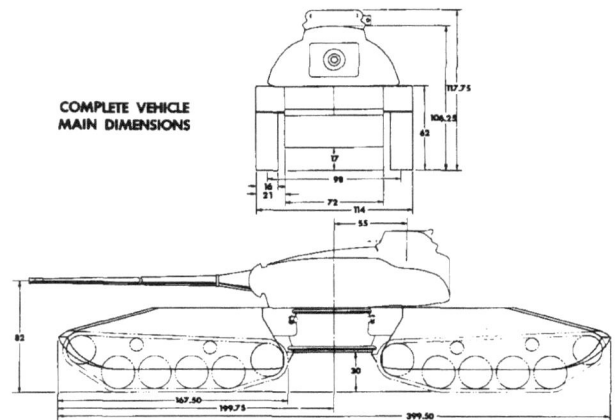

COMPLETE VEHICLE
MAIN DIMENSIONS

revolutions per minute. The engine was coupled to the final drives by a new design transmission consisting of a hydrokinetic torque converter and a planetary gear train. Wobble stick steering control was used similar to that on the M46 and M47 tanks. Each propulsion unit was fitted with a torsion bar suspension using tracks either 16 inches or 22 inches in width. The two identical propulsion units supported the fighting compartment between them. The forward propulsion unit was rigidly connected to the fighting compartment at three points by mechanically operated coupling jaws that engaged mating pins on the fighting compartment. The rear propulsion unit also was attached at three points. However, the rear connecting pins on the fighting compartment were fixed to a bracket attached to two bands encircling the hull structure. The rear bracket and bands rotated through an arc of 60 degrees permitting the vehicle to make smooth turns with a radius of 17 feet. The coupling jaws were operated from within the propulsion units so that the separation or assembly of the three element vehicle could be carried out under combat conditions. The fighting compartment was supported on three self-contained jacks when it was detached from the propulsion units. These jacks were controlled from within the compartment to permit alignment with the propulsion units. A driver could be carried in each of the two propulsion units, although only one was required to operate the complete vehicle.

The turret was mounted on the fighting compartment using a 79 inch inside diameter ring. It provided space for the tank commander, gunner, and loader, but the latter two were reversed from the usual arrangement with the gunner on the left side of the cannon and the loader on the

right. The tank commander was located under the cupola in the turret bustle. This cupola was fitted with a 30 inch coincidence type optical range finder and armed with a .50 caliber machine gun. The 90mm liquid propellant gun was stabilized in both azimuth and elevation. Ammunition proposed for this 90mm gun included a ram jet assisted armor piercing projectile as well as high explosive rounds. A .30 caliber machine gun and a gunner's telescopic sight were mounted coaxially with the cannon on the right and left sides respectively. Another .30 caliber machine gun was available for use in a ball mount in the turret side wall. Armor on the front of the propulsion units and the fighting compartment was equal to six inches at zero degrees obliquity and was reduced to a minimum of one inch at the rear.

The assembly of the Hen fighting compartment and propulsion units in the field is illustrated in the sketch at the right.

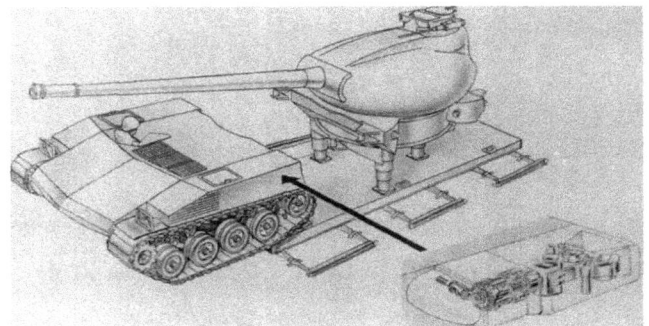

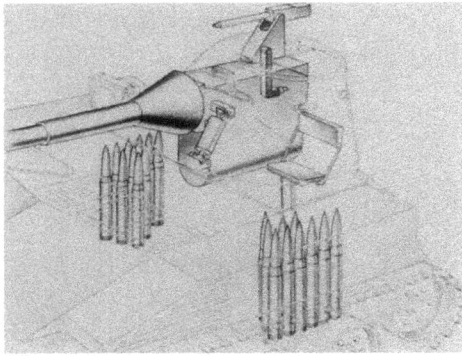

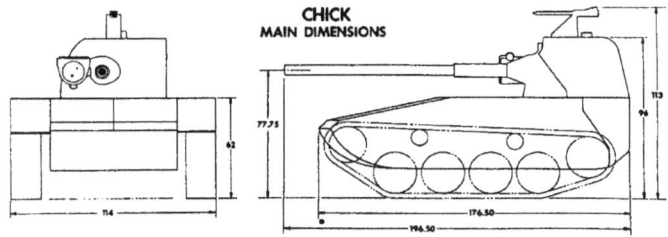

CHICK
MAIN DIMENSIONS

Above, the interior arrangement of the Chick can be seen in the cutaway drawing at the left. The dimensions of the vehicle are shown in the sketch at the right.

The secondary vehicle or Chick weighed about 21 tons and was manned by a crew of two. The driver remained in the same location in the standard propulsion unit. The commander-gunner rode in the left side of the turret fitted on the demountable fighting compartment installed at the rear. The 105mm rocket boosted gun was rigidly mounted in the turret, but it was offset to the right of the center line to provide space for the commander-gunner. Armor on the turret front equaled 3¾ inches at zero obliquity decreasing to one inch at the rear. A .30 caliber remote controlled machine gun was installed on a rotating plate in the turret roof. This weapon was operated independently of the 105mm gun and had a 360 degree traverse. An Optar pulsed light range finder was attached to the right side of the gun mount.

It also was proposed to use single propulsion units as the basis for several ancillary vehicles. These included an antiaircraft vehicle armed with twin 30mm liquid propellant machine guns and two types of missile launchers. Equipped with a tow bar and a trailer, the basic propulsion unit could be used as a tank retriever or transporter.

The concept of tactical employment proposed for the Hen and Chick envisiged the use of the former against long range, heavily armored targets. The Chick was intended to provide a high volume of fire against infantry and light armor or against heavy armor at short ranges. Except for the articulated rear propulsion unit which permitted smooth turns, the Hen was similar to the four track tank concepts examined earlier. As such, it also would have suffered from the same disadvantages. For example, it was not possible to apply full vehicle power to the rear tracks when crossing difficult terrain or if the forward tracks were damaged. Although the Hen and Chick concept provided a new approach to armored operations, its advantages were not considered sufficient to warrant further development.

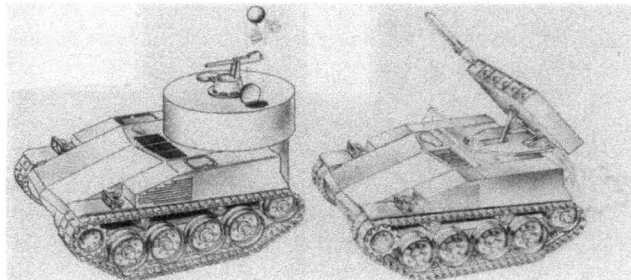

At the right, the Chick chassis carries the launcher for the D-40 (left) or a new ram jet (right) missile. Below, the Chick serves as an antiaircraft vehicle with two 30mm liquid propellant guns (left) or in the retriever role with a trailer (right).

30

These three views show the X-Weapon proposed by General Motors for the Astron Project. A four view drawing of the vehicle is at the bottom of the page.

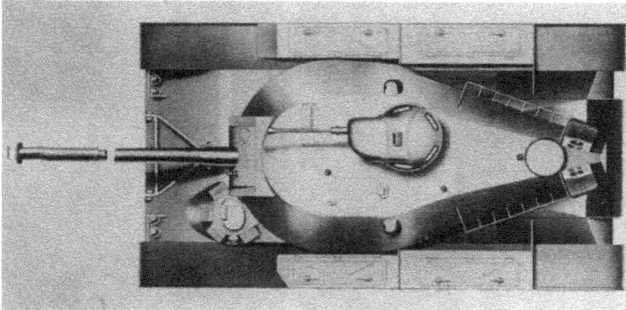

General Motors approached the ASTRON project from a different direction. Their analysis concluded that high mobility provided greater protection than any amount of armor. Thus the armor on their entry was limited to that required for protection against small arms fire and shell fragments. The resulting vehicle, using the original designation of X-Weapon, was actually a heavily armed light tank or a lightly armored, highly mobile tank destroyer. Armed with a modified 90mm T208 smooth bore gun in a recoiling mount, it was estimated to weigh almost 26 tons. The cannon was not stabilized. The welded homogeneous

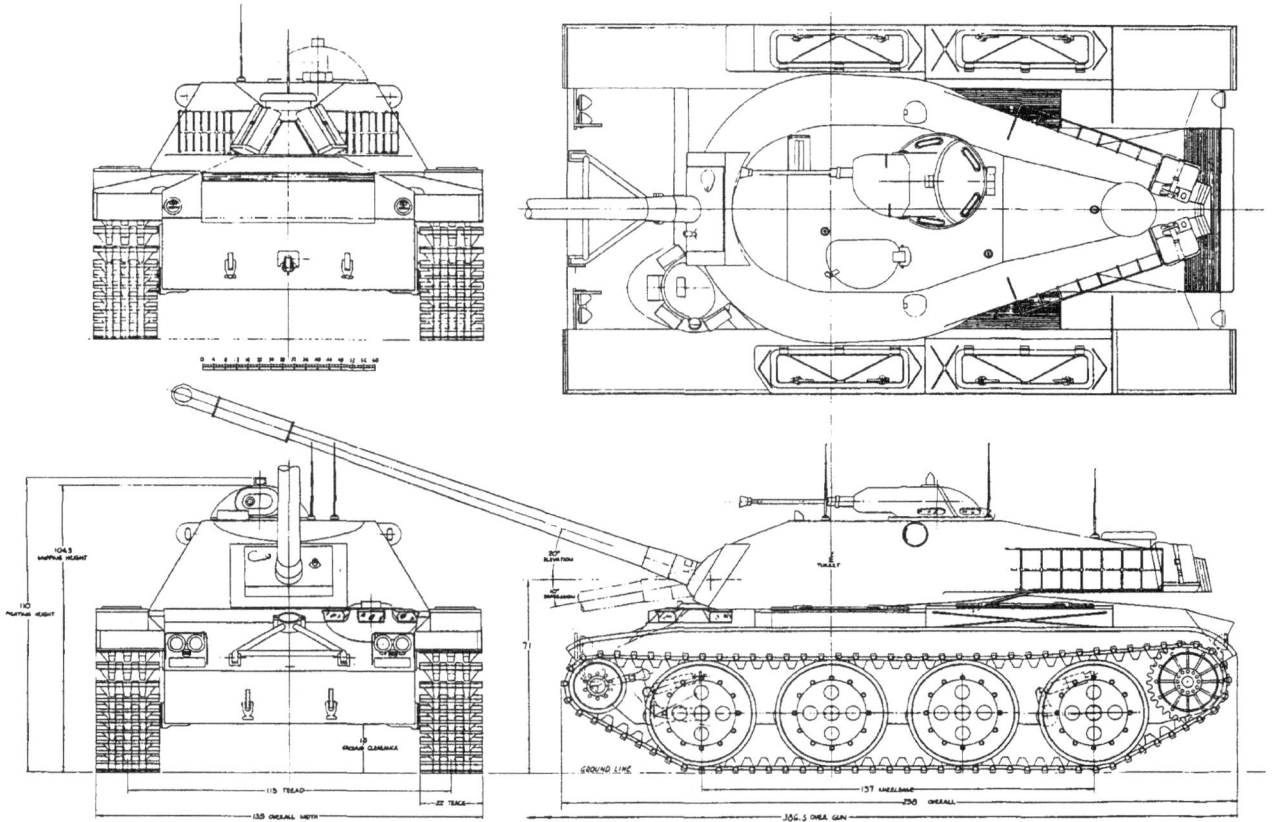

31

Details of the General Motors X-Weapon and its interior arrangement can be seen in these drawings. Note that all of the 90mm ammunition is stowed in the hull below the turret ring.

armor plate on the front and sides of the turret was only 1 1/8 inches thick at an angle of 30 degrees from the vertical. The upper front hull was ¾ inches thick at 67 degrees from the vertical.

The three man turret of the X-Weapon was mounted on an 85 inch inside diameter ring with the gunner and the tank commander in their conventional positions on the right side of the cannon. The loader was on the left. Two .50 caliber machine guns were provided with one mounted coaxially to the left of the cannon and one installed in the commander's cupola. The range finder and other fire control equipment was the same as on the M48 tank. A high speed turret was proposed with only eight seconds required for a full 360 degree rotation.

The driver was located in the left front hull of the X-Weapon with a stowage rack for 90mm ammunition on his right. The vehicle was powered by a Continental AOI-1195 engine with the XT-500GS transmission. The engine was expected to develop 590 gross horsepower at 2800 revolutions per minute when using 100 octane gasoline. The four large road wheels on each side of the flat track suspension were sprung using half-width solid torsion bars surrounded by concentric torsion tubes. Since these springs extended only to the center line of the hull, the road wheels on the opposite sides of the vehicle did not have to be offset from each other. Shock absorbers were fitted on the first and last road wheels on each side. The single pin, cast steel tracks were 22 inches wide. The maximum estimated speed for the X-Weapon on level ground was over 50 miles/hour.

The concept of employing a lightly armored, highly mobile combat vehicle with powerful armament was similar to the U.S. tank destroyer doctrine developed during World War II. Although these self-propelled antitank guns proved effective in certain situations, the lack of protection made them extremely vulnerable in others, particularly during offensive operations. It is interesting to note that the General Motors X-Weapon closely followed their previous experience in developing the M18 tank destroyer during World War II and their later production of the M41 light tank series. Both were highly mobile combat vehicles with a minimum of armor protection. During the study of the X-Weapon, consideration was given to the development of a new version of the T101 self-propelled gun with the original 90mm M54 cannon replaced by the modified 90mm T208 gun. Such a vehicle was estimated to weigh about 10½ tons and was referred to as the T101A.

At the right is a model of the T101A self-propelled gun concept armed with the 90mm T208 smooth bore cannon.

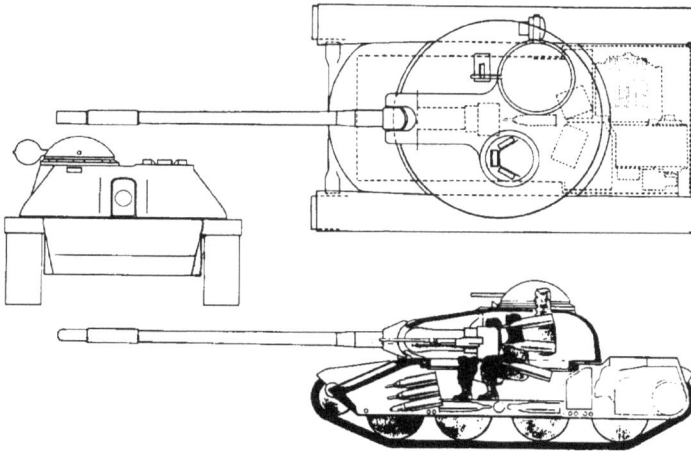

A model of the 90mm gun Rex tank is shown above. At the left are sectional drawings of the same vehicle. This was the version of the Rex originally presented on 17 May 1955. Note the very large overhang of the long barreled T208 cannon when it was in the forward position.

The Ordnance Tank Automotive Command (OTAC) presented two programs in response to the requirements of the ASTRON project. The first was based on technology that was only slightly advanced beyond what was available at that time, while the second program utilized a level of technology that could be expected by 1962, provided that high priorities were placed on the development of certain components.

The first program consisted of proposals for a new medium gun tank and two specialized combat vehicles. The latter were a ten ton airborne light tank and a one man armored reconnaisance aircraft with a tracked carrier. As presented on 17 May 1955, the medium gun tank was referred to as the Rex. Later, it was redesignated as the Rex I after the development of subsequent versions.

The Rex tank incorporated many of the features included in some of the earlier proposals. The driver was located in the turret with the gunner and the tank commander. The turret ring was only 80 inches in diameter, but this provided ample space since an automatic loading device replaced the loader reducing the crew to three men. The gunner and tank commander were in their usual positions on the right side of the rigidly mounted modified 90mm gun T208 and the tank commander was provided with a cupola armed with a .30 caliber machine gun. This was part of an integrated commander's station including an Optar range finder, a sight, and an automatic target designation system. The latter held the cupola and its sight on a target while the turret and gun were traversed and brought to bear on the same target. The correct superelevation was automatically applied to the main armament during this operation. The 90mm gun T208 was modified to use combustible case ammunition eliminating the need for a case ejection mechanism and simplifying the design of the automatic loader. This loader was supplied by a 14 round rotary magazine underneath the turret basket floor which could be reloaded when needed by the tank commander or driver. The estimated maximum firing rate with the automatic loader was 28 to 30 rounds per minute. A coaxial .50 caliber machine gun was mounted on the left side of the cannon and an articulated telescopic sight was located on the right side for the gunner. No stabilization was provided for the main armament.

The driver was seated on a platform suspended from the roof in the left front of the turret. His seat was hydraulically adjustable between a low position for use when the hatch was closed to a high position for operation with the driver's head exposed. A gear train rotated the driver's platform so that he always faced forward regardless of the turret position. The linkages for the driver's controls were actuated by electrical signals transmitted through slip rings. The turret location provided the driver with a greatly increased field of vision allowing him to maneuver the tank even in reverse without aid from another observer. He also was less vulnerable to dirt and water splash as well as mine explosions. Protection for the entire crew compartment against chemical, biological, and radiological (CBR) attack was provided by a single unit so that the wearing of individual masks was not required.

The front of the turret and hull was protected from attack in the 60 degree frontal arc by siliceous cored armor. This armor was effective against shaped charge rounds up to 100mm in diameter. Against armor piercing kinetic energy projectiles, it was equivalent to 4.8 inches of homogeneous steel armor at 60 degrees from the vertical. The combat loaded weight of the Rex was estimated to be 31 tons and it was to be powered by a transversely mounted Continental AOI-628 engine with the XTG-400 transmission. This version of the engine was expected to deliver a maximum of 350 gross horsepower. The flat track torsion bar suspension had four dual road wheels per side with 19 inch wide tracks. The driver-in-turret design resulted in an overall height of only 95 inches for the Rex. The width was 126 inches and the overall length with the gun forward was 348 inches.

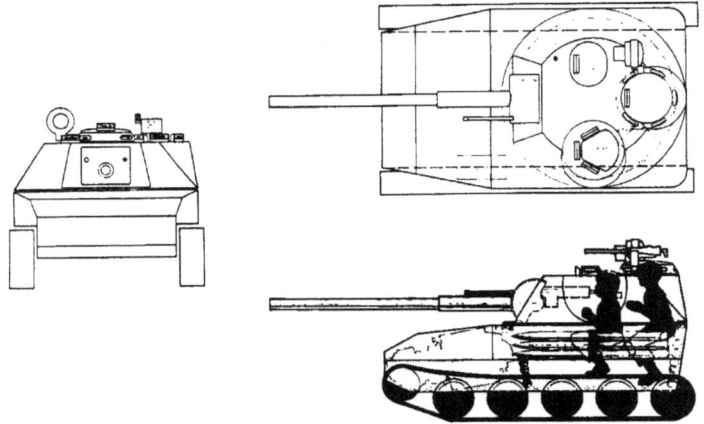

Above is a photograph of a model (left) and a three view drawing (right) of the Wasp light tank.

The ten ton light tank proposed under the first program was dubbed the Wasp and it was armed with a 4.7 inch rocket boosted cannon using combustible case ammunition. A seven round automatic loader was fitted and stabilization was provided in both azimuth and elevation. The driver was located in the turret and the entire three man crew was protected against CBR attack by a single system as on the Rex. Secondary armament consisted of a .50 caliber coaxial machine gun and a .30 caliber machine gun on the commander's cupola. Fire control equipment included the T53 Optar range finder, an electronic computer, and M16E1 periscopes for the gunner and tank commander. An articulated telescopic sight also was provided for the gunner.

Armor on the Wasp was limited to ½ inch at 60 degrees from the vertical on the upper front hull and ½ inch of vertical armor on the sides. The vehicle was powered by an AOI-470 engine with an XT-90 transmission. This power plant developed a maximum of 250 gross horsepower. The Wasp was a compact vehicle 210 inches in length with the gun in the forward position. Overall height and width were 91 inches and 95 inches respectively.

The most radical of all of the proposals under the first OTAC program was the one man armored reconnaisance aircraft. A vertical take off and landing vehicle with the flying characteristics of a helicopter, it was designed to be carried by and launched from a tracked carrier based on the Rex tank. The aircraft portion of this combination was called Falcon and the tracked vehicle was appropriately named Aerie.

Powered by an aircraft radial engine, the 113 inch diameter flying vehicle was protected by one inch thick titanium armor. Armed with ten 4.5 inch free flight rockets and a .30 caliber machine gun, the estimated weight of

Below, the model of the Falcon is depicted in flight at the left and a three view drawing is at the right.

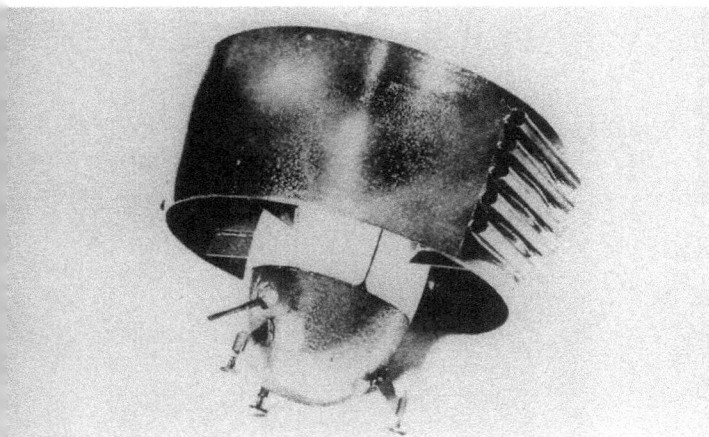

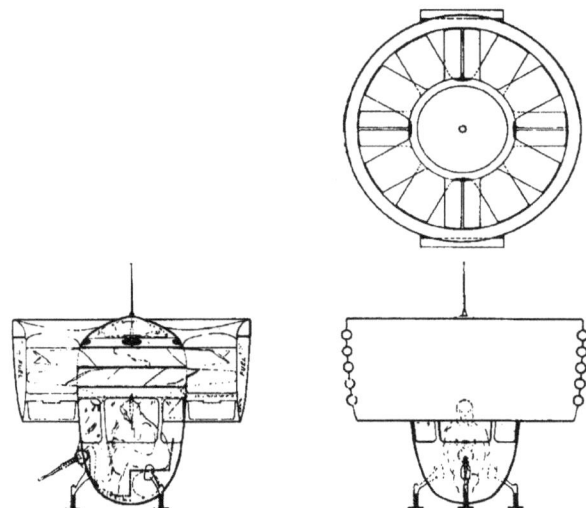

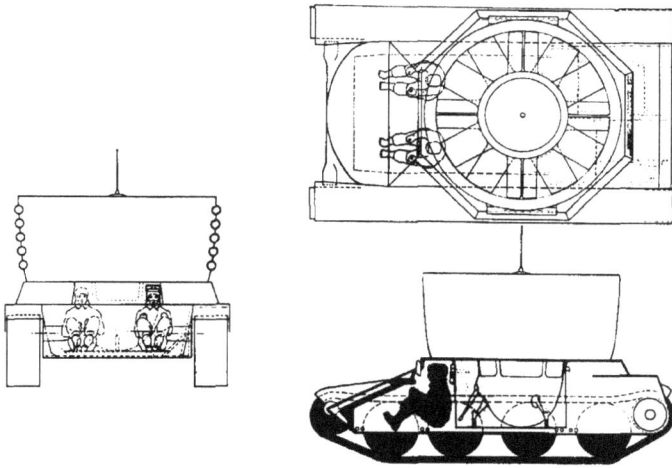

Above, the Falcon is mounted on the Aerie carrier in the drawing at the left and in the photograph of the model at the right.

the Falcon was 3000 pounds. Its speed range was calculated to vary from 0 (hover) to 100 miles per hour with an endurance of about one hour. An emergency parachute installed at the top center of the aircraft was expected to be capable of providing a safe landing from an altitude as low as 100 feet.

The Aerie carrier alone was estimated to weigh 20 tons with space in the front hull for a two man crew. It was fitted with a hydraulically retractable take off platform for the Falcon.

The second or long range program presented by OTAC under the ASTRON project proposed a siliceous armor, rocket and hypervelocity (SARAH) gun tank based on advanced technology expected to be available by 1962. The SARAH tank featured two guns rigidly mounted side by

side in the turret. The gun on the left was a hypervelocity small bore weapon expected to have a muzzle velocity of 9000 feet per second. Several such weapons were under study at that time using gas, liquid, or solid propellants. A 4.7 inch rocket boosted cannon was mounted on the right. Both guns used combustible cartridge cases. As on the Rex, the entire three man crew was located in the turret and protected from CBR attack by a single system.

The SARAH tank illustrated in the proposal was designed around a supercharged, four cylinder X engine which was expected to deliver 300 net horsepower. However, if the vehicle was selected for 1962 production, the Rex power train would be used. A lightweight band type track was proposed for the new tank. Its dimensions were similar to the Rex except for the overall length. With the gun forward, the latter was 312 inches.

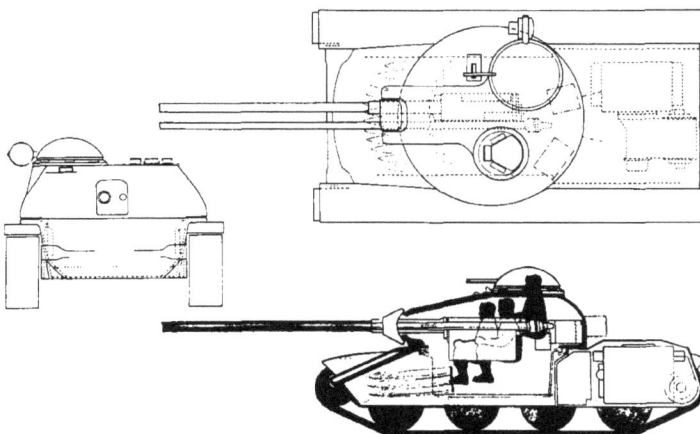

The concept model of the SARAH tank is shown below and a three view drawing is at the left.

Chrysler's TV-8 proposal is illustrated in the photographs above by the model at the left and the full size mock-up at the right.

Subsequent to the ASTRON meeting on 17-18 May, Chrysler Corporation presented a separate proposal for an unusual tank designated as the TV-8. This design located the entire crew, armament, and power plant in a pod shaped turret mounted above a lightweight chassis. The total weight was estimated to be 25 tons with about 15 tons in the turret and 10 tons in the chassis. The two were separable for shipment by air.

The TV-8 was armed with the 90mm gun T208 rigidly mounted in the turret and fitted with an hydraulic ramming device. The 90mm ammunition stowage was in the rear of the turret separated from the crew by a steel bulkhead. Secondary armament consisted of two coaxial .30 caliber machine guns and one remote controlled .50 caliber machine gun on the turret top operated by the tank commander. Closed circuit television was provided to protect the crew from the flash of tactical nuclear weapons and to increase the field of vision.

On the phase I TV-8, a Chrysler V-8 engine developing 300 gross horsepower was coupled to an electric generator in the rear of the turret. This generator supplied power to the two electric motors in the front hull. One motor drove each of the two 28 inch wide tracks. Other power plants were considered for later development including a gas turbine electric drive, a vapor cycle power plant with hydrocarbon fuels, and finally a vapor cycle power plant with nuclear fuel. The fuel tanks for the phase I vehicle were located in the hull separating them from the crew in the turret.

Space was provided in the heavily armored inner turret for a crew of four, although only two were required to operate the tank, the gunner and the driver. These two were located in the front at the right and left of the cannon respectively. The driver could operate fully protected inside the turret or with his head and shoulders exposed above the roof. The tank commander was at the right rear with the loader on his left. The heavily armored inner turret was surrounded by a light outer shell that gave the turret its podlike appearance. This shell was watertight creating sufficient displacement to allow the vehicle to float. Propulsion in the water was by means of a water jet pump installed in the bottom rear of the turret. The outer turret shell was of sufficient thickness to detonate shaped charge rounds and it acted as spaced armor to help protect the inner turret. The turret was supported by an assembly which rotated in a ring in the hull roof and it was moved in elevation by two large hydraulic cylinders. The TV-8 was 352 inches long with the gun forward, 134 inches wide, and 115 inches high over the remote controlled machine gun.

The three ASTRON proposals, as well as the TV-8 design, were reviewed and it was concluded that they did not offer sufficient advantages over the conventional medium gun tank to justify further development. This was confirmed by OTCM 36225, dated 23 April 1956, which terminated the ASTRON program. However, the OTCM indicated that consideration would be given to the novel features of the ASTRON proposals and the TV-8 in the design of future tanks.

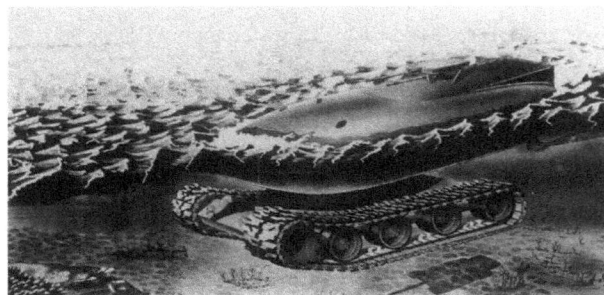

The artist's concept above at the right shows the Chrysler TV-8 swimming supported by the buoyancy of the large turret envelope. The sectional drawings below show the proposed interior arrangement of the TV-8.

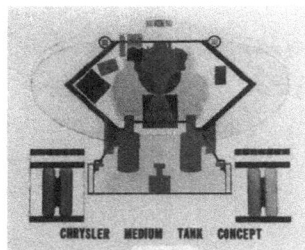

36

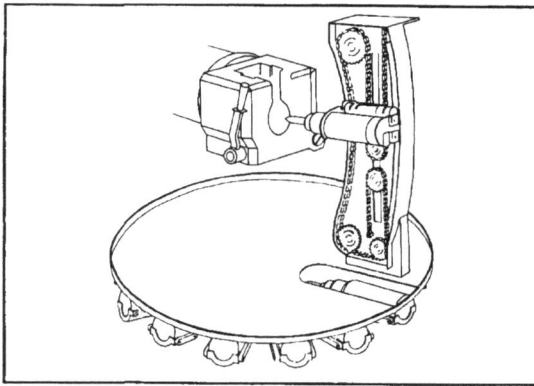

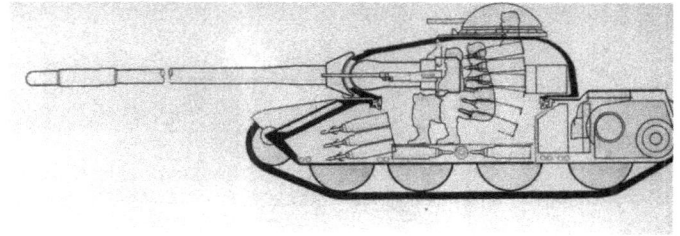

At the left is a drawing of the automatic loader proposed for installation in the Rex tank. The sectional view above shows the R-12a version of the Rex as presented during the Questionmark III conference. Below is a comparison of the front silhouettes of the Rex and the 90mm gun tank M48A1.

The Questionmark IV conference was held in Detroit during August 1955 and the OTAC ASTRON proposals were included among the concepts presented. However, the Rex appeared with several modifications. The original Rex, with the siliceous cored armor, was listed as concept R-12a. On R-12b and R-12c, the siliceous core was eliminated and the front armor was reduced to four inches and one inch respectively at 60 degrees from the vertical. The thinner armor reduced the estimated weights of the latter vehicles to 28 and 21 tons and their mobility increased accordingly. Three additional versions of the Rex were presented powered by a 400 horsepower gas turbine with a Jered transmission. Designated as the R-13, R-14, and R-15, they all retained the siliceous cored front armor. The side armor also was increased on the R-14. The R-13, R-14, and R-15 were estimated to weigh 28.5 tons, 31 tons, and 28 tons respectively. Like the original Rex, the R-13 retained the 90mm T208 smooth bore gun modified to use combustible case ammunition with a total round length of 30 inches. In R-14, the T208 cannon was redesigned to use a pivot breech. With this arrangement, the gun separated at the chamber which then rotated ¼ turn, moved backward and pivoted down to the loading position. The gun was then loaded from front to rear, just the opposite of conventional guns. This design permitted the gun to extend deep into the turret allowing it to be balanced and stabilized. The pivot breech version of the T208 gun also was intended to use combustible case ammunition.

The proposed armament for the R-15 concept was a 90mm experimental gas gun using a mixture of hydrogen, helium, and oxygen as the propellant. It was expected to defeat five inches of armor at 60 degrees obliquity at a range of 2000 yards using a 30mm penetrator. It also could fire standard 90mm chemical energy rounds at lower velocities. The estimated weight of the weapon was 1900 pounds and the installation was balanced and stabilized. For safety, the propellant tanks were located in the turret bustle isolated from the crew by a steel bulkhead.

The hull and suspension were redesigned on the R-14 and R-15 shortening the vehicle. These changes also would appear in later versions of the Rex tank.

The R-15 concept appears in the sectional drawing at the right.

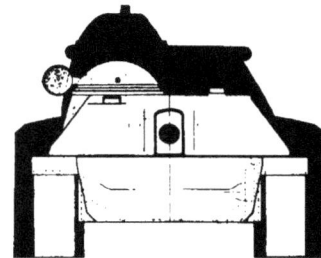

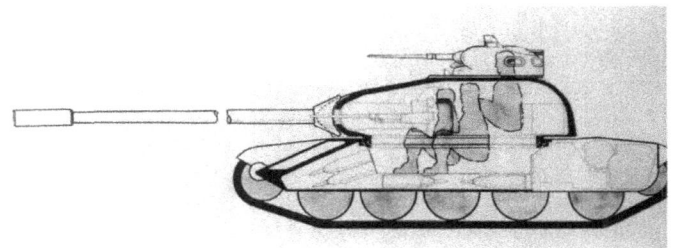

The R-13 concept as presented at Questionmark III is illustrated above.

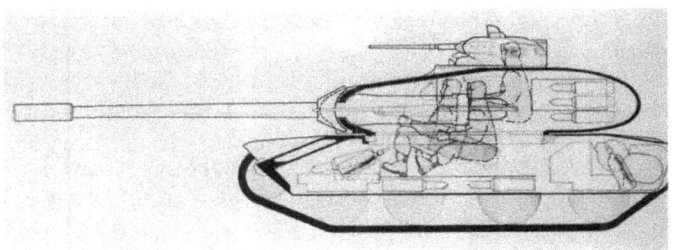

The sectional drawing above shows the R-14 concept from Questionmark III.

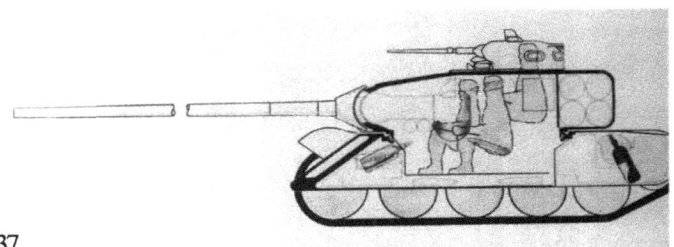

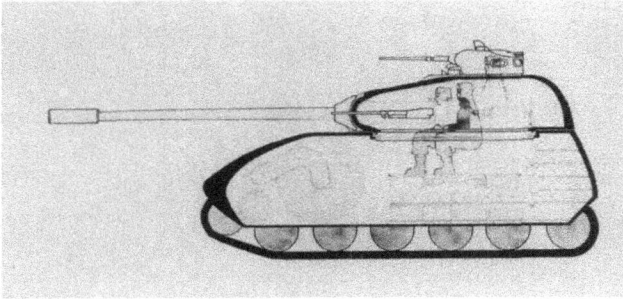

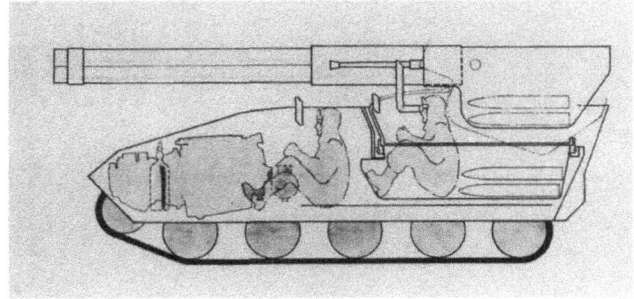

The nuclear powered R-32 concept is at the top left. At the top right is a sectional drawing of the R-34 and below are two views of the R-35.

The ASTRON Wasp proposal also appeared in Questionmark IV under the designation R-31. Concept R-32 presented a design for a nuclear powered tank. This vehicle was lighter in weight than previous OTAC nuclear powered studies and it was proposed as a possible replacement for the M48 tank series. Armed with the modified 90mm T208 gun, its estimated weight was 50 tons. Powered by a turbine using a nuclear reactor as a heat source, the estimated range of the R-32 was 4000+ miles. Armor on the front equaled 4.8 inches at 60 degrees from the vertical and the overall length of the tank with the gun forward was 220 inches. Obviously, such a tank would have been extremely expensive and the radiation hazard would have required crew changes at periodic intervals.

Concepts R-34 and R-35 were designed around two versions of the 105mm rocket boosted gun under development by the American Machine and Foundry Company. Armor protection on both vehicles was limited to ½ inch and each was estimated to weigh 10 tons. In both cases, power was supplied by a Continental AOI-470 engine with an XT-90 transmission.

A 105mm rocket boosted gun with two barrels, one above the other, was installed on the R-34 in an automatic turret. Each barrel of the weapon was fed independently from two concentric ammunition drums and they were expected to be capable of firing at a combined rate of 60 rounds per minute.

The 105mm rocket boosted gun in the R-35 concept diverted part of the combustion gas to the rear making the weapon recoilless. This gun also was fed by an automatic loader producing an estimated maximum firing rate of 300 rounds per minute. Both vehicles were only 80 inches in height and 98 inches in width. The length without the gun was 192 inches for the R-34 compared to 190 inches for the R-35.

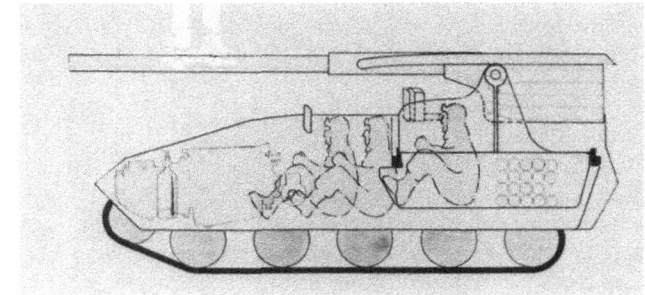

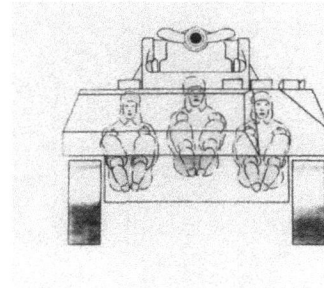

The proposed Wabnitz leveling system is illustrated in these two views. Two turret bearing rings at an angle with each other permitted compensation for sloped terrain and extended the elevation range of the gun.

Two deficiencies noted in the original Rex I concept were the lack of stabilization for the main armament and a ground pressure of 13.3 psi. This ground pressure exceeded that of both the M48 and the T95 tanks, the latter then under development. To correct both of the problems, the vehicle was redesigned to the Rex II configuration. This tank had the cannon stabilized in azimuth and elevation and the flat track torsion bar suspension was lengthened by adding one dual road wheel to each side increasing the ground contact length from 123 inches to 142 inches. This reduced the ground pressure to 11.5 psi with the 19 inch wide tracks.

An analysis of the new tank was carried out by the Engineering Staff of the Ford Motor Company and their report concluded that the Rex II was a great advance in tank design. Other changes in the Rex II compared to the Rex I included the use of the AOI-628-2 engine with the XTG-350 transmission. The new power plant developed a maximum of 400 gross horsepower by means of a higher engine speed and compression ratio. The power package still retained the transverse mounting. The front turret armor remained the same as on the Rex I, but the front hull was modified to the equivalent of four inches of solid homogeneous steel armor at 65 degrees from the vertical. Actually, it consisted of one inch of steel, four inches of siliceous core, and two inches of steel, all at 65 degrees from the vertical. The Optar range finder was relocated from the commander's cupola to the turret right side wall. Although the hull was lengthened to 227½ inches and the width increased to 128 inches, redesign of the armor maintained the estimated combat weight of the Rex II at 31 tons. The estimated cruising range on roads increased from 191 miles to 210 miles with the new power package.

The Ford report recommended additional modifications to the Rex II. These included the replacement of the torsion bar springs with a hydropneumatic suspension system. This would have reduced the height of the tank by four to five inches by eliminating the space required for the torsion bars in the bottom of the hull. It would, of course, also reduced the weight. A gas turbine power plant and a pivot breech gun also were proposed for future use. The report concluded that the Rex II was superior to any tank previously developed and recommended its production as is or in modified form at the earliest possible date.

A late concept of the Rex I (top left) can be compared with a similar view of the Rex II (above) and a sectional drawing, also of the Rex II (below). Top and front view drawings of the Rex II are at the bottom.

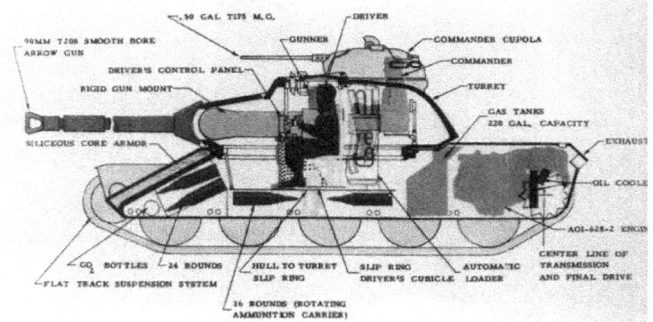

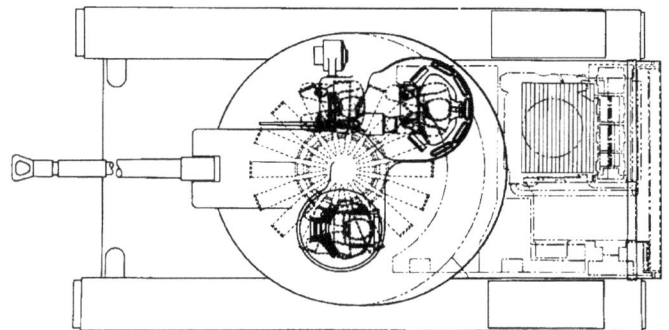

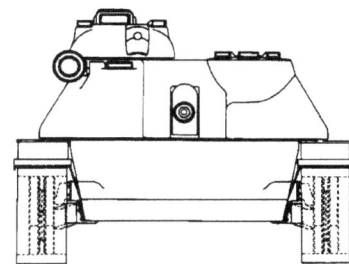

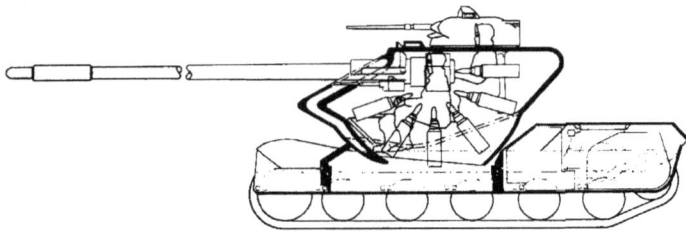

The artist's concepts and sectional drawings are shown here for the Rex III (above) and the Rex IV (below) proposals. Note the extremely large diameter turret on the latter and the leading idler design of the suspension on both tanks.

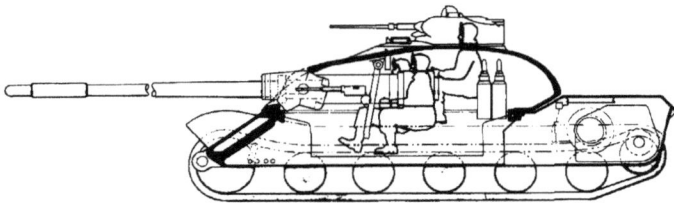

Studies continued at OTAC of various modifications to the Rex tank. One of these evaluated the possibility of developing a tank in the 30 ton weight class with the T208 cannon in an oscillating turret. This concept was designated as the Rex III. However, it was concluded that the armor overlap of the heavy turret castings, as well as the heavy trunnions required for the oscillating turret, would raise the estimated weight to 35½ tons. To minimize the increase in ground pressure, the ground contact length was extended by redesigning the flat track torsion bar suspension. The new design replaced the usual compensating idler with a large front road wheel which also served the same function as the idler in maintaining track tension. The same 19 inch wide tracks were retained as on the earlier Rex tanks. The armor protection on the turret was equivalent to the Rex I and that on the hull equaled the Rex II.

The results of earlier studies of 104 inch diameter turret rings were incorporated into the Rex program using components of the Rex II. With ballistic protection equal to that on the Rex II, the estimated combat weight was 33½ tons. Designated as the Rex IV, this tank had a suspension similar to that on the Rex III. Like the other Rex tanks, the Rex IV was armed with the 90mm T208 gun modified for use with combustible case ammunition.

The Rex V design appears below. Note the six road wheels in the artist's concept and the five road wheels in the sectional drawing.

A characteristic feature of all Rex tanks was the mounting of a very long cannon on a small chassis. The resulting large overhang of the gun barrel would have made the vehicle difficult to maneuver in rough terrain or confined areas. An effort to alleviate this problem resulted in the Rex V. This concept was based on keeping the gun muzzle in approximately the same location relative to the ground as that of the weapon on the M48 tank. The Rex V was armed with the pivot breech version of the 90mm gun T208 mounted in a cleft turret. With this design, the gun, automatic loader, and tank commander occupied a small central portion of the turret which elevated as a unit. Thus not only the gun, but the tank commander and his cupola were stabilized. The gunner and the driver rode in the conventional part of the turret on the right and left sides of the cannon respectively. Each crew member was provided with a hatch. This design moved the cannon to the rear deeper into the turret reducing the barrel overhang. This was possible because of the pivot breech loading system. The front hull also was sloped upward from the bottom reducing the tendency to dig in when crossing rough terrain. The estimated weight of the Rex V was 28 tons and the length of the hull was reduced to 204 inches. All of the experimental concepts were constantly being changed as the designs evolved and occasionally this resulted in some confusion. Note that the artist's concept of the Rex V was equipped with six road wheels while the side sectional drawing shows five. Apparently the latter number was correct.

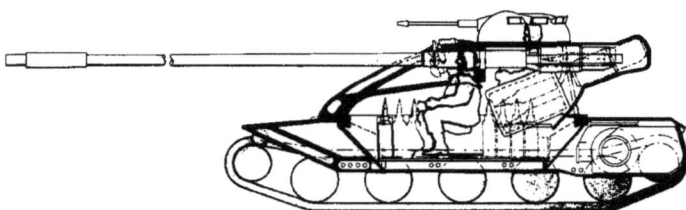

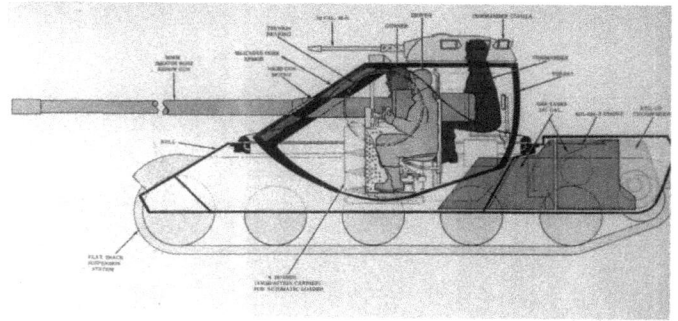

Above are the artist's concept and the sectional drawing of the Rex VI. Top and front view drawings are shown below. Note the extremely compact design of this tank.

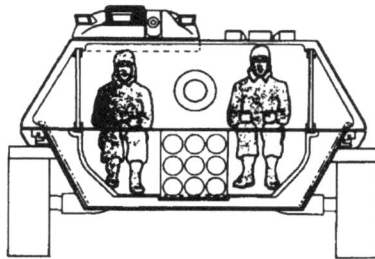

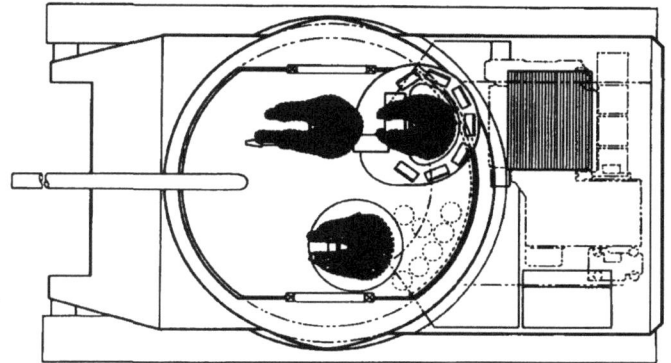

During this period, an important development objective at OTAC was to provide combat vehicles which could be transported by air. Although the Rex tanks were much lighter than the M48 series then in service, they still did not satisfy this objective with the aircraft then available. To achieve an even greater weight reduction, a new design was prepared and designated as the Rex VI. By utilizing a new concept of ballistic protection and armor distribution, the estimated weight of the new vehicle was reduced to 25 tons. This new approach divided the tank into three basic elements. These were the turret fighting compartment, the engine compartment, and the hull support structure. The first consisted of a heavily armored oscillating or ball turret containing all of the items requiring the greatest protection, such as the main armament, ammunition, fire and gun control equipment, and the crew. The

engine compartment in the rear hull also was provided with sufficient armor to protect the engine, transmission, and fuel tanks. The forward hull was a lightly armored structure supporting the turret and suspension system. Components and stowage contained in this area were not critical to the survivability of the vehicle and the tank could still fight even if they were damaged by enemy action. With this arrangement, the crew could have the heavy armor on the turret front facing the threat regardless of the relationship of the turret to the hull. The armor on the turret bottom combined with that of the hull floor greatly increased the mine protection for the crew. The spaced armor effect of the combined hull and turret armor provided protection equal to the Rex I on the front and rear and equivalent to the Rex II on the sides. The turret armor itself was equal to the Rex I.

Below, the fording capability of the Rex VI is illustrated at the left and its silhouette is compared with that of the 90mm gun tanks T95 and M48A1 at the right.

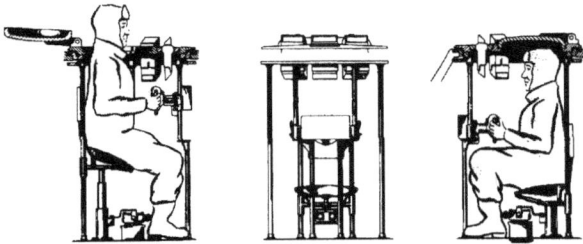

The Rex VI was armed with the pivot breech version of the 90mm gun T208 modified for the use of combustible case ammunition. As on all of the Rex tanks, the cannon was rigidly mounted in the turret without a recoil mechanism. This rigid mount in the oscillating turret eliminated the relative movement between the gun breech and the automatic loader greatly simplifying its design. It was estimated that the number of moving parts would be reduced by 50 per cent compared to earlier automatic loaders thus reducing maintenance requirements. The entire turret was stabilized in both azimuth and elevation reducing the exposure of the crew to shocks when moving across rough terrain and improving their ability to fire while moving. The gunner and the tank commander were seated on the right side of the cannon and the tank commander's cupola was armed with a .50 caliber machine gun. The driver's station was located on the left side of the turret and, as on the other Rex tanks, it counterrotated to keep the driver facing forward at all times.

As on the Rex II through V, the power package on the Rex VI consisted of the transversely mounted AOI-628-2 engine with the XTG-350 transmission. As mentioned earlier, the engine was expected to develop a maximum of 350 net horsepower or 400 gross horsepower. With a usable fuel capacity of 185 gallons, the maximum cruising range on roads was estimated to be about 230 miles.

To minimize the ground pressure with the 14 inch wide tracks, the Rex VI used the suspension with the large front road wheel-idler combination as on the Rex III and IV. This design resulted in a ground contact length of 151 inches and a ground pressure of 11.8 psi. Despite its lighter weight,

the overall dimensions of the Rex VI chassis were similar to those of the Rex II with a hull length of 109½ inches and an overall width of 128 inches.

The effort to design a 25 ton medium tank continued with the presentation of a proposal, dated 21 December 1956, by the Cadillac Division of General Motors Corporation. This proposal described two design concepts, one armed with the 90mm gun T208 and the other fitted with a guided missile launcher. The hull and chassis were identical for both vehicles and the flat track torsion bar suspension was fitted with four dual road wheels per side. As described in the proposal, both versions were powered by the AOI-628 engine with the XTG-350 transmission. However, it was recommended that the AOI-628 be replaced by the General Motors free piston turbine power plant. The latter was expected to deliver 350 net horsepower with afterburning and it had the ability to operate on fuels ranging from gasoline to crude oil. With its lighter weight and low fuel consumption, the free piston engine was expected to reduce the total vehicle weight by about 400 pounds, half in the engine weight and half in the fuel supply.

An oscillating turret on the gun armed tank was fitted with the pivot breech version of the 90mm gun T208 in a rigid mount with a coaxial .30 caliber machine gun. The provision of an automatic loader reduced the crew to three men. The gunner was located in the lower right front of the turret with the tank commander and the driver seated side by side in the upper rear. The tank commander's cupola was armed with a .50 caliber machine gun and the driver's station counterrotated so that he faced forward at all times. Dual controls were provided for the tank commander so that he could take over the duties of the gunner or the driver if required.

Concept drawings for the General Motors 25 ton tank proposal are shown below armed with the 90mm gun T208 (left) and a guided missile launcher (right).

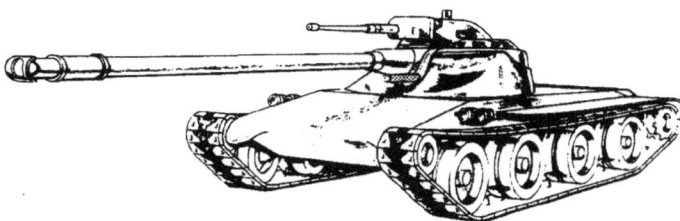

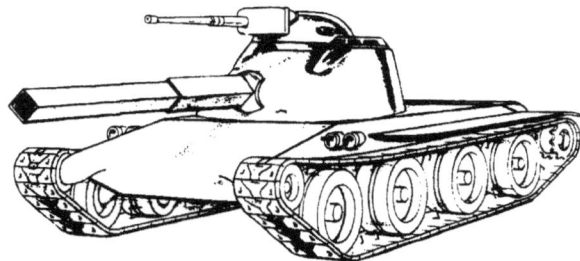

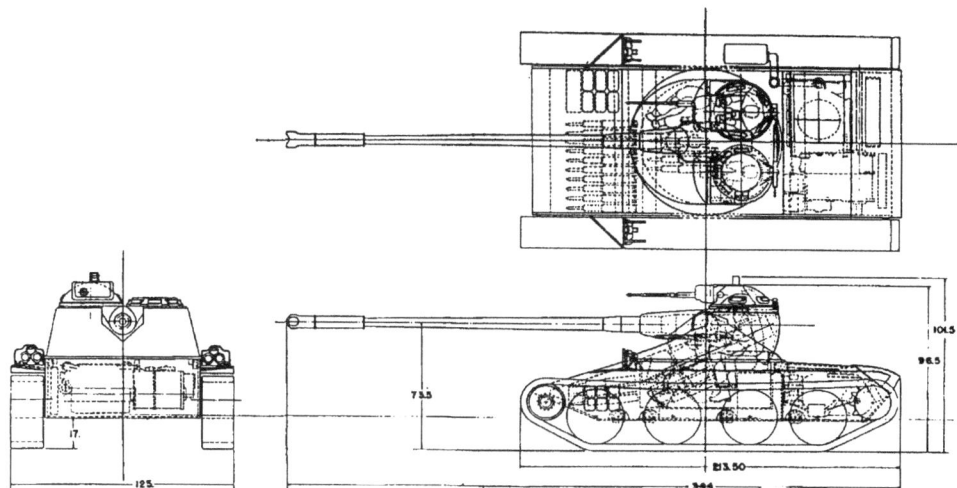

Three view drawings for the proposed General Motors 25 ton tank are shown here armed with the 90mm gun T208 (above) and a guided missile launcher (right).

A conventional turret on the second version of the tank was fitted with a lightweight launching tube for a radio controlled guided missile. The crew locations in this turret were the same as in the gun armed tank. The missile proposed by the Bell Aircraft Corporation featured a solid rocket for propulsion and a line of sight guidance system. With this arrangement, the gunner had only to keep the sight on the target. An infrared sensor measured the deviation of the missile from the line of sight and correction commands were automatically transmitted to the missile by a radio.

The front hull of both vehicles was protected by 3½ inches of homogeneous armor at 70 degrees from the vertical. This was equivalent to 10½ inches of armor at zero obliquity and the same level of protection was specified for the front of both turrets. In the proposal, this was reduced to 1½ inches of vertical armor on the sides and rear. However, analysis of the design indicated that it would not be possible to hold the vehicle weight to 25 tons with this level of protection, particularly on the gun armed tank.

In January 1957, the Assistant Secretary of the Army for Research and Development received a request from the Joint Coordinating Committee on Ordnance to establish a panel to review the status of the tank development program. At this time, questions were being raised regarding the progress of the Army's main tank development program, that for the 90mm gun tank T95. The M48 series, then in front line service, was rapidly falling behind the state of the art and apparently, the T95 would not be available for several years. Also, there was considerable controversy regarding the type of armament required for future tanks. A congressional investigation of the entire defense program during this same period further emphasized the need to reassess the tank development effort. To review the situation, the Army Chief of Staff, in February, established the Ad Hoc Group on Armament for Future Tanks. Dubbed ARCOVE, the objective of this group was to determine the appropriate weapon systems for tank use after 1965. The effect of atomic weapons also was to be considered. ARCOVE submitted its recommendations in May

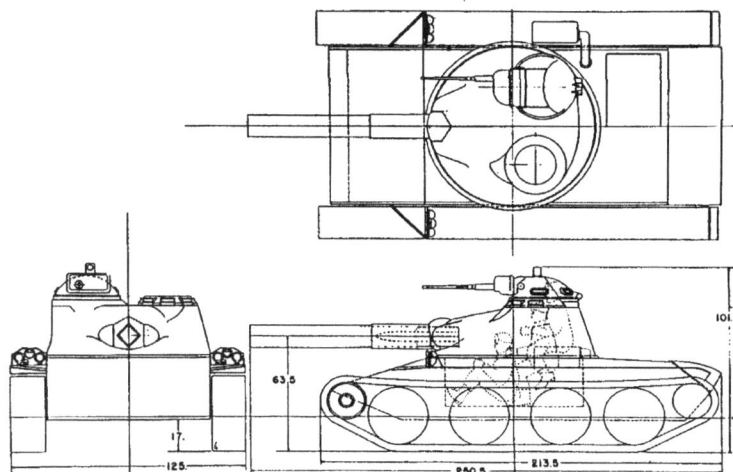

1957 although the official report did not appear until January 1958. ARCOVE concluded that the most effective weapon system for future tanks would be a missile using line of sight guidance. They recommended that such a system be developed for use in tanks after 1965. If necessary, it was suggested that the development of high velocity guns and ammunition be curtailed to provide the required funds. However, it was expected that research on shaped charge warheads would continue. No doubt this was essential, since if subsonic missile systems completely replaced the kinetic energy weapons, the shaped charge or other chemical energy warheads would be the only means available to defeat enemy armor.

The wide diversity in the research and development program was reflected in the variety of concept proposals presented at the Questionmark V conference in March 1958. These included several types of tanks, some amphibious, armed with either guns or missiles, and a tank chassis with interchangeable turrets. It also featured a concept of using radar to detect incoming armor piercing rounds and the use of shaped charges to destroy them. A radiological tank was proposed for use in high radiation environments. Eventually, this wide range of ideas would be organized into a program to select the most promising directions for future tank development. However, these studies will be reviewed in a later section after consideration of the T95 development program.

Above are the concept models of the 105mm gun tank T96 (left) and the 90mm gun tank T95 (right).

THE T95 TANK PROGRAM

As mentioned previously, the TL-1 concept presented at the Questionmark III conference was recommended for development as the future medium gun tank. On 23 September 1954, the Army Chief of Staff approved development programs for the TL-1 as well as the TL-4 and TS-31. The latter two were long and short range concepts of a future heavy gun tank proposed as replacements for the 120mm gun tank T43 series.

The TL-1 and TL-4 were to be armed with new smooth bore 90mm and 105mm guns respectively. These weapons resulted from experimental work indicating greatly improved penetration performance using fin stabilized arrow type projectiles. The length to diameter ratio of these long rod penetrators was too great for satisfactory spin stabilization resulting in the fin stabilized smooth bore design. On 12 August 1954, OTCM 35432 assigned designations for the 90mm guns T208 and T209 and the 105mm gun T210. These were essentially smooth bore versions of

the 90mm guns T54 and M41 and the 105mm gun T140 respectively. The T209 was intended only for early low velocity tests in the ammunition development program.

The greatly increased performance of the smooth bore cannon was expected to result in lighter weight medium and heavy gun tanks. However, if the development of the smooth bore weapons was unsuccessful, the TL-4 could be fitted with the rifled 105mm gun T140 to serve as the new medium gun tank and the TS-31, with the conventional 120mm gun T123E1, would be available as a lighter weight replacement for the heavy tank T43 series. To further define the technical characteristics of all three vehicles, a conference was held at Detroit Arsenal on 1 October 1954. This resulted in several modifications to the original concepts.

The TL-1 as presented at the Questionmark III conference was armed with the smooth bore 90mm gun T208 modified by shortening the barrel by three feet. The revised

The 105mm gun tank T96 is shown in the drawing below and in the sectional view and the artist's concept at the right.

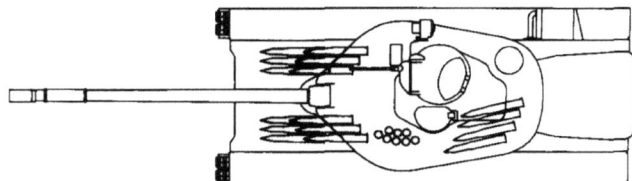

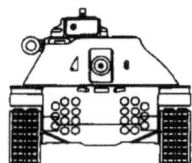

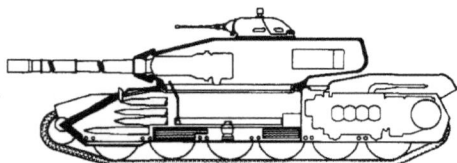

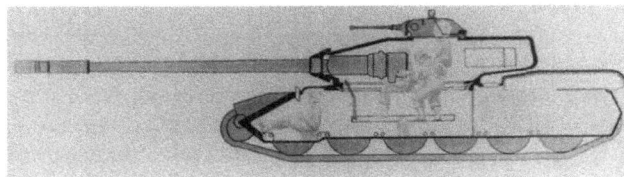

44

The layout of the 90mm gun tank T95 is sketched at the right.

characteristics specified the full size T208 cannon increasing the overall length of the tank with the gun forward from 354 to 378 inches. This increase was minimized by redesigning the hull and reducing its length without the gun from 265 to 246 inches. At the same time, the overall height of the vehicle was reduced from 112 to 109½ inches, but the 78 inch inside diameter turret ring was retained from the original proposal. This was then considered to be the optimum size ring for use with the 90mm gun. It was expected that a considerable weight reduction could be realized by designing the new tank around a smaller turret ring compared to the 85 inch ring on the M48 series. OTCM 35667, dated 27 January 1955, assigned the designations 90mm gun tank T95, 105mm gun tank T96, and 120mm gun tank T110 to the TL-1, TL-4, and TS-31 projects respectively. Two pilots were authorized for each tank at this time subject to the completion and approval of wooden mock-ups. When OTCM 35667 was published, it still indicated that a minimum turret ring diameter would be used on the T95 as a means of reducing the size and weight of the vehicle. However, this was soon increased to 85 inches identical to that on the T96 and the M48 series tanks.

The original proposals for the TL-1 and TL-4 were powered by the 12 cylinder version of the General Motors X type diesel engine which then was estimated to develop 750 gross horsepower. An XT-500 transmission was to be utilized with this power plant. However, since the X type engine was in an early stage of development, it was proposed that the TL-1 and TL-4 initially be powered by the Continental AOI-1195 and AOI-1490 engines respectively. Later, in order to use as many common components as possible in the two tanks, the AOI-1490 was dropped and both vehicles were designed around the AOI-1195 using a modified version of the XT-270 transmission. It was expected that this power train could be replaced by the X type diesel when it became available with only a minimum of modification. The H type engine also was considered as a possible future power plant. An infrared suppression rear deck

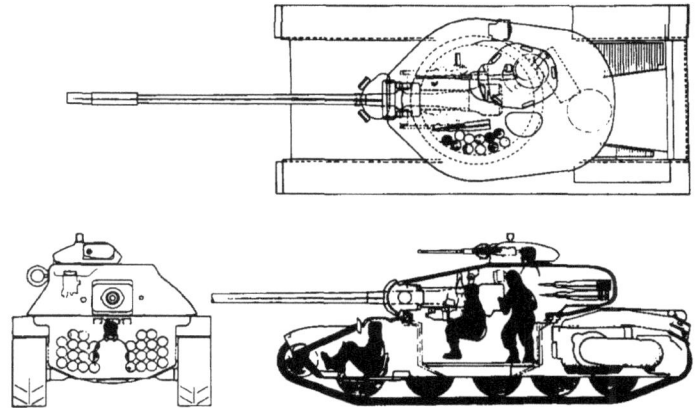

was incorporated in the design regardless of which engine was selected.

The modified transmission used with the transversely mounted AOI-1195 engine was designated as the XTG-410 and it utilized clutch-brake and geared steering. Originally, the transmission for the T95 was intended to have six forward speeds, but tests at the General Motors Proving Ground with a six speed XT-500 revealed difficulty in steering. As a result, the XTG-410 as installed in the T95 only had four forward speeds.

Development of the T95 was carried out at Detroit Arsenal, but the T96 program was contracted out to the Ford Motor Company. However, the two projects were closely coordinated to utilize as many common components as possible. The number of pilot vehicles also was increased to four for each tank. The T95 and T96 were similar in appearance with the main armament mounted rigidly in the turret without a recoil system. The 90mm gun T208 was balanced in the T95 turret and it was fitted with a stabilizer. However, the 105mm gun T210 could not be balanced in the T96 turret and it was not stabilized. Both tanks used the same components in the low silhouette flat track suspension, but the shorter T95 only had five road wheels per side compared to six on the T96.

Early in the design study at Ford, it was noted that the hull armor thickness on the T96 would have to be reduced below that on the T95 in order to maintain the

Below are the engine compartment arrangements for the 90mm gun tank T95 when powered by the AOI-1195 engine (left) and the proposed X engine (right).

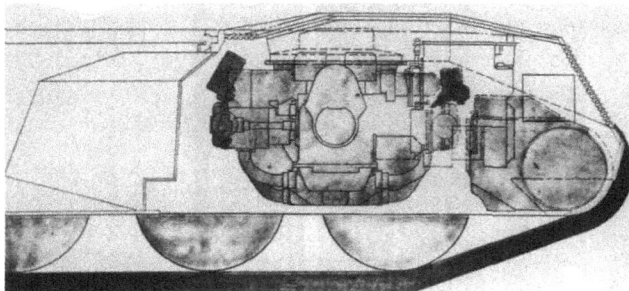

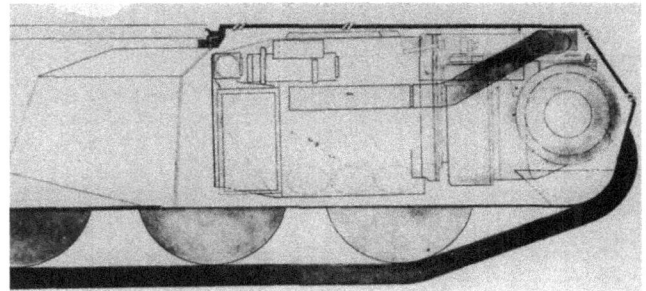

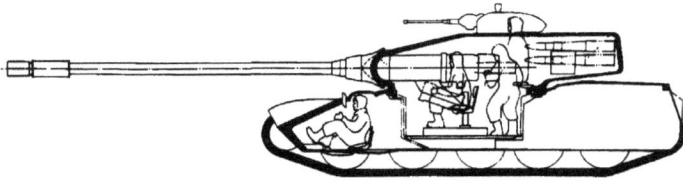

The 120mm gun T123E1 (Study A above) and the lightweight version of the same weapon (Study B below) are shown mounted in the T96 turret on the T95 chassis.

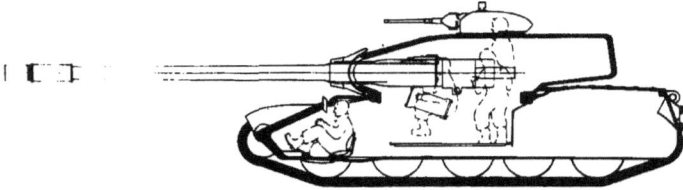

45 ton weight limit. For example, the upper front hull was equal to homogeneous steel armor 3.8 inches thick at an angle of 60 degrees from the vertical compared to the equivalent of 4.4 inches at 60 degrees on the T95. Since this was not considered desirable in a heavy gun tank, Ford proposed two variations of the basic vehicle designated as the T96-1 and the T96-2 with the upper front armor increased to 4.1 and 4.8 inches respectively, both at 60 degrees from the vertical. The estimated combat weight was about 47.8 tons for the T96-1 and 49.3 tons for the T96-2. However, the T96 as presented in the final report by Ford on phase one of the project had an upper front hull 3.2 inches thick at an angle of 65 degrees from the vertical providing essentially the same protection as 3.8 inches of armor at 60 degrees. The combat weight was then estimated as about 46 tons.

By this time, further analysis of the power train had indicated that the T95 chassis with its 85 inch diameter

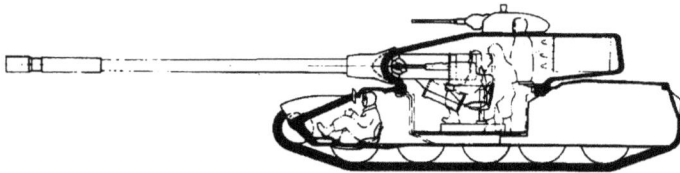

Above, the 105mm gun T140E3 is installed in a T96 turret on the T95 hull for Study C. Below in Study D, the 105mm gun turret from the T54E2 tank is mounted on the T95 chassis at the left and a model of this combination is at the right.

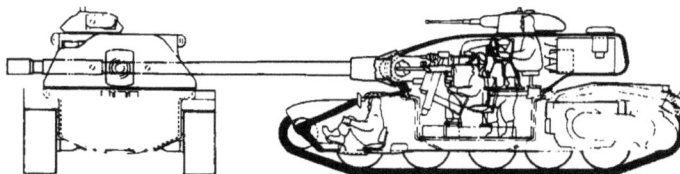

turret ring could be fitted with the T96 turret and still maintain adequate performance for a heavy gun tank. This was the course of action selected and the program was reoriented cancelling the T96 chassis and authorizing procurement of four additional T95 chassis to mount the T96 turrets.

The reorientation of the development program was outlined in OTCM 36383 dated 29 November 1956. The project now included nine T95 chassis. They consisted of four vehicles mounting the basic T95 turret with the 90mm gun T208 and four armed with the 105mm gun T210 in the T96 turret. The weapons in all eight of these turrets were to be rigidly installed without a recoil system in the combination gun mounts T191(T95) and T193(T96). An additional T95 chassis (serial number 5) was fitted with a T95 turret armed with a modified 90mm gun T208 in the combination gun mount T192 incorporating a recoil system. This vehicle was designated as the 90mm gun tank T95E1 and it was equipped with a simplified fire control system eliminating the range finder. Because of early progress in the development of the smooth bore guns and the successful completion of the 120mm gun tank T43E2 modification program, OTCM 36383 also terminated the development of the 120mm gun tank T110.

During the latter part of 1956 and early 1957, a series of studies evaluated a wide variety of armament and turret combinations for the T95 tank series. Labeled A through P, these studies considered just about every type of weapon and turret configuration readily available at that time. Study A mounted the 120mm gun T123E1 with a recoil system in the T96 turret on the T95 chassis. This was the same weapon used in the heavy tank T43 series and it resulted in an estimated total vehicle weight of over 45 tons. Study B utilized a lightweight version of the 120mm gun with a recoil mount in the T96 turret. This weapon was later designated as the 120mm gun T123E6. The estimated total vehicle weight for this combination was between 43 and 44 tons. Unlike Study A, the analysis indicated that the lighter weapon could be balanced and stabilized in the turret.

The concept in Study C was armed with the 105mm gun T140E3 in a T96 turret with a recoil system. This arrangement resulted in an estimated total vehicle weight of a little over 44 tons. A balanced gun and stabilization were not considered feasible with this combination. Study D installed the turret from the T54E2 tank with its 105mm gun T140E3 on the T95 chassis. The estimated total vehicle weight was slightly under 45 tons, but the armor protection of the turret was not equal to that of the T95 tank.

5.0 DIA — PENETRATION 2.9" @ 60° AT 2000 YARDS
MUZZLE VELOCITY-3000 FT/SEC.
MUZZLE ENERGY-3,368,000 FT LBS

5.8 DIA — PENETRATION 4.7" @ 60° AT 2000 YARDS
MUZZLE VELOCITY-4,800 FT/SEC.
MUZZLE ENERGY-4,427,800 FT LBS

5.8 DIA — PENETRATION 4.7" @ 60° AT 2000 YARDS
MUZZLE VELOCITY-4,800 FT/SEC.
MUZZLE ENERGY-4,427,800 FT LBS

7.0 DIA — PENETRATION 5" @ 60° AT 2000 YARDS
MUZZLE VELOCITY-4,730 FT/SEC.
MUZZLE ENERGY-6,683,900 FT/LBS

6.5 DIA — PENETRATION 5" @ 60° AT 2000 YARDS
MUZZLE VELOCITY-5,200 FT/SEC.
MUZZLE ENERGY-4,198,700 FT/LBS

3.0 DIA — PENETRATION 5.3" @ 60° AT 2000 YARDS
MUZZLE VELOCITY-4,800 FT/SEC.
MUZZLE ENERGY-6,883,400 FT/LBS

6.2 DIA — PENETRATION 4.8" @ 60° AT 2000 YARDS
MUZZLE VELOCITY-3,500 FT/SEC
MUZZLE ENERGY-9,511,000 FT LBS

7.5 DIA — PENETRATION 4.8" @ 60° AT 2000 YARDS
MUZZLE VELOCITY-3500 FT/SEC.
MUZZLE ENERGY-9,511,000 FT LBS.

6.3 DIA — PENETRATION 5" @ 60° AT 2000 YARDS
MUZZLE VELOCITY-5700 FT/SEC.
MUZZLE ENERGY-6,305,960 FT LBS.

5.2 DIA — PENETRATION 3.5" @ 60° AT 2000 YARDS
MUZZLE VELOCITY-3500 FT/SEC.
MUZZLE ENERGY-6,657,000 FT LBS.

7.5 DIA — PENETRATION 4.8" @ 60° AT 2000 YARDS
MUZZLE VELOCITY-3500 FT/SEC.
MUZZLE ENERGY-9,511,000 FT/LBS

PENETRATION	HEAT	AP
@2000 yards	6"/55°	5.5"/55°
MUZZLE VELOCITY-1200 ft/sec		1300 ft/sec
BURN-OUT VELOCITY-3450 ft./sec		4840 ft/sec

PENETRATION 6"/64° 500-6000 yards
VELOCITY 350 ft/sec
OPERATING TIME FLIGHT, 52 sec

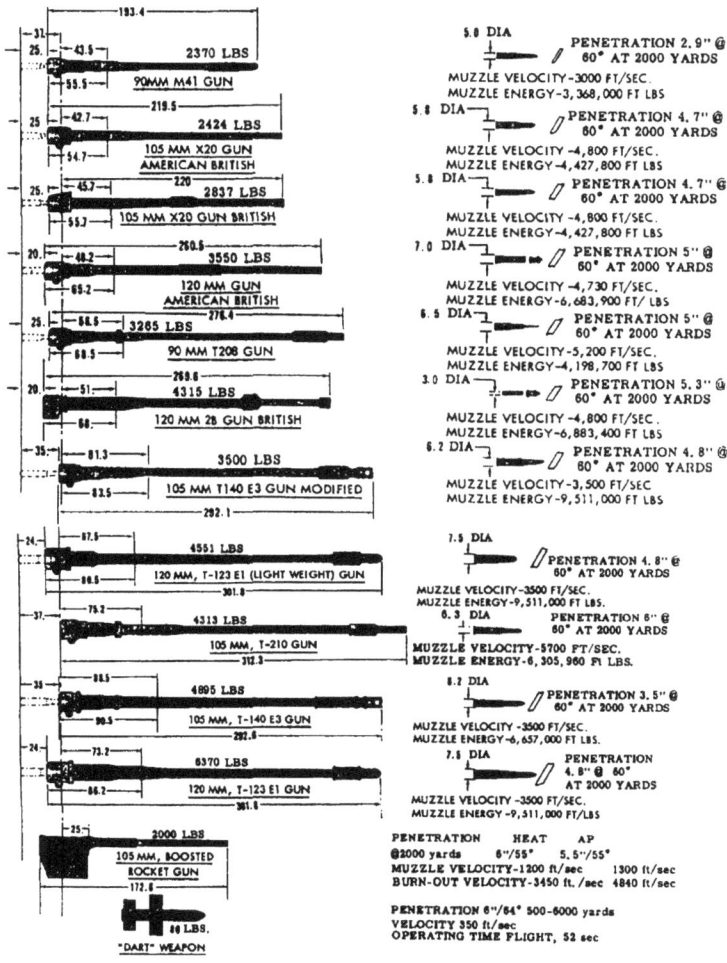

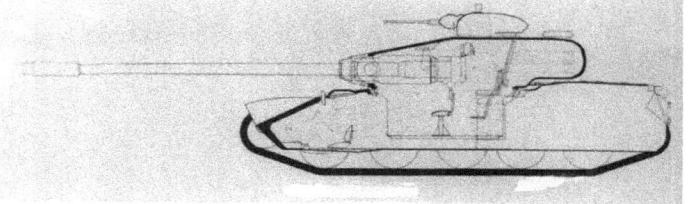

The 105mm gun T140E4 in the T95 turret (Study E above) and the British 120mm gun in the T96 turret (Study F below) are shown fitted to the T95 hull. The chart at the left shows the characteristics of the various weapons under consideration as tank armament.

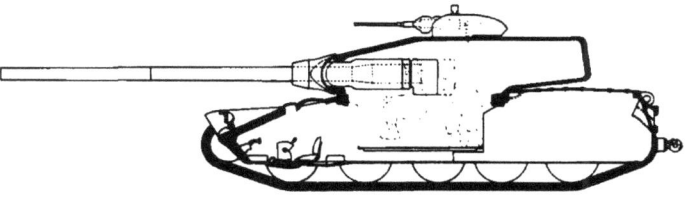

The 105mm gun T210 rigidly mounted in the T96 turret on the T95 chassis was covered in Study H. The estimated total weight exceeded 43 tons and it was not considered possible to balance and stabilize the weapon with this mount, turret, and hull combination. Study I covered the basic T95 tank. This was, of course, the 90mm T208 gun rigidly mounted in the T95 turret on the T95 chassis. The mount could be both balanced and stabilized. At this time, the estimated total vehicle weight was a little under 42 tons.

A lightweight version of the 105mm gun T140E3 was installed in a T95 turret in Study E. The weapon was tentatively designated as the 105mm gun T140E4. This tank with its recoil mounted weapon had an estimated total weight of slightly over 42 tons, but balancing and stabilizing the gun was not considered practical.

The British 120mm tank gun using a bag charge was recoil mounted in a T96 turret in Study F. The weapon was balanced and could be stabilized. The estimated total weight of the vehicle was about 43 tons. Since the bag charge used with the British 120mm gun was not a popular feature, an American version of the weapon was proposed using a new breech and combustible stub case ammunition. Study G featured this modified gun mounted in a T95 turret with a recoil system. The estimated total weight of the tank was under 42 tons. The concept had a balanced gun mount which would permit stabilization.

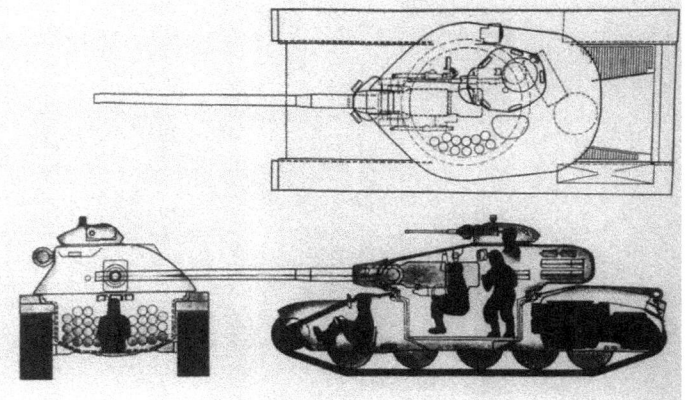

Above, the American version of the British 120mm gun is installed in the T95 for Study G. Study H below shows a sectional drawing and a model of the T96 turret with the 105mm gun T210 on the T95 chassis.

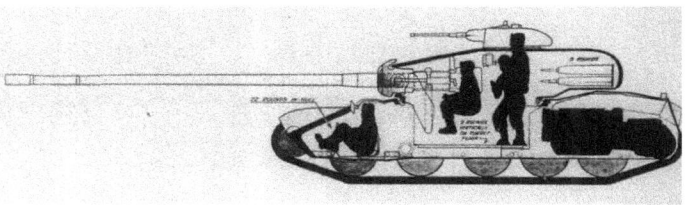

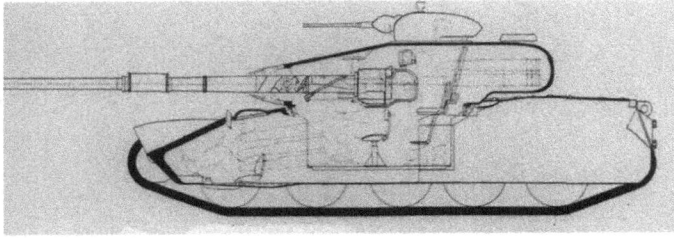

At the left, the British 105mm gun is mounted in the T95 for Study J.

Below, Study K shows the British 105mm gun in an M48A2 turret on a T95 hull.

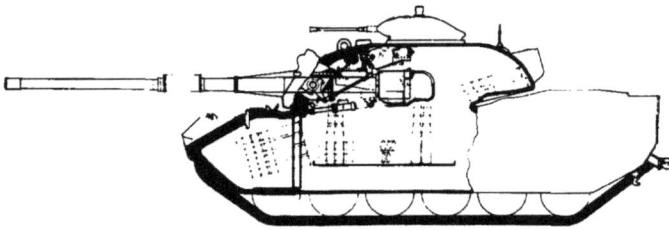

Study J explored the possibility of mounting the British 105mm gun in the T95 turret. At that time, this weapon was frequently called the Ex-20 Pounder or X20 Pounder referring to the gun from which it was developed. The use of this cannon reduced the estimated total vehicle weight of the T95 to just a little over 41 tons and the mount could be both balanced and stabilized. The application of this same weapon to the M48A2 tank was covered in Study K. The promising results foreshadowed its eventual application in the M60 tank. An American version of the British

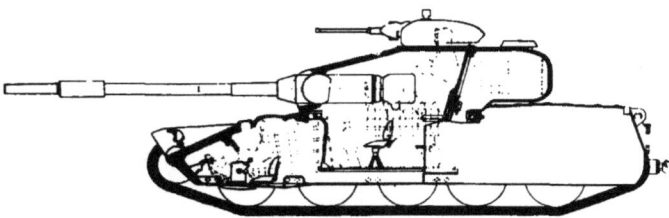

Above, Study L mounts the U. S. version (T254) of the British 105mm gun in the T95 tank. Below, the 105mm T147 rocket boosted gun is mounted on the T95 for Study M.

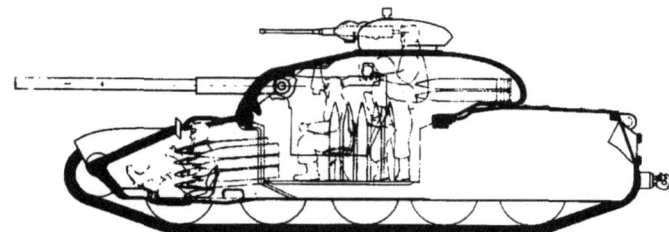

105mm gun was designed which was somewhat lighter than the original weapon. Study L considered this new gun mounted with a recoil system in a T95 type turret. The American version of the British cannon was later designated as the 105mm gun T254. The estimated total vehicle weight was between 40 and 41 tons and the mount was balanced and could be stabilized.

Study M replaced the conventional cannon with the 105mm boosted rocket launcher T147 mounted in a turret similar to that on the T95. The trunnions were moved up 2 3/4 inches and 18 1/4 inches to the rear to permit adequate internal clearance for the magazine when the launcher was elevated and traversed. This modification also resulted in a new turret nose configuration. The launcher was balanced with a full magazine load of 11 rounds. Each rocket boosted 105mm round weighed 38.1 pounds and was about 38 inches long. The 29.1 pound kinetic energy armor piercing projectile had a burn-out weight of 17.1 pounds and a muzzle velocity of 1300 feet per second. Burn-out velocity was 4840 feet per second occurring at a range of about 717 yards. At this time, penetration performance was estimated as 5.5 inches of homogeneous armor at 55 degrees obliquity and ranges of 500 and 2000 yards. Defeat of about six inches of armor at this angle was expected at 1000 yards and a slightly greater amount at the maximum velocity range of 717 yards. A shaped charge, high explosive anti-tank (HEAT) round also was proposed for this weapon. The estimated total vehicle weight was slightly over 40 tons.

Study N considered an even more radical approach to solve the armament problem. A special turret was proposed for the T95 chassis armed with the Dart missile. The optically tracked, wire guided Dart was a powerful weapon which could destroy any armored vehicle at that time. Since the seven inch diameter warhead easily overmatched any known tank armor, its performance was

Below, Study N fits a special turret armed with the Dart missile on the T95 chassis.

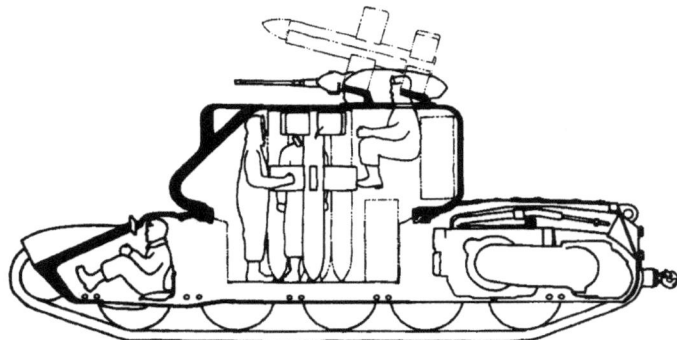

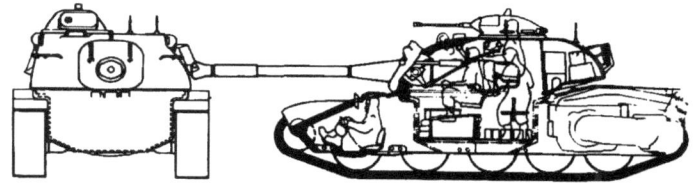

Study O above mounts the standard M48A2 turret on the T95 hull. This was the version that later emerged as the T95E2.

At the right, Study P shows a 120mm gun converted for liquid propellant installed in the T95.

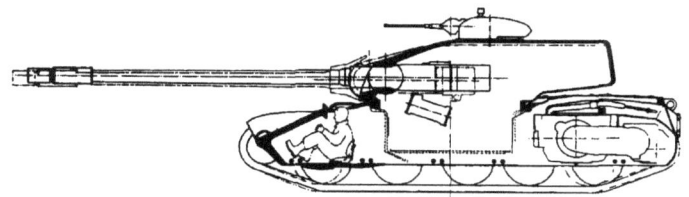

optimized not for penetration, but to obtain the maximum destructive effect behind the armor. Thus the conical copper liner used to achieve maximum penetration with small warheads, was replaced by an aluminum liner and the charge was designed to produce the maximum spall effect and high overpressure behind the armor. Because of the large size of the missile, only a few rounds could be carried. With an estimated total vehicle weight of about 45 tons, it really was not a tank, but an armored antitank missile carrier.

Study O considered the feasibility of mounting the complete turret from the M48A2 tank on the T95 chassis. This was estimated to produce a vehicle with a total weight of slightly over 40 tons having an increase in automotive performance and cruising range compared to the M48A2 tank. Liquid propellant guns also were considered as possible future armament and Study P showed a 120mm gun modified as a liquid propellant weapon. It was expected that the reduced space required to load only the projectile would permit moving the trunnions to the rear resulting in a modified turret design with lighter weight. The overall length and gun overhang also would be reduced.

The conference reviewing the various gun, ammunition, and turret combinations concluded that there were two outstanding weapon systems for tank installation. These were the American versions of the British 105mm gun and the British 120mm gun. These two weapons and the original British guns were superior for tank use because of their lethality combined with light weight, relatively short tubes, and short rounds requiring less loading space. The shorter length of the guns resulted in less tube distortion reducing dispersion and the compact weapons were more easily balanced and stabilized in the turret. The only drawback to the use of the 120mm gun in the T95 turret was the necessity for a single loader to handle the two piece ammunition. However, at that time it was considered that the development of suitable loading assist equipment would

result in satisfactory performance. The review also noted that the smooth bore 105mm gun T210 was the only weapon which could meet the heavy tank armor penetration requirement of six inches at 60 degrees obliquity at a range of 2000 yards.

Several of the configurations studied were recommended for application to the various T95 pilots. Since the T96 turrets were not expected to be available for some time, it was recommended that their four chassis be used initially to mount two M48A2 tank turrets and two T54E2 tank turrets armed respectively with the 90mm gun M41 and the 105mm gun T140E3. This would permit early engineering and user tests and would provide insurance against the failure of the smooth bore gun development program. The nomenclature for the T95 series vehicles was revised by OTCM 36543, which was approved on 13 June 1957. This item designated the T95 chassis with the M48A2 turret as the 90mm gun tank T95E2. The T95 chassis with the T54E2 turret became the 105mm gun tank T95E3. When these chassis were refitted with the T96 turrets as originally intended, their designation was to become the 105mm gun tank T95E4. Thus the nine chassis for the program were allocated 13 turrets as follows:

90mm gun tank T95 (chassis numbers 4, 7, 8, and 9)
90mm gun tank T95E1 (chassis number 5)
90mm gun tank T95E2 (chassis numbers 1 and 3)
105mm gun tank T95E3 (chassis numbers 2 and 6)
105mm gun tank T95E4 (chassis numbers 1, 2, 3, and 6)

Subsequent developments resulted in the cancellation of the T95E4 and chassis 1, 2, 3, and 6 were never converted to that configuration.

The sketches below show, from top to bottom, the W, X, Y, and Z fire control systems for the T95 tank.

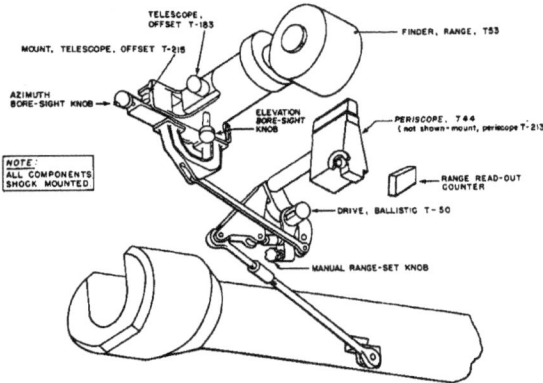

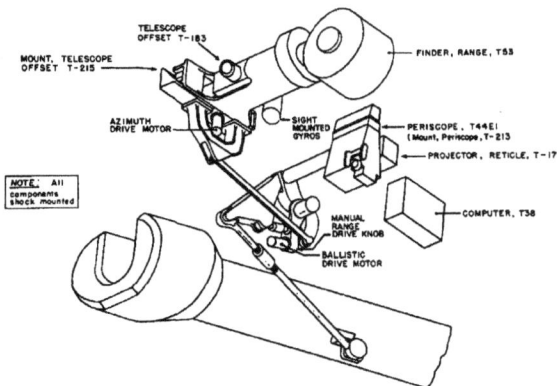

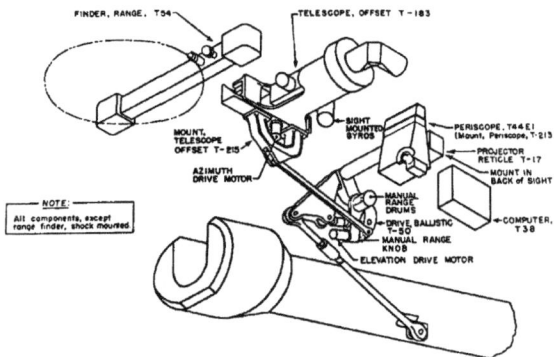

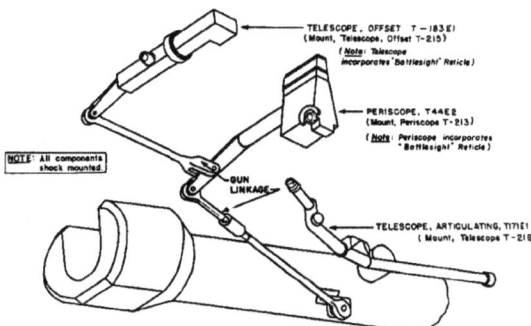

TELESCOPE, OFFSET T-183
MOUNT, TELESCOPE, OFFSET T-215
FINDER, RANGE, T53
AZIMUTH BORE-SIGHT KNOB
ELEVATION BORE-SIGHT KNOB
PERISCOPE, T44 (not shown-mount, periscope T-213)
RANGE READ-OUT COUNTER
NOTE: ALL COMPONENTS SHOCK MOUNTED
DRIVE, BALLISTIC T-50
MANUAL RANGE-SET KNOB

TELESCOPE OFFSET T-183
MOUNT, TELESCOPE OFFSET T-215
FINDER, RANGE, T53
AZIMUTH DRIVE MOTOR
SIGHT MOUNTED GYROS
PERISCOPE, T44E1 (Mount, Periscope, T-213)
PROJECTOR, RETICLE, T-17
NOTE: All components shock mounted
COMPUTER, T38
MANUAL RANGE DRIVE KNOB
BALLISTIC DRIVE MOTOR

FINDER, RANGE, T54
TELESCOPE, OFFSET T-183
MOUNT, TELESCOPE OFFSET T-215
AZIMUTH DRIVE MOTOR
SIGHT MOUNTED GYROS
PERISCOPE, T44E1 (Mount, Periscope, T-213)
PROJECTOR RETICLE T-17
MOUNT IN BACK of SIGHT
MANUAL RANGE DRUMS
DRIVE, BALLISTIC T-50
COMPUTER, T38
MANUAL RANGE KNOB
ELEVATION DRIVE MOTOR
NOTE: All components, except range finder, shock mounted.

TELESCOPE, OFFSET T-183E1 (Mount, Telescope, Offset T-215) (Note: Telescope incorporates 'Battlesight' Reticle)
PERISCOPE, T44E2 (Mount, Periscope T-213) (Note: Periscope incorporates 'Battlesight' Reticle)
NOTE: All components shock mounted.
GUN LINKAGE
TELESCOPE, ARTICULATING, T171E1 (Mount, Telescope T-215)

The fire control equipment in the T95E2 and T95E3 was the same as in the 90mm gun tank M48A2 and the 105mm gun tank T54E2 respectively. Both were equipped with a commander operated stereoscopic range finder, a mechanical "superelevation only" computer, and a gunner's periscope. A straight tube coaxial telescope also was provided for emergency use by the gunner. The standard two man mode of operation required the commander to determine the range which was then automatically transmitted through the computer determining the superelevation applied to the central lay reticle in the gunner's periscope. In the T54E2 turret, a ballistic shock mount structure was installed to isolate the fire control equipment from the turret walls.

For the new turrets of the T95 series, several fire control systems were under development. These were designated as the W, X, Y, and Z systems and they were intended to take advantage of the extremely high velocity and flat trajectory characteristics of the main armament to reduce the time required for a hit on an enemy tank. System W utilized the Optar pulsed light range finder T53 with an offset telescope T183, a T37 "superelevation only" electrical computer, and a gunner's periscope T44. A T171E1 articulated telescope for emergency use was provided for the gunner. This system featured one man operation and it could be operated by either the gunner or the tank commander.

System X replaced the T37 "superelevation only" computer with the T38 electrical computer which provided corrections for jump, parallax, cant, drift, and lead as well as superelevation. The computer corrections were automatically introduced into the gun directing system where sight mounted gyros maintained the line of sight on the target at all times. The gunner's periscope was fitted with a reticle projector and redesignated as the T44E1. Like system W, it could be operated by either the gunner or the tank commander.

System Y differed from W and X in that it reverted to a two man operation. The T53 Optar range finder was replaced by a short base length optical range finder mounted on the commander's cupola. This transmitted the range information to the full solution T38 electrical computer which then positioned the central lay reticle in the gunner's T44E1 periscope. Early documents designate the commander's range finder as the T54. This became the T57 in later versions.

System Z was the simplified arrangement installed in the 90mm gun tank T95E1. The ballistic drive, computer, and range finder in the more complex systems were eliminated. The gunner's T44E2 periscope and the commander's T183E1 offset telescope were mechanically linked directly to the gun trunnions. Either man could perform the firing problem by visually estimating the range and selecting the proper reticle marking in the battlesight of their sighting device. Like the other three, system Z retained a T171E1 articulated telescope for emergency use by the gunner. The fire control equipment in all four systems was isolated from the turret walls by a shock mount structure attached at the turret ring.

Above is the mock-up of the 90mm gun tank T95. Note the bulldozer mounted on the vehicle.

The first pilot vehicle of the T95 series, a T95E2 with the M48A2 turret, was completed at Detroit Arsenal during May 1957 and shipped to Aberdeen Proving Ground on 7 June. Its registration number was 9B1048 (chassis number 1). The second pilot was a T95E3, registration number 9B1050 (chassis number 2), with the turret and armament from the 105mm gun tank T54E2. It was completed in July and shipped to the Yuma, Arizona Test Station in August for engineering tests. Chassis number three, the second pilot T95E2 with registration number 9B1049, was shipped to Fort Knox, Kentucky in December 1957 for evaluation by the Armor Board. The first actual T95 armed with the 90mm T208 smooth bore gun in the rigid mount was the fourth vehicle completed. It carried registration number 9B1043 and was shipped to Aberdeen Proving Ground on 31 March 1958. Chassis number 5 was the single T95E1 with the recoiling gun mount and simplified fire control system. It was shipped to Fort Benning, Georgia on 7 March 1958 for display and at the end of the month, it was transferred to Aberdeen. Its registration number was 9B1047. The second T95E3, chassis number 6, registration number 9B1051, was shipped to the Armor Board in mid 1958 along with the second pilot T95. The registration number for the latter was 9B1044, chassis number 7. The remaining two T95s, chassis numbers 8 and 9, carried registration numbers 9B1045 and 9B1046 respectively. They were used for various tests at Detroit Arsenal and chassis number 8 later was shipped to Aberdeen for stabilizer tests. Chassis number 9 was retained at Detroit, but the turret was transferred later to the Aeroneutronics Division of the Ford Motor Company for use in the development of the Shillelagh weapon system.

These four photographs show the first pilot 90mm gun tank T95, registration number 9B1043, at Detroit Arsenal on 4 February 1958. Note the lightweight road wheels on this vehicle.

Above, the first pilot T95 is shown with its cannon locked in the travel position. Note the counterweight added to the gun tube just behind the bore evacuator to balance the weapon in the mount. In the front views of the tank below, the turret is shown in both the forward and travel positions.

Below is the second pilot 90mm gun tank T95, registration number 9B1044, at Fort Knox. This tank is armed with the 90mm gun T208E13 with the sleeve added to the tube to balance the gun. Note that no counterweight is installed behind the bore evacuator. Additional views of this tank are shown on the opposite page. Unlike the first pilot, 9B1044 is fitted with solid road wheels.

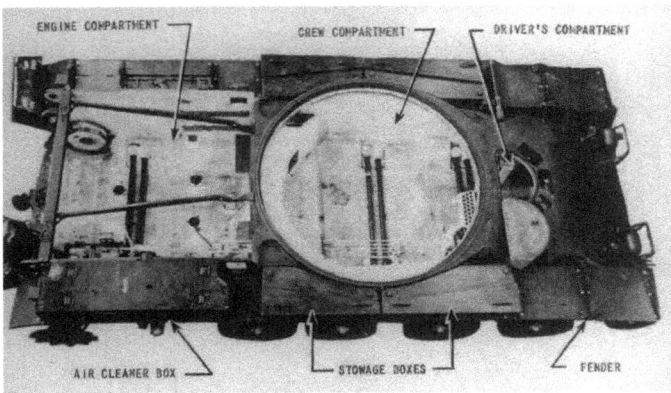

Above, the T95 hull appears at the left with the turret and power train removed. Details at the right rear of the tank can be seen in the photograph above at the right.

The hull of the T95 chassis was an armored steel structure used to mount the turret, power train, suspension, and other components of the vehicle. It was divided into two sections with the crew compartment in the front and the engine compartment in the rear. A homogeneous steel armor casting comprised the front, top, and sides of the hull extending to the rear of the crew compartment. Rolled homogeneous steel armor plates were assembled by welding to form the sides and rear of the engine compartment as well as the floor of the tank. The crew was separated from the engine compartment by a watertight bulkhead welded in place across the full width of the hull. The rear deck over the engine compartment consisted of cast armor steel grill doors on each side, an aluminum alloy top deck plate in the center, and two steel support cross beams. A traveling lock for the main gun was attached to the top deck plate just in front of the rear cross beam.

The air-cooled, eight-cylinder, opposed, Continental AOI-1195-5 engine was mounted transversely in the engine compartment with the General Motors Allison XTG-410-1 transmission. As indicated by its designation, the engine utilized fuel injection equipment instead of a carburetor. Three fuel tanks fabricated from fiber glass reinforced epoxy resin were shaped to fit the available space in the compartment with one on each side at the front of the engine and the third at the right rear. A 300 ampere auxiliary generator connected directly to a single cylinder air-cooled gasoline engine was installed on the right side of the engine compartment between the right front and right rear fuel tanks.

Combustion air for the engine, drawn from the crew compartment, passed through the dry type filters installed in the armor plate box welded to the right side wall of the engine compartment and then to the engine air intake. The engine cooling fan pulled air through the intake grills on each side of the rear deck. It passed over the fuel tanks to below the engine, up through the cylinder cooling fins and oil coolers, through the fan, and out through the insulated passage under the center top deck plate to the two grill doors in the rear wall of the engine compartment. The engine exhaust, after passing through the mufflers, was mixed with this cooling air to lower the temperature and reduce the infrared signature of the vehicle. To aid field maintenance, an hydraulic pump and a vertical ram were installed in the engine compartment to raise the front end of the power plant about 12½ degrees, pivoting about the final drives as an axis. This permitted access to the fuel filters and the spark plugs for cylinders 1, 3, 5, and 7.

Below are right front (left) and top (right) views of the T95 power pack consisting of the AOI-1195-5 engine and the XTG-410-1 transmission. In the top view, the insulated exhaust air passageway has been removed revealing the engine driven cooling fan.

Above, the muffler can be seen inside the exhaust air passageway in the rear view of the power pack at the left and the power pack is installed in the hull at the right. Below are rear views of the T95 with the exhaust doors open (left) and closed (right).

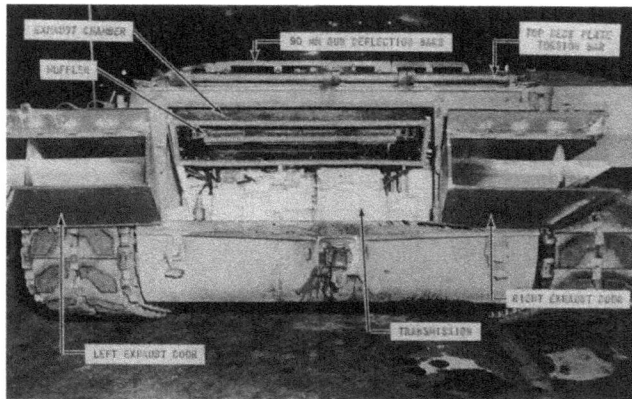

Above are photographs of the gun travel lock on the rear of the top deck (left) and the fuel tank filler cap (right). Below, the driver's hatch in the front hull is at the left and details of the personnel heater exhaust can be seen at the right.

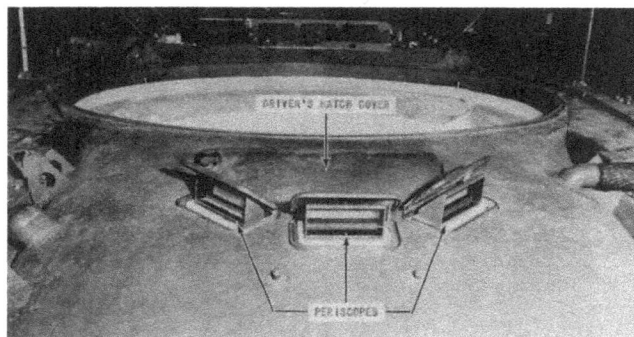

The headlight and taillight groups installed on the T95 tank are pictured above and below respectively.

The XTG-410-1 transmission consisted of an hydraulic torque converter, a planetary gear set, and an automatic lockup clutch with a combination of clutch-brake and geared steering. Manual shifting was required between the four forward and one reverse speeds and neutral. The forward speeds were designated as low, 1st, 2nd, and 3rd. Only geared steering was available in 2nd and 3rd while low, 1st, and reverse were restricted to clutch-brake steering. Unlike the CD-850 transmission used in the M48 series tanks, the XTG-410-1 would not steer in neutral. Thus the vehicle would not pivot with the tracks moving in opposite directions. However, with the clutch-brake steering in low, 1st, or reverse, the tank could be made to pivot with one track stopped and the other moving.

The final drives were mounted in holes on each side at the rear of the engine compartment and coupled directly to the transmission. A ten tooth sprocket was installed on the hub of each final drive. These sprockets engaged the 21 inch wide T114 double pin track provided on the original T95 pilots. As mentioned previously, the T95 series tanks featured a flat track suspension without track support rollers. By supporting the top run of the track on the road wheels, the weight of the track support rollers was eliminated and the lower suspension reduced the height of the vehicle. The torsion bar suspension supported the tank on five dual road wheels per side with hydraulic shock absorbers fitted on the first and last road wheel arms. Both solid disc wheels and wheels with ten lightening holes were installed on the pilot tanks. An example of the latter was on the first pilot T95 (chassis number 4). Occasionally, both types were mounted on the same vehicle. Track tension was adjusted by the dual idler at the front of each track.

Below, the T95 suspension is shown with and without the T114 tracks installed. Note the lightweight road wheels in the right photograph.

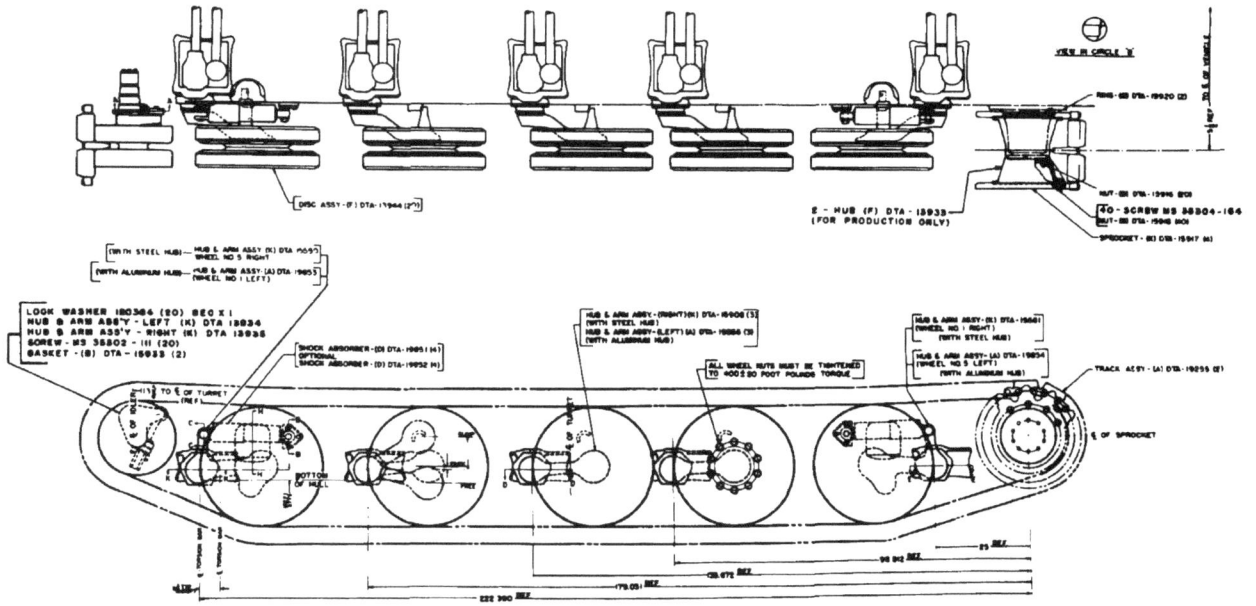

Above is a drawing of the T95 suspension system. At the right, the top deck plate, cross beam, and rear exhaust doors are being removed as a unit.

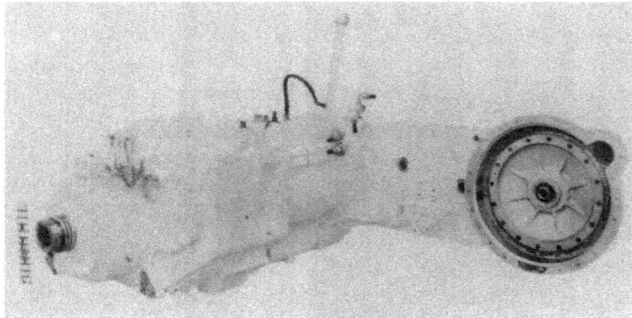

Above, the XTG-410-1 transmission is shown separate from the engine. Below are details of the right sprocket (left) and the towing pintle on the lower rear hull (right).

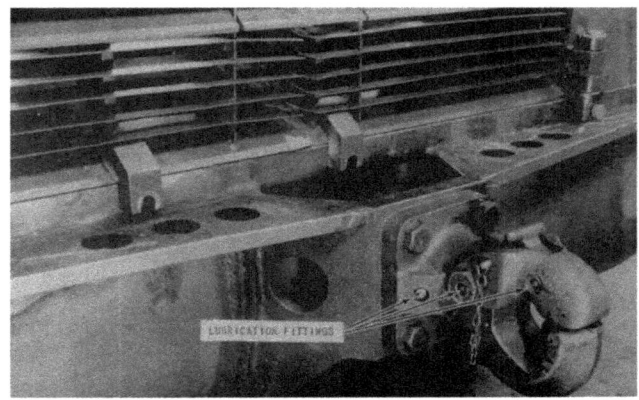

The left and right sides of the driver's compartment appear at the left and right above. The 90mm ammunition racks on each side of the driver can be seen below at the left.

The driver's seat was located in the center of the front hull. The seat was adjustable in height to permit operation with the driver's head exposed or lowered for use when the hatch was closed. A slight bulge in the floor beneath his seat provided increased headroom when the hatch was secured. This hatch cover was mounted on a torsion bar located inside the hull. When closed, it was locked in place by an exterior cam controlled by a handle from inside the hull. When released, the hatch cover would rise about 23 degrees and then could be slid directly sideways to the right. Three retractable T48 periscopes were installed in the hull armor in front of the driver's hatch. It was necessary to retract the periscopes before opening the hatch. A steering handle bar was located directly in front of the driver and the manual transmission shift lever was at his right side. Accelerator and brake pedals were mounted on the

floor of the driving compartment and two instrument panels were located one on each side. A circular emergency escape hatch was installed in the floor directly in front of the driver's seat. A rack for main armament ammunition was fitted on each side of the driver and a gasoline fueled personnel heater was located to his left front. This provided hot air outlets to the driver's compartment as well as to the turret basket. The tank was equipped with a fixed fire extinguisher system which could be operated from the driver's position or by an exterior control handle. A portable fire extinguisher was provided in the turret. An inflatable seal was installed in a ring groove under the turret to provide a watertight seal for the crew compartment during fording operations. The turret could not be rotated when the seal was inflated.

The T95's cast homogeneous steel armor turret was mounted on an 85 inch inside diameter race ring over the center of the hull crew compartment. The gunner was located on the right side in front of the tank commander with the loader in his usual position on the left side of the cannon. The seats for the tank commander and the loader were suspended from the turret ring and the gunner's seat was pedestal mounted on the turret basket platform. A T6

At the right is a top view of the T95 turret assembly. The armament and fire control equipment have not been installed.

These views show details of the T95 turret with the armament and fire control equipment in place. The Optar range finder is shown with its armored cover open.

cupola with a .50 caliber machine gun was specified for the tank commander. However, an Aircraft Armaments, Inc. Model 108 cupola was installed on the pilot tanks pending the availability of the T6 which corresponded to Aircraft Armaments, Inc. Model 232. The two cupolas differed only in the thickness of the armor. On the Model 232, the maximum armor thickness was increased to about two inches compared to one inch for the Model 108. Five vision blocks, two on each side and one in the rear, as well as the M28 periscopic sight in the cupola roof, provided all round protected vision for the tank commander. The M28 periscope was linked to the M2 .50 caliber machine gun following it through the elevation range of +60 to –10 degrees. The .50 caliber ammunition was fed from a 210 round storage box on the left side of the cupola through

a feed chute to the gun. In addition to the commander's cupola hatch, a hatch with a flat hinged cover was provided over the loader's position.

The 90mm gun T208 in the rigid combination mount T191 was installed in the T95 turret with a coaxial .30 caliber M37 machine gun to the left of the cannon. The .30 caliber ammunition was fed to the machine gun from the 1500 round storage box on the turret basket platform. Some changes were required to balance the cannon for stabilization. This resulted in a number of differences between the four T95 pilot tanks. It was intended to balance the weapon on the first two tanks (chassis numbers 4 and 7) by adding a 360 pound sleeve to the tube and redesignating it as the 90mm gun T208E13. However, in order to meet the shipping date, a temporary fix was applied to the first

Below is a drawing of the 90mm gun T208.

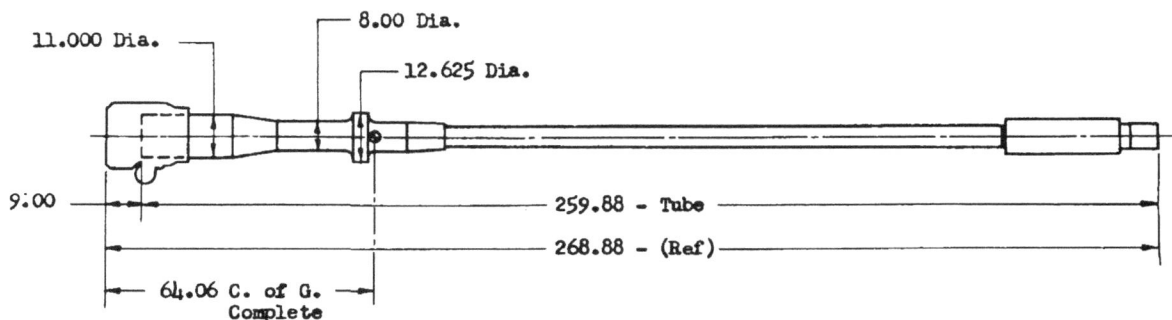

59

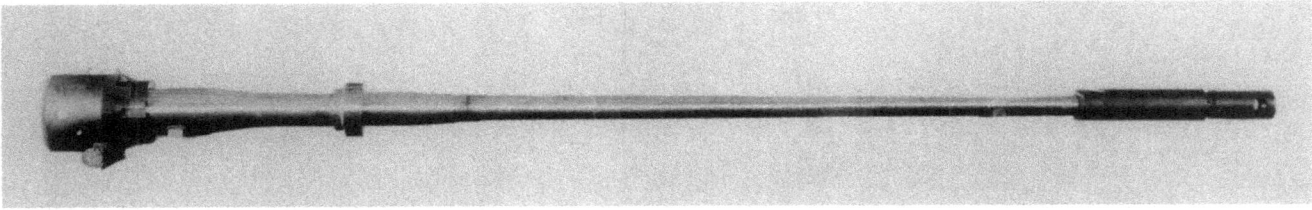

Details of the 90mm gun T208 can be seen above and at the left.

pilot (chassis number 4). This consisted of leaving the original T208 gun installed, but adding a 124 pound weight to the muzzle of the tube just behind the bore evacuator. It was intended that the T208E13 gun with the sleeve added would replace the original weapon when it became available. The T208E13 gun was installed in the second pilot (chassis number 7) before it was shipped to Fort Knox. A different solution was applied to the third and fourth T95s (chassis numbers 8 and 9). On these tanks, the gun was moved forward three inches in the mount and the blast deflector was replaced with a 150 to 200 pound weight. All of the T95 turrets measured 47 inches from the center line of the gun mount trunnions to the turret center line. On the first and second pilots (chassis numbers 4 and 7), the distance from the trunnion center line to the rear face

of the gun breech was 63 inches or 16 inches behind the turret center line. However, with the cannon shifted forward three inches on pilots three and four (chassis numbers 8 and 9), these dimensions were decreased to 60 inches and 13 inches respectively. These changes were reflected, of course, in the overall length of the vehicles.

An external shoulder ring near the chamber end of the T208 tube was used to retain the cannon in the rigid T191 combination mount. This mount consisted of a shield cap, a combination armor gun shield and front cradle, and a rear armor cradle welded to a cradle extension. The front and rear cradles separated at the trunnion axis and each contained a semicircular cavity which enclosed the trunnion bearings when the cradles were assembled in the turret. The cast gun shield and front cradle had a threaded cylindrical bore which enclosed the ring shoulder on the gun tube and it was attached to the rear cradle by four large bolts from inside the turret. An externally threaded locking nut was screwed into the front cradle clamping the ring shoulder on the tube against the front face of the rear cradle. The shield cap was then attached to the front face of the gun shield by four cap screws. Since the quick change gun tube was screwed into the breech ring with an interrupted thread, the tube could be replaced from outside the turret by removing only the shield cap and the locking nut without disturbing the remainder of the mount.

Below, the 90mm gun T208 is installed in the T95 tank turret. Some of the controls for the gunner and the tank commander are visible in the view at the right from below the gun breech.

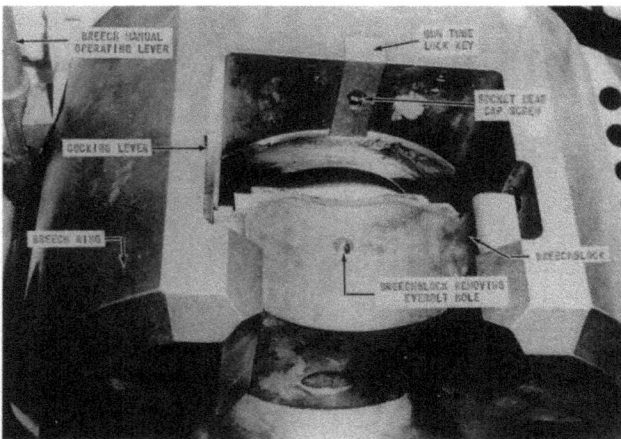

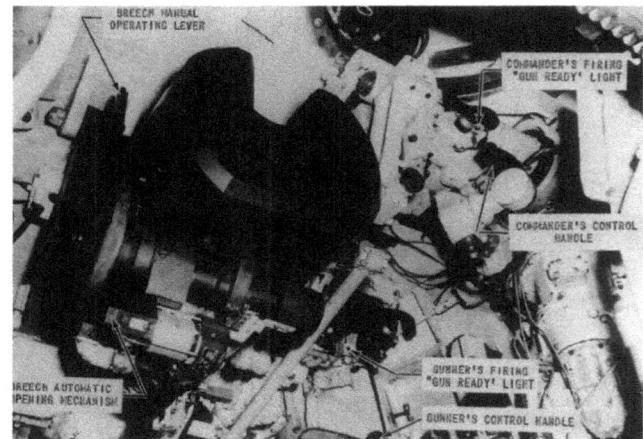

60

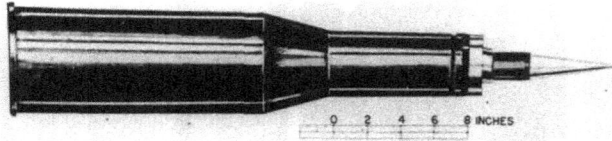

Above is the 90mm T320E60 shot. The drawings at the right show the left (upper) and right (lower) sides of the rigid non-recoiling gun mount in the T95 tank.

The smooth bore 90mm gun T208 had a muzzle velocity of over 5000 feet per second with the T320 armor piercing, fin stabilized, discarding sabot (APFSDS) ammunition. This round with its 40mm diameter tungsten carbide penetrator was expected to defeat five inches of homogeneous steel armor at 60 degrees obliquity at a range of 2000 yards. High explosive and white phosphorus rounds also were under development for this cannon. Since there was no recoil movement of the rigidly mounted gun to open the breech after firing, an electric motor was installed to perform this function. A sensor detected the firing shock and, after a timed delay, actuated the electric motor to open the breech. A precise operating sequence was developed to provide the proper delay interval for the effective operation of the bore evacuator and to prevent accidental opening of the breech due to road shocks or other impacts. Power traverse and elevation were provided in the turret by an electric-hydraulic system manufactured by the Minneapolis-Honeywell Regulator Company. It was stabilized in both azimuth and elevation.

The W fire control system was installed in all four of the T95 pilots. However, the gunner's periscope, originally from the T44 series, was replaced by one from the later model T50 series. The two periscopes were similar, but the T50 had a magnification of 8x compared to 6x for the

At the right is a top view drawing of the T95 gun mount. The .30 caliber coaxial machine gun and the gunner's telescope can be seen on the left and right sides respectively. Some of the hydraulic system components are visible in the photograph of the turret basket below. At the lower right is a schematic diagram of the T95 fire control system.

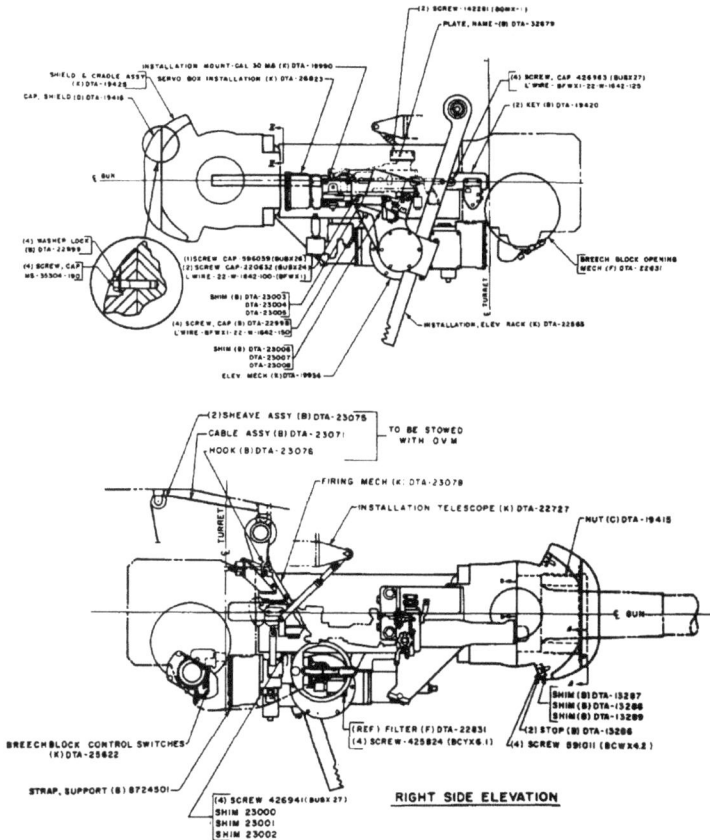

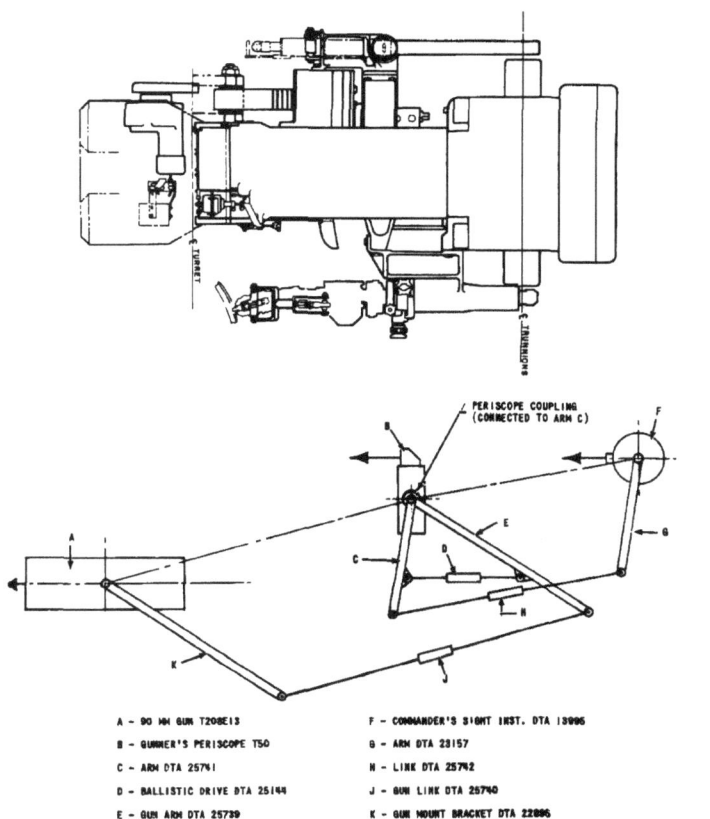

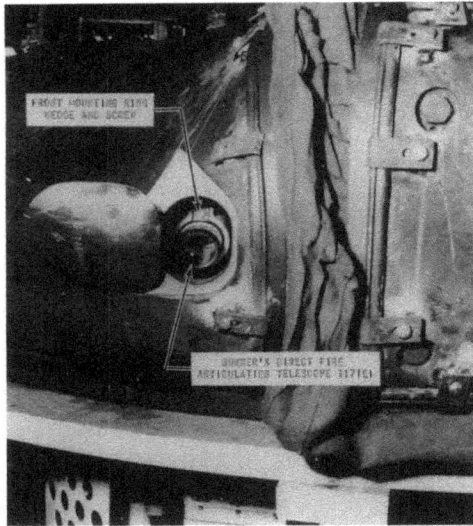

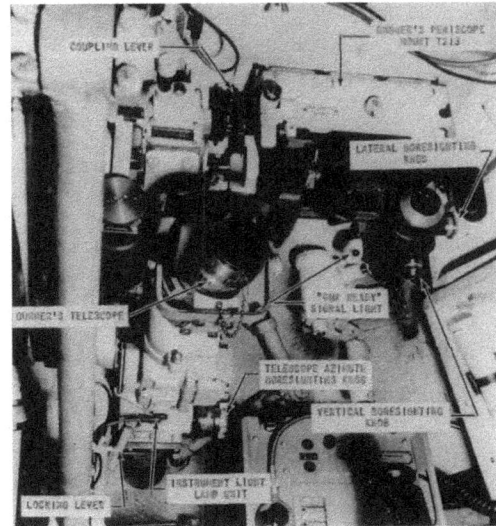

Above, the gunner's telescope is seen from the outside (left) and from the inside (right) of the turret. At the left is an outside view of the gunner's periscopic sight.

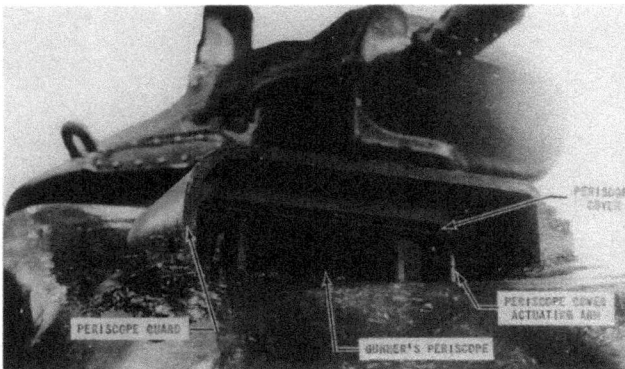

T44. The T171E1 articulated telescope was retained on the right side of the T191 combination mount for emergency use by the gunner and its port in the turret nose was protected by an armor cover when it was not in use. Except for this telescope, all of the fire control system components were installed on a ballistic shock mount attached to the turret ring. This mount isolated the fire control equipment from the turret walls reducing the effect of any impacts. Also, since the turret no longer formed part of the fire control linkage, the equipment was shielded from the effect of thermal expansion or contraction of the armor. The T53 Optar range finder and the commander's T183 offset telescope were enclosed in a spherical armor hood on the right side of the turret. A moveable shield on the front of the hood could be opened from within the turret to expose the instruments for use. It was intended to replace the W fire control system with the X system in T95 pilots two and four (chassis numbers 7 and 9) by early 1959. Pilots one and three (chassis numbers 4 and 8) were to be fitted with the Y fire control system at approximately the same time. However, by 1959, interest was shifting away from the T95 project and these changes were never carried out.

The tank commander's telescopic sight is shown below from both the outside and the inside of the turret. The outside photograph also provides a clear view of the Optar range finder with the armor cover in the open position.

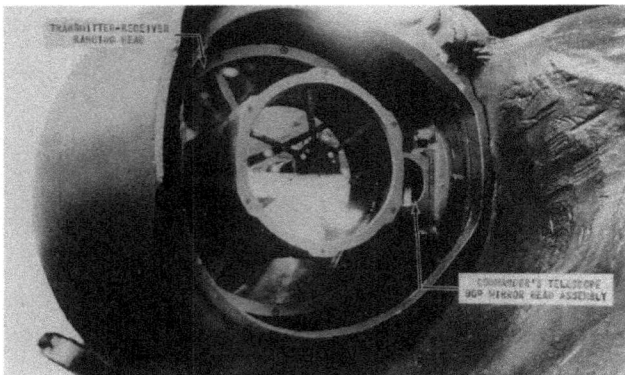

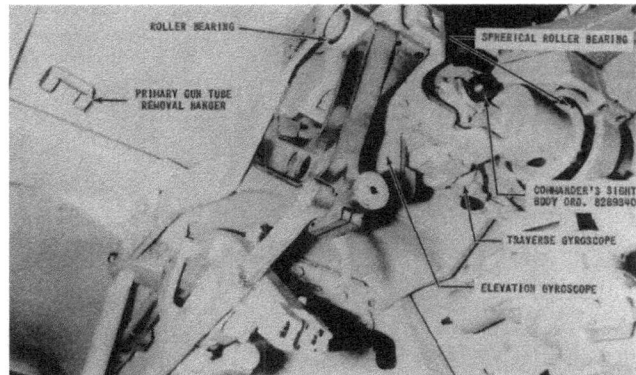

62

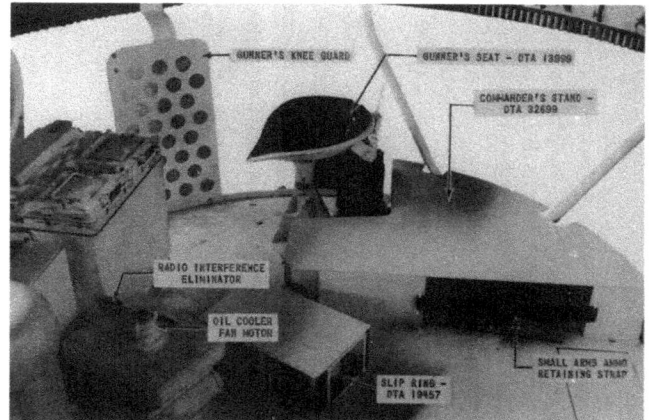

Above, the gunner's controls and the azimuth indicator are at the left and the gunner's seat as well as the tank commander's stand are visible in the right side of the turret basket at the right. At the left is the mount for the coaxial machine gun, but the weapon has not been installed.

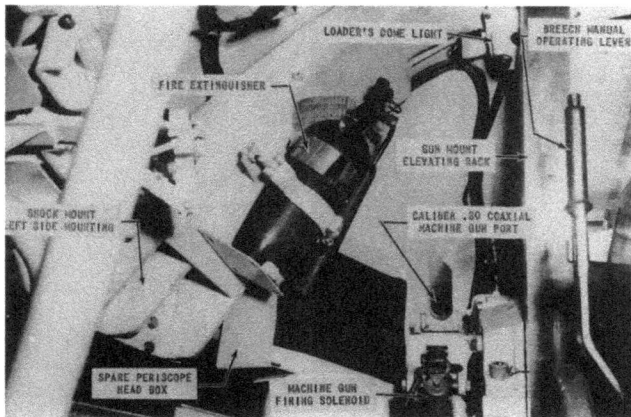

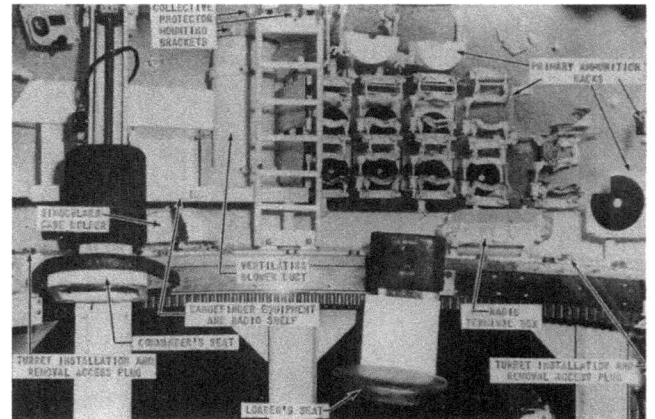

Fifty rounds of 90mm ammunition were stowed in the T95. Twentyeight of these were in the front hull with 14 rounds on each side of the driver. The remaining 22 were located in the turret with eight in a vertical stand on the turret platform near the loader's position. Three rounds were suspended in a rack on the left turret rim and 11 rounds were stowed in the turret bustle rack. The turret bustle also contained the radio, the M8A1 collective gas protection equipment, and the ventilation blower.

The turret bustle stowage rack can be seen above. Below, the interior of the tank commander's cupola is at the left and the lower rear of the turret basket appears at the right. In the latter view, the gunner's seat and the tank commander's stand are visible as well as part of a 90mm ammunition rack.

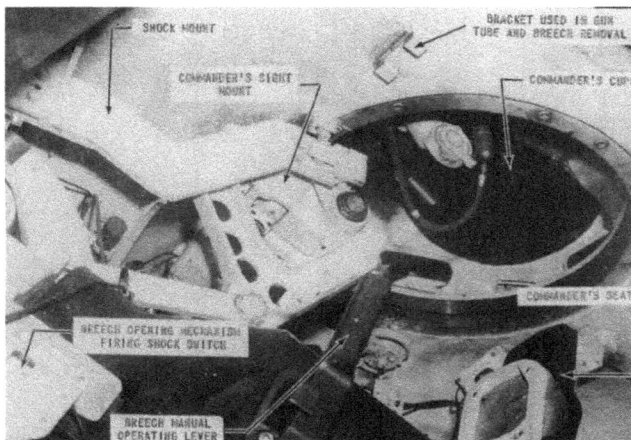

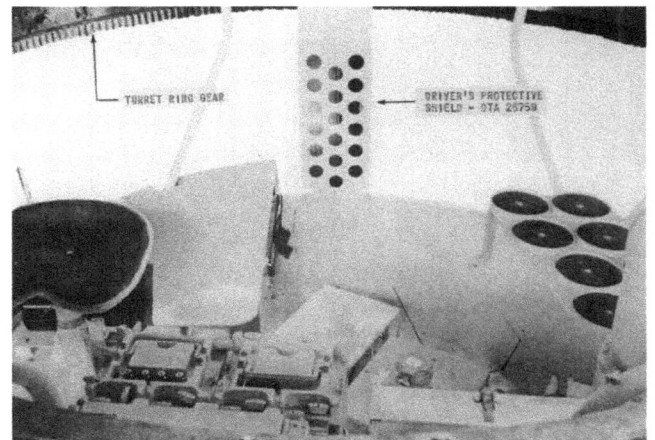

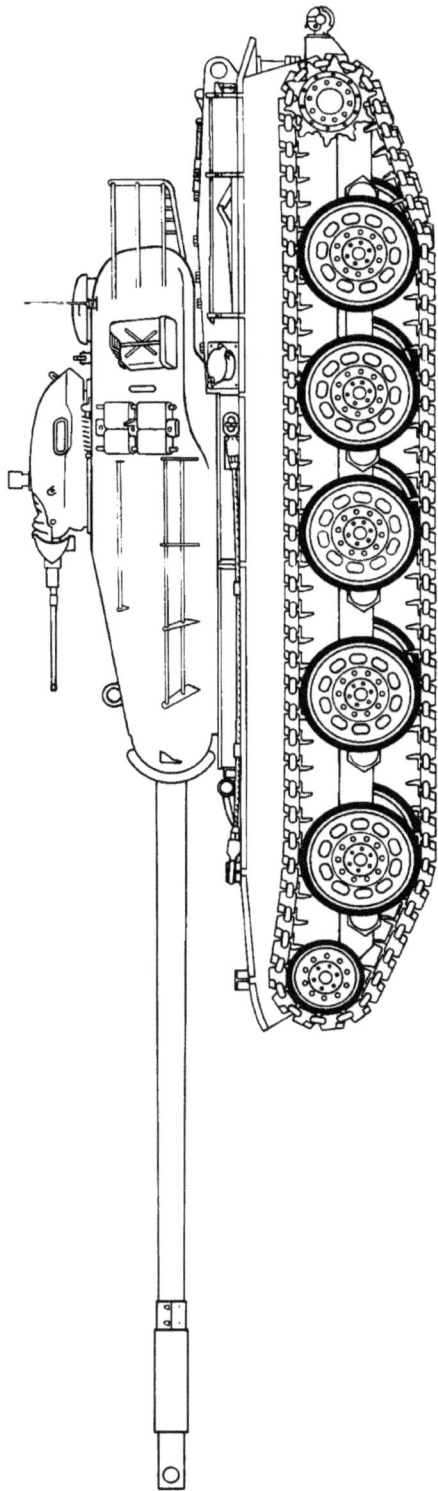

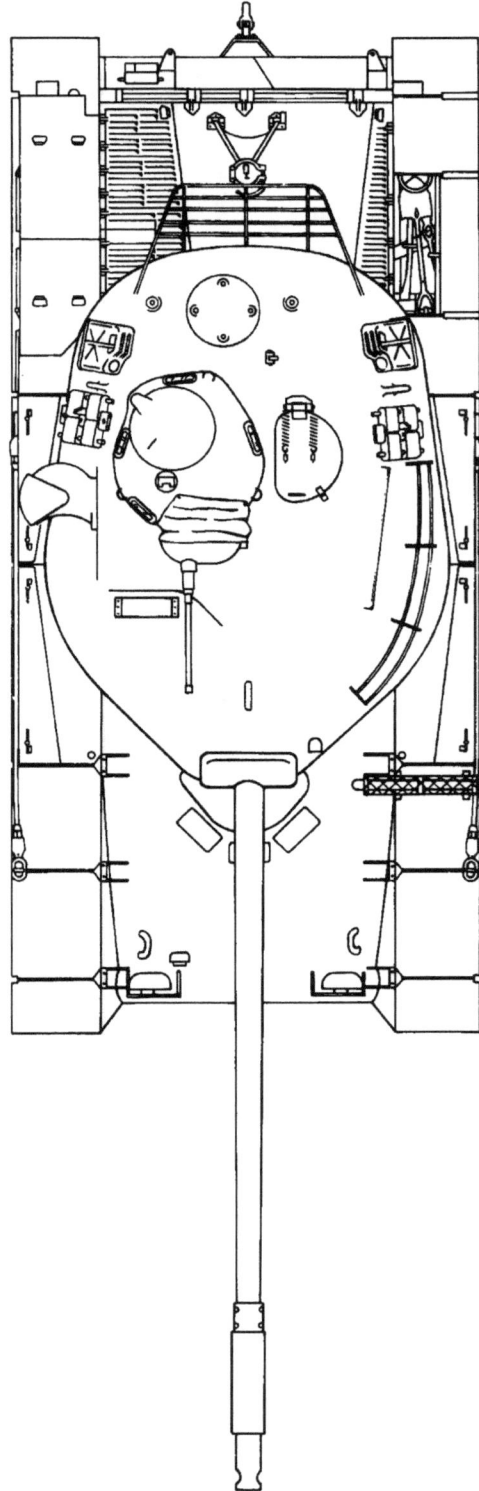

Scale 1:48

© D.P. Dyer

90mm Gun Tank T95

64

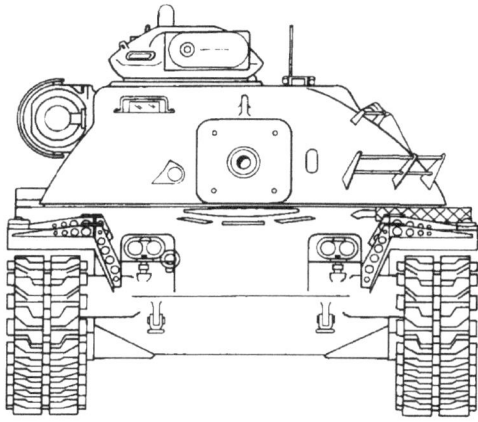

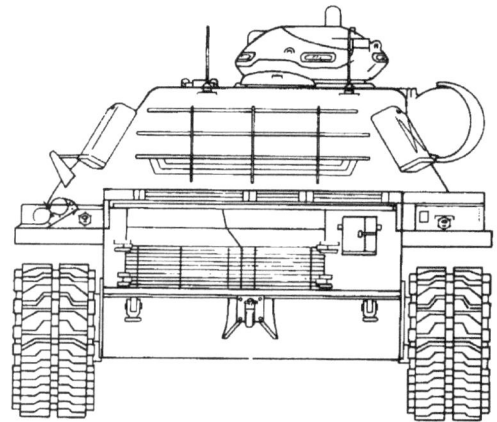

Above is another photograph of the second pilot T95, registration number 9B1044, during the evaluation at Fort Knox. The tank below, registration number 9B1049, was labeled as the T95 Special. However, it actually was the second pilot T95E2 fitted with a T95 turret in October 1957 prior to its shipment to Fort Knox.

The views above and below show the pilot 90mm gun tank T95E1, registration number 9B1047, after completion at Detroit. Note that this tank was fitted with the solid road wheels.

The chassis of the 90mm gun tank T95E1 was essentially the same as the other T95 series tanks. However, the single pilot vehicle (chassis number 5) was not fitted with the gun deflection bars across the top rear deck on the hull. The T95E1's turret casting was identical with that of the T95, but, as mentioned previously, it was armed with a modified version of the 90mm smooth bore gun in the combination mount T192 which featured a concentric recoil mechanism. Ballistically, the cannon was identical to the T208 gun from which it was converted. Designated as the

Both sides of the T95E1 pilot can be seen below. With the Z fire control system, the Optar range finder has been omitted from the right side of the turret.

66

These additional photographs of the T95E1 pilot show further details of the tank after its completion at Detroit Arsenal.

90mm gun T208E9, it was modified to fit the recoil system in the mount and had a cam operated semiautomatic breechblock. The T192 combination mount consisted of a shield cap, a gun shield, and a combination gun cradle and concentric recoil mechanism. The latter was a constant recoil hydrospring type with a recoil distance of eight inches. As in the T95 turret, the mount was supported on two self-aligning roller trunnion bearings with a center line 47 inches forward of the turret center line. To allow space for the recoil, the gun was moved forward in the T192 mount so that the distance from the trunnion center line

to the rear face of the breech ring was 58 inches or 11 inches behind the turret center line. Like the T208, the T208E9 was screwed into the breech ring with an interrupted thread and the tube could be changed by removing the shield cap, the follower, and the locking key. The tube then could be rotated 1/8 of a turn and extracted from the front without disturbing the remainder of the mount.

A sectional view of the recoil system for the 90mm gun T208E9 appears at the right. The dimensions of the 90mm gun T208E9 are shown in the sketch below.

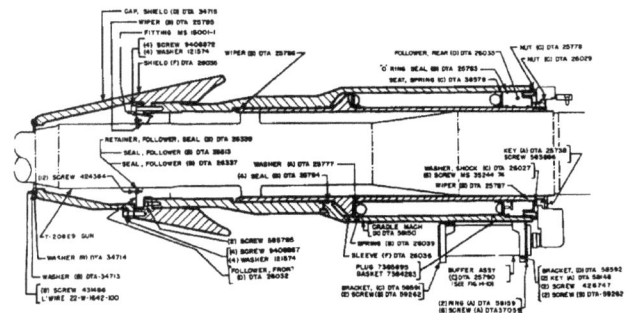

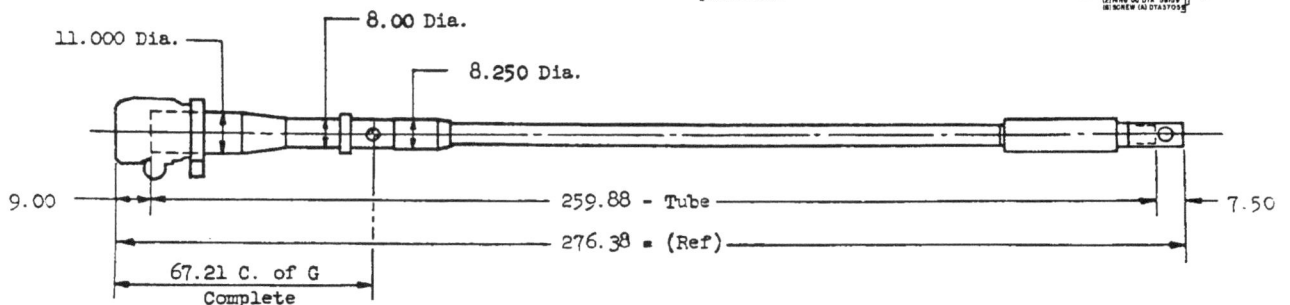

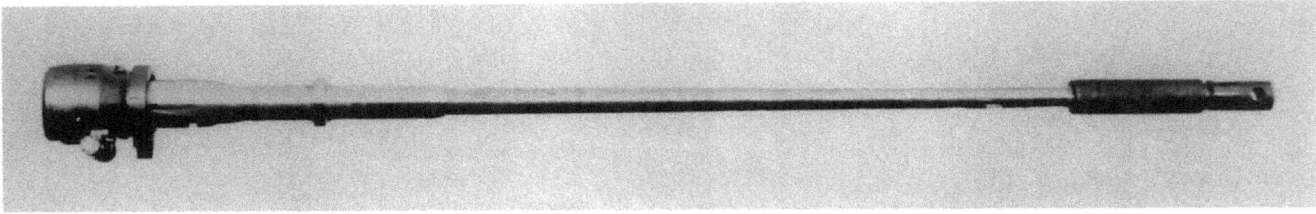

Above and to the right are photographs of the 90mm gun T208E9. Below are drawings of the recoiling gun mount installed in the T95E1 tank.

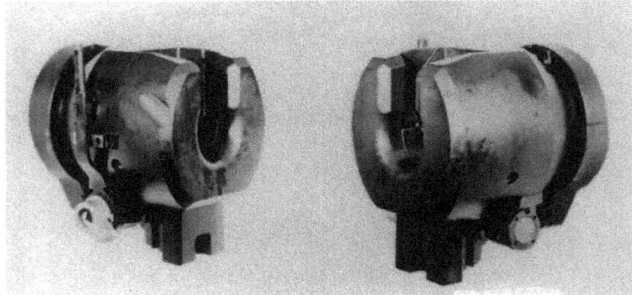

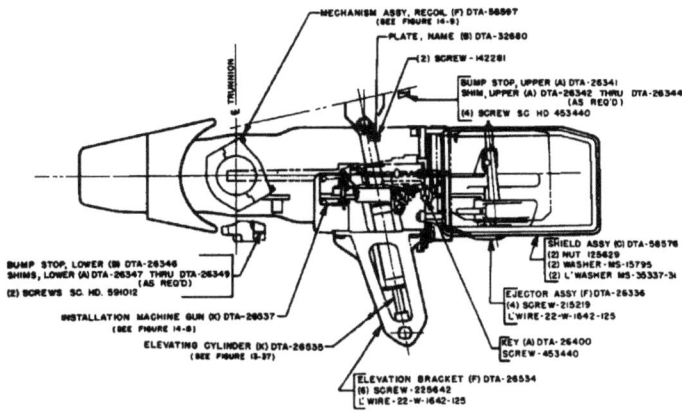

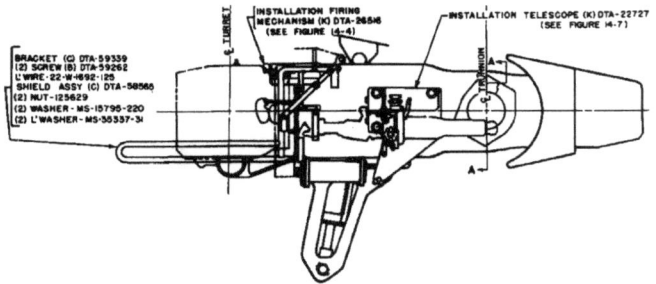

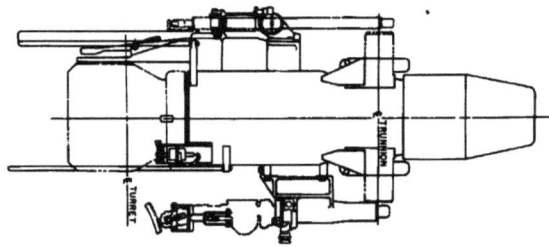

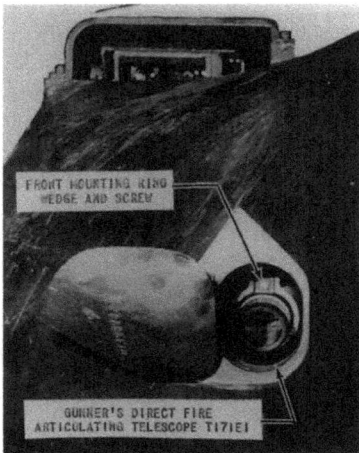

MECHANISM ASSY, RECOIL (F) DTA-58597
(SEE FIGURE 14-9)
PLATE, NAME (B) DTA-32680
(2) SCREW - 142281
BUMP STOP, UPPER (A) DTA-26341
SHIM, UPPER (A) DTA-26342 THRU DTA-26344
(AS REQ'D)
(4) SCREW SC HD 453440

SHIELD ASSY (C) DTA-58576
(2) NUT 125629
(2) WASHER MS-15795
(2) L'WASHER MS-35337-34

EJECTOR ASSY (F) DTA-26336
(4) SCREW-215219
L'WIRE-22-W-1642-125

KEY (A) DTA-26400
SCREW-453440

BUMP STOP, LOWER (B) DTA-26346
SHIMS, LOWER (A) DTA-26347 THRU DTA-26349
(AS REQ'D)
(2) SCREWS SC HD. 59IOI2
INSTALLATION MACHINE GUN (X) DTA-26637
(SEE FIGURE 14-8)
ELEVATING CYLINDER (X) DTA-26635
(SEE FIGURE 8-27)

ELEVATION BRACKET (F) DTA-26534
(6) SCREW-225442
L'WIRE-22-W-1642-125

BRACKET (C) DTA-59339
(2) SCREW (B) DTA-59262
L'WIRE-22-W-4692-125
SHIELD ASSY (C) DTA-58566
(2) NUT-125629
(2) WASHER-MS-15795-220
(2) L'WASHER-MS-35337-34

INSTALLATION FIRING
MECHANISM (K) DTA-26546
(SEE FIGURE 14-4)

INSTALLATION TELESCOPE (K) DTA-22727
(SEE FIGURE 14-7)

FRONT MOUNTING RING
WEDGE AND SCREW

GUNNER'S DIRECT FIRE
ARTICULATING TELESCOPE T171E1

The armor cover is open at the left exposing the gunner's telescope. The right and left sides of the T95E1 turret front can be seen below. Note the differences compared to the T95 on page 59.

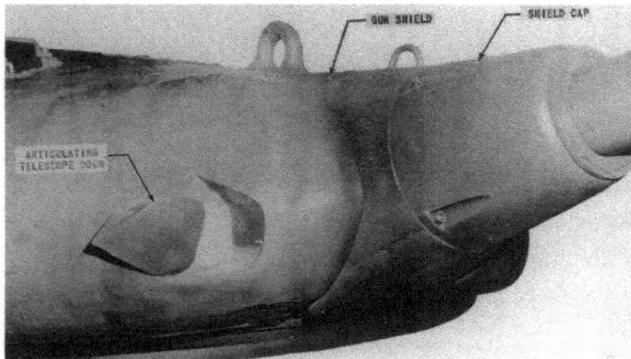

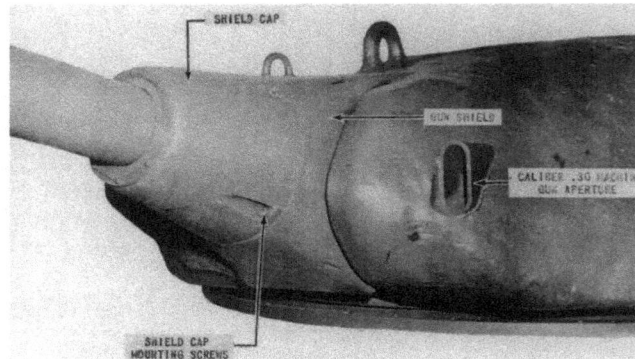

GUN SHIELD
SHIELD CAP
ARTICULATING
TELESCOPE DOOR

SHIELD CAP
GUN SHIELD
CALIBER .30 MACHINE
GUN APERTURE
SHIELD CAP
MOUNTING SCREWS

68

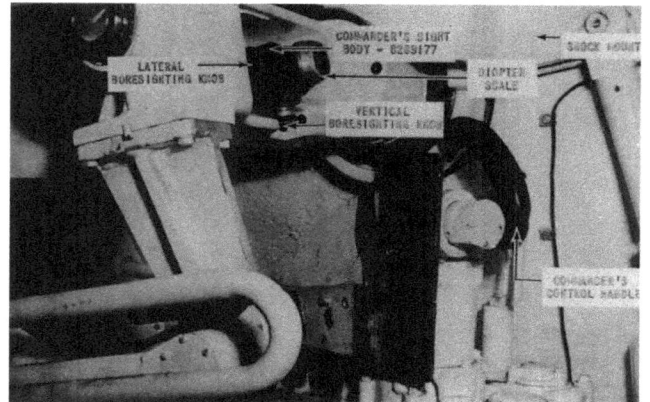

Above and at the left are inside and outside views of the tank commander's telescopic sight on the T95E1. Note the absence of the Optar range finder. Below, the fire control equipment can be seen through the tank commander's hatch.

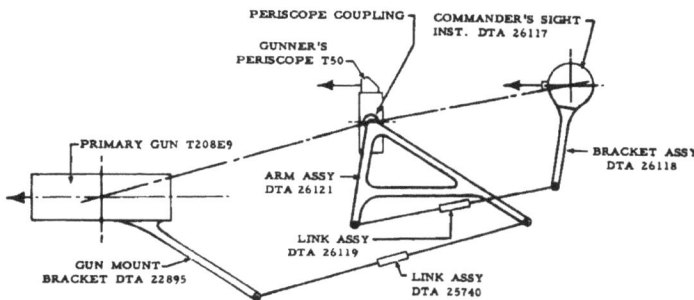

Above is a schematic diagram of the fire control system in the T95E1. Below, the gunner's sighting equipment is at the left and the interior of the tank commander's cupola is at the right.

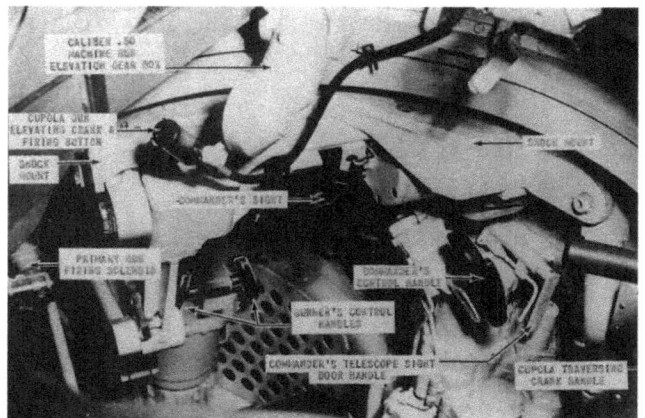

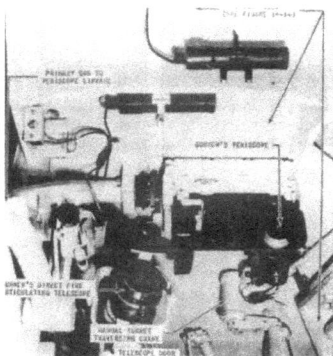

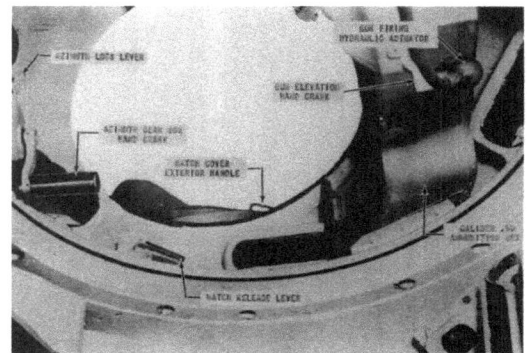

The cannon was elevated and traversed both manually and by a Cadillac Gage constant pressure hydraulic system. The latter equipment was the same as on the M48A2 tank except that with the simplified fire control system, no provision was made for superelevation to be applied to the mount. The Type Z fire control system eliminated the range finder and coupled the gunner's T50E2 periscope and the commander's T183E1 offset telescope directly to the gun mount. Without the Optar range finder, only a small armor cover was required to protect the offset telescope on the right side of the turret. The range was estimated visually and use was made of the proper reticle in either of the battlesights. The gun mount was not stabilized in either azimuth or elevation. The crew arrangement and stowage in the T95E1 was essentially the same as in the T95.

69

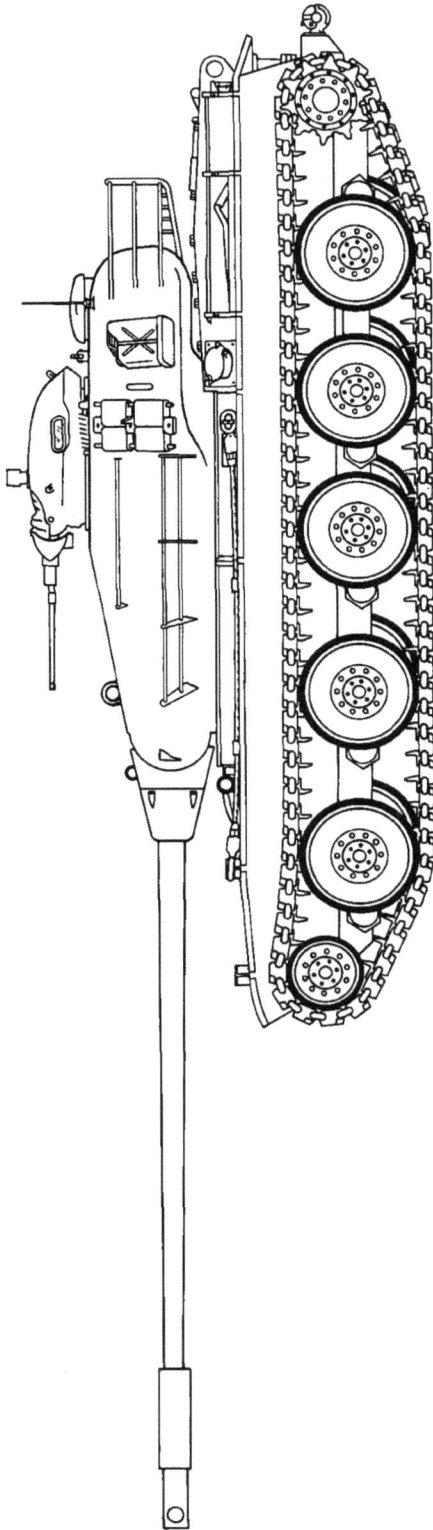

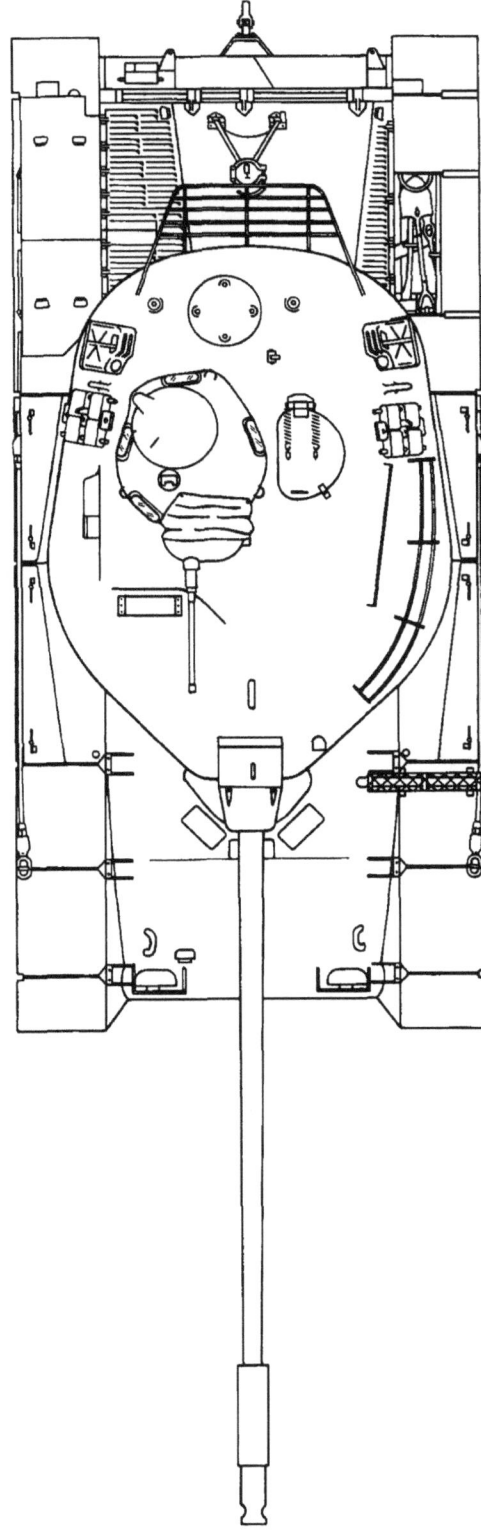

© D.P. Dyer

90mm Gun Tank T95E1

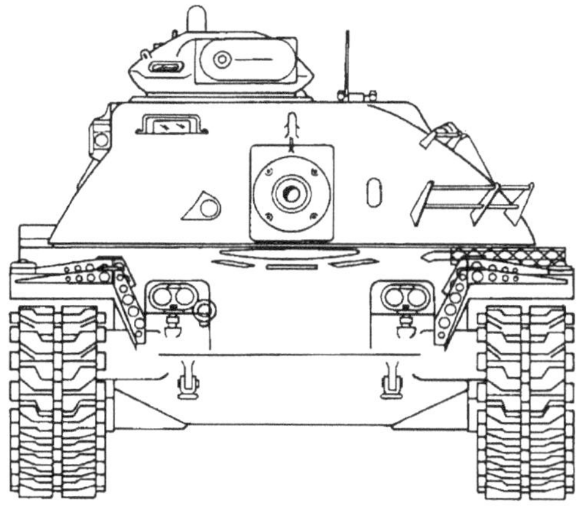

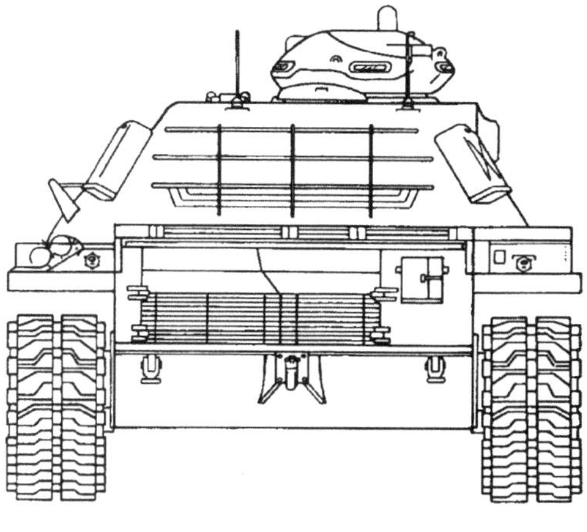

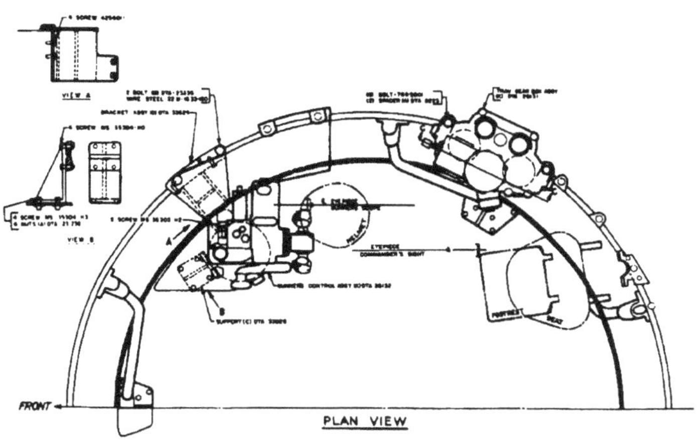

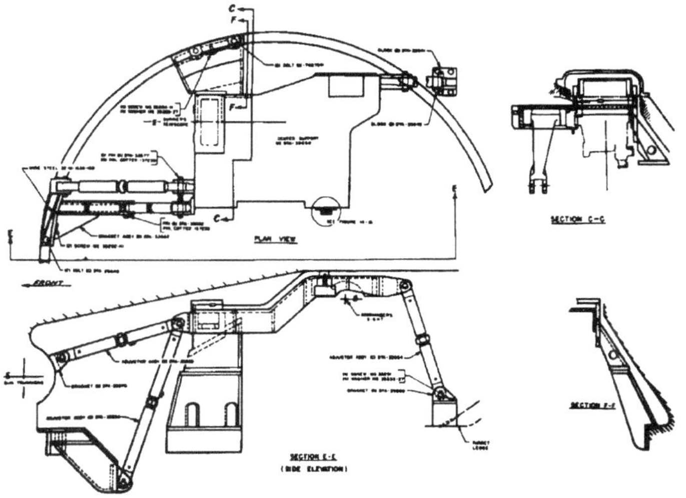

The right side of the turret basket is sketched above with drawings of the fire control equipment shock mount at the right. Below, the T95E1 turret basket is at the left and part of the fire control equipment shock mount is visible at the right.

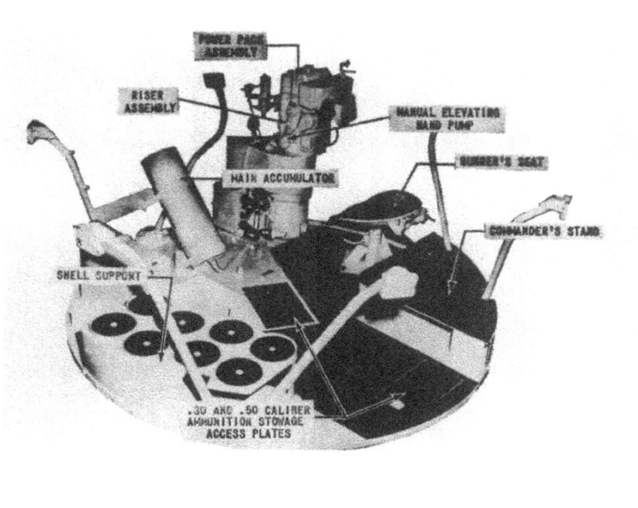

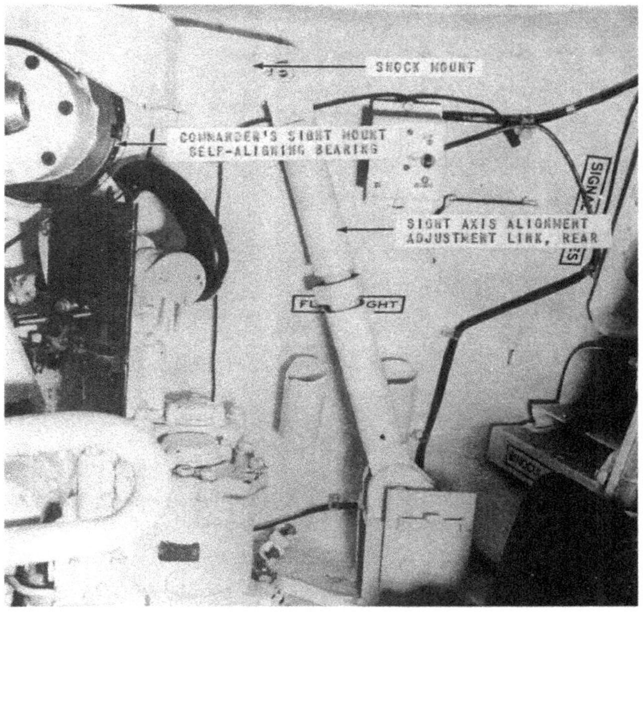

The first pilot 90mm gun tank T95E2, registration number 9B1048, appears above and below. This was the first tank completed in the T95 series.

The T95E2 and T95E3 tanks were utilized primarily as automotive test vehicles. Although they also originally were intended to provide insurance against the failure of the smooth bore gun development program, the appearance of other more effective weapons eliminated that requirement before they were completed. The T95E3 was particularly valuable in evaluating the performance of the vehicle at a weight of about 45 tons thus providing useful information for the T95E4 development project. During the various test programs, turrets were frequently switched from one chassis to another. At one time, an M48A2 turret was installed on the T95E1 chassis (serial number 5) and the T95E1 turret was mounted on an M48A2 hull. This latter combination was used during the trials in October 1958 to select the main armament for the XM60 tank.

In the photographs of the first T95E2 below, the gun shield cover and the .50 caliber machine gun have been installed in the top view.

These are additional views of the first pilot T95E2, dated 4 June 1957, after completion at Detroit Arsenal. Note the lightweight road wheels installed in the number 1 position on each side of the tank.

The second pilot 90mm gun tank T95E2, registration number 9B1049, is shown in these photgraphs after its arrival at Fort Knox. The vehicle has full stowage and the secondary armament is installed.

Below are additional views of the second T95E2 at Fort Knox. Note that the solid road wheels are installed at all positions on this tank.

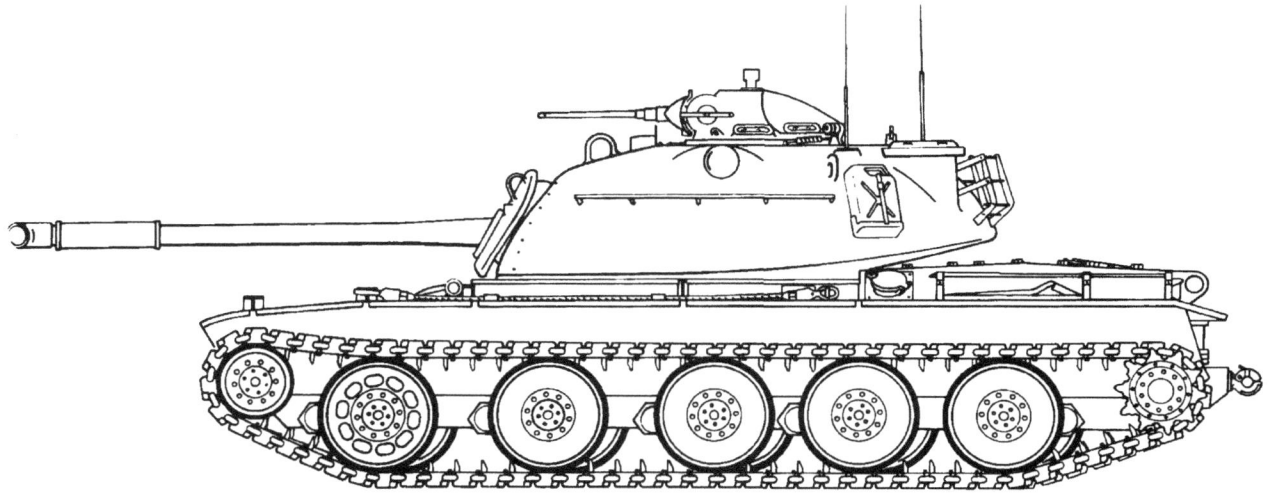

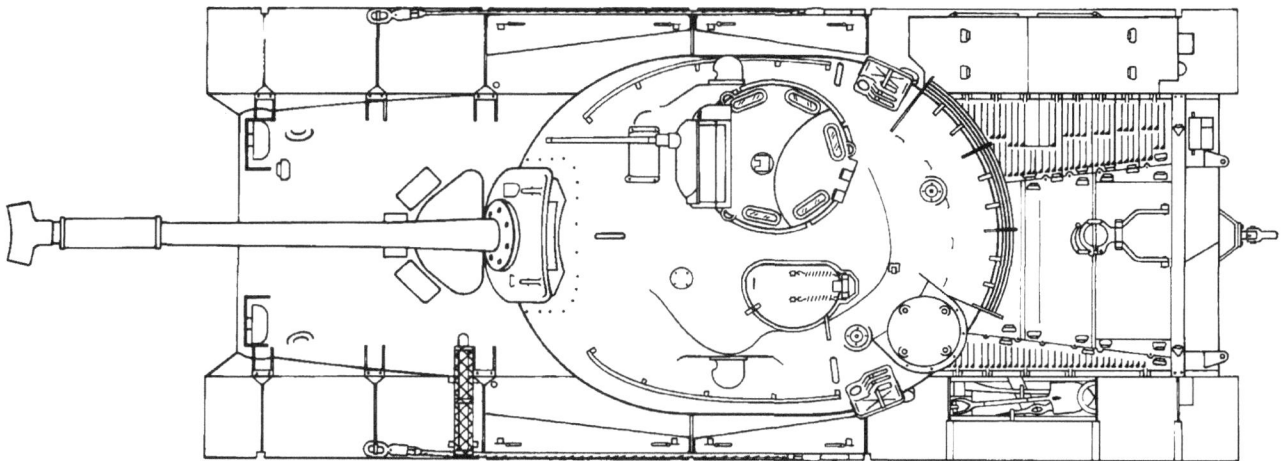

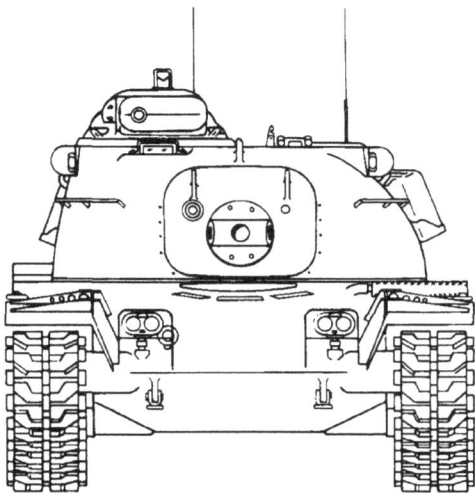

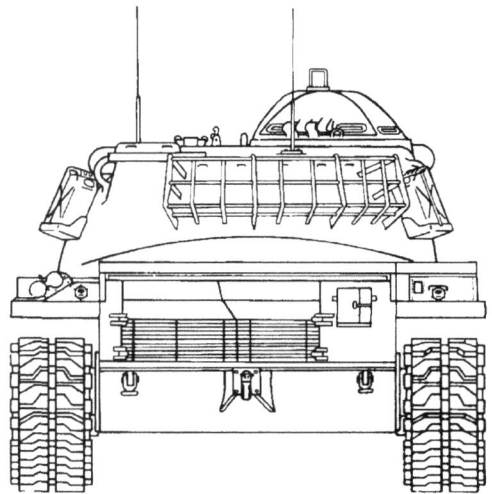

Scale 1:48

© D.P. Dyer

90mm Gun Tank T95E2

The first pilot 105mm gun tank T95E3, registration number 9B1050, appears above and at the right after assembly at Detroit Arsenal. These photographs were dated 9 August 1957. Note that the chassis is fitted with the lightweight road wheels.

The long barrel of the 105mm gun T140E3 is particularly obvious in these views of the first pilot T95E3. Note the new gun travel lock for this weapon. The secondary armament and the vehicle stowage have not yet been installed.

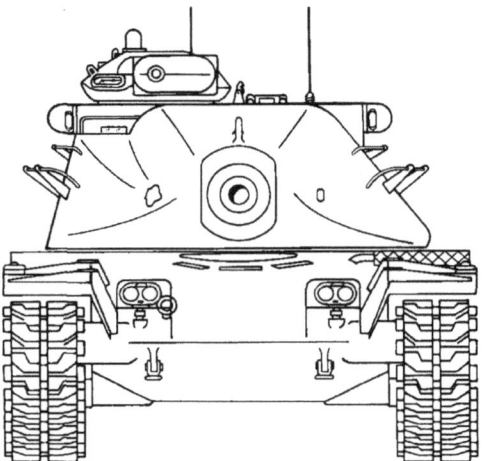

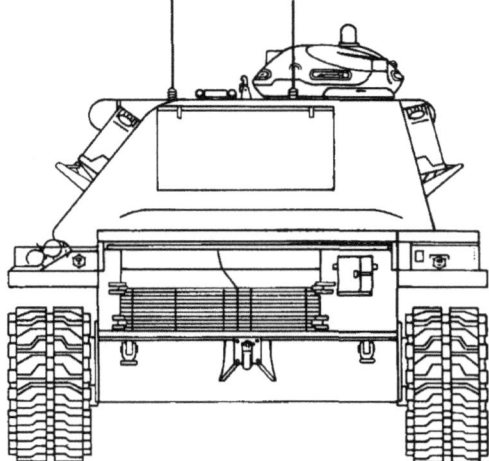

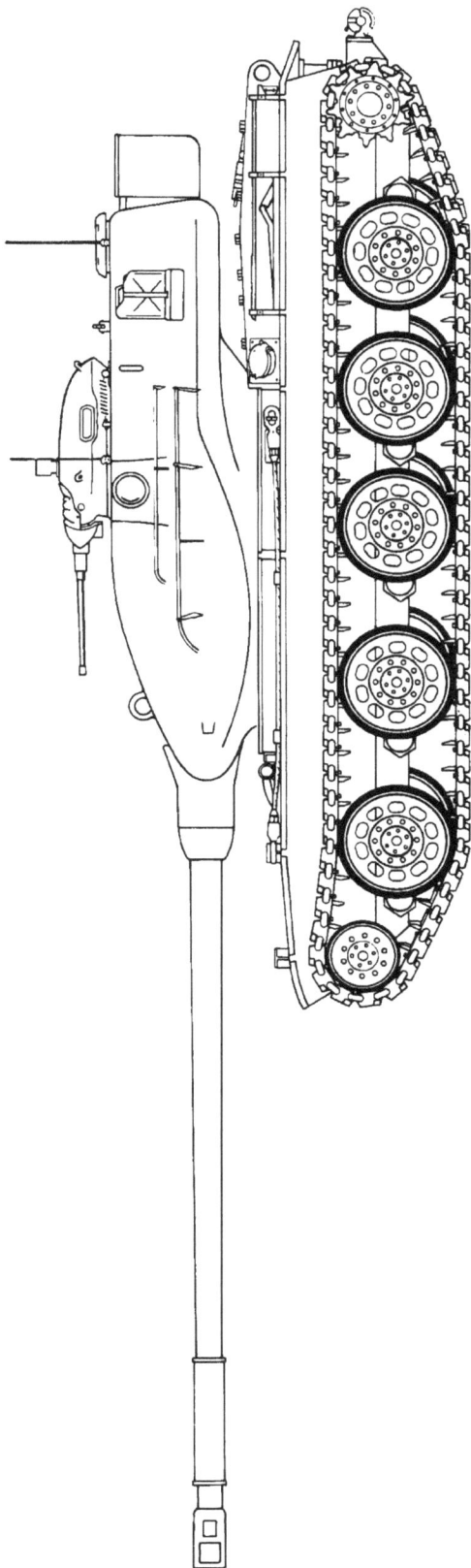

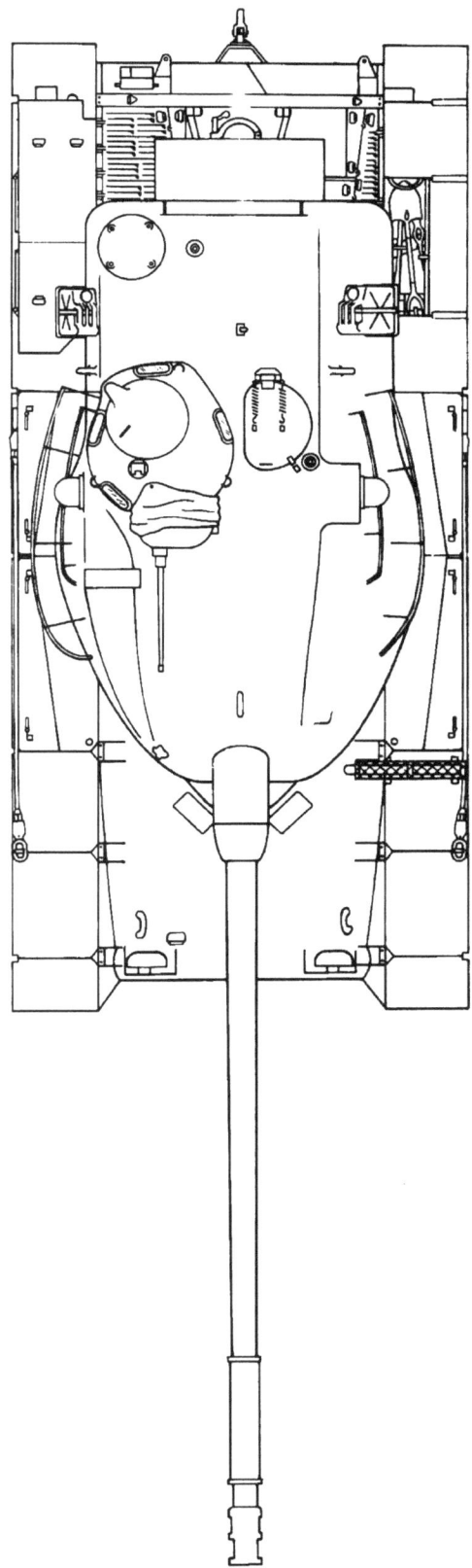

Scale 1:48

105mm Gun Tank T95E3

77

Above and below are the second pilots T95 and T95E2 respectively during the automotive tests at Fort Knox. Note the sectioned mine exploder discs added to the T95 to simulate the weight of a fully stowed tank.

The automotive tests revealed a number of deficiencies. The report from Fort Knox indicated that the T95 was inferior to the M48A2 in cross-country mobility. This was particularly true in deep mud. To lower the ground pressure and increase the traction, several types of tracks were tested as possible replacements for the T114. These included the T95 and the T127, both of which were 24 inches wide compared to the 21 inch wide T114. The T95 was the track originally used on the 90mm gun tank T42 and sufficient quantities remained from that program for test purposes. Both the T95 and the T127 were single pin steel tracks with detachable rubber pads. However, the grouser height was 1¾ inches on the T127 compared to 1¼ inches

on the T95 which gave it slightly better traction in soft earth. The T127 proved to be vulnerable to breakage during the tests and it was strengthened by adding reinforcing members to the body of the track shoe. This modified version was designated as the T127E1. Tests of the T95E2 tank in mud indicated that both the T95 and the T127 tracks were superior to the T114 and the mobility in most cases equaled that of the M48A2 tank with the T97E2 track. This was particularly true if the rubber pads were removed. However, the clutch-brake steering in the lower speed ranges of the XTG-410-1 transmission caused problems in deep mud. Since the inner track came to a complete stop during a turn, the agility of the vehicle was less than that of

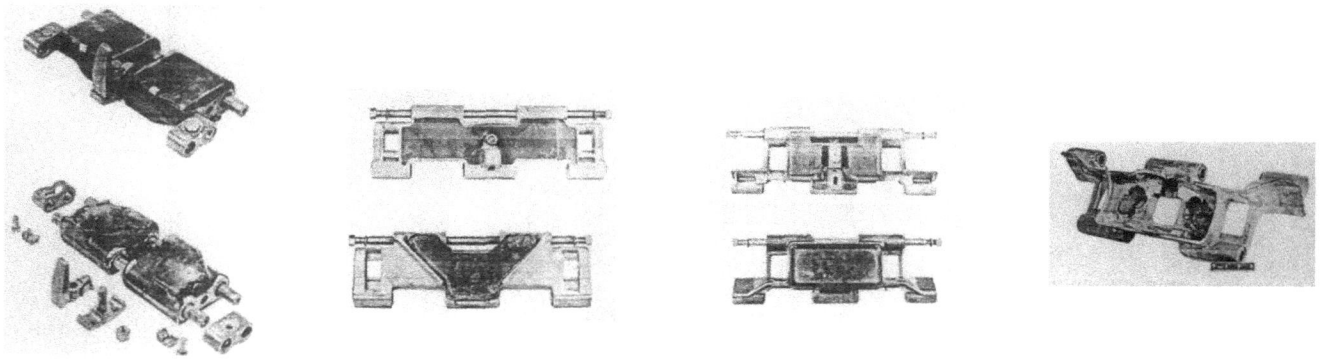

Above from left to right are the T114, T95, T127, and the T127E1 track links. At the right is a T95 test rig fitted with the T127 track.

the M48A2 with its CD-850 transmission. In the latter tank, power was transmitted to the inner track at all times. During the test program, the XTG-410-1 transmission was modified to permit geared steering in a lower gear improving the performance in deep mud.

As a result of the various armament studies, procurement was approved for two conversion kits to install the British 105mm gun in the M48A2 turret. Later, two additional kits were authorized to install the American version of this weapon, the 105mm gun T254, in the M48A2 turret on the T95E2. At that time, the vehicle was to be redesignated as the 105mm gun tank T95E5. Since the armor protection provided by the M48A2 turret was inferior to that of the T95 type turret, another variation was proposed. The T95E1 (chassis number 5) and the second T95 (chassis number 7) were to be modified to carry the 105mm gun T254E2 in a recoiling mount. These vehicles were to be designated as the 105mm gun tank T95E7. Unfortunately, time ran out for the T95 project before either the T95E5 or the T95E7 were completed. However, a conversion kit was used to install the 105mm gun T254 and later the British X15E8 105mm gun in an M48A2 turret for the October 1958 trials to select the gun for the XM60 tank.

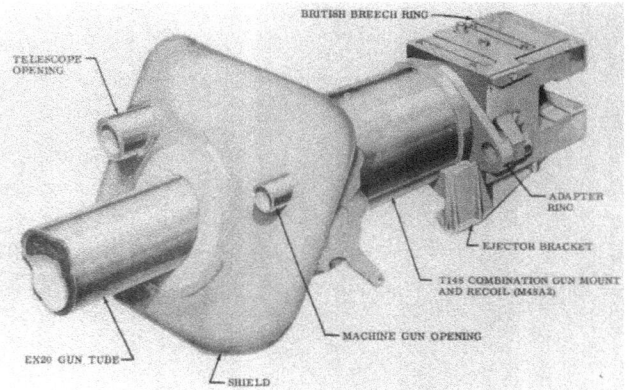

The kit for mounting the British 105mm gun in the M48A2 tank turret is depicted at the right. Below, the dimensions are given for the British 105mm gun (upper) and the U.S. 105mm gun T254 (lower).

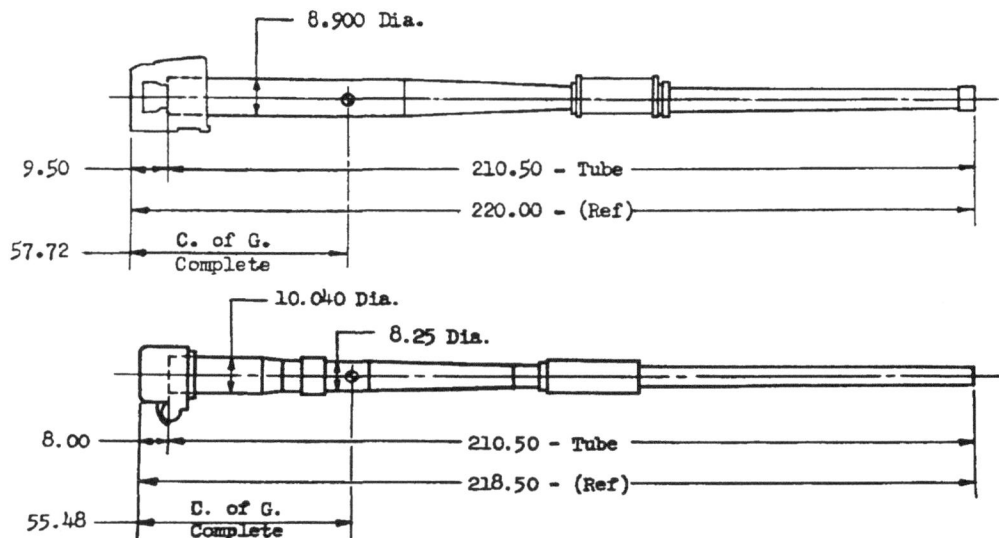

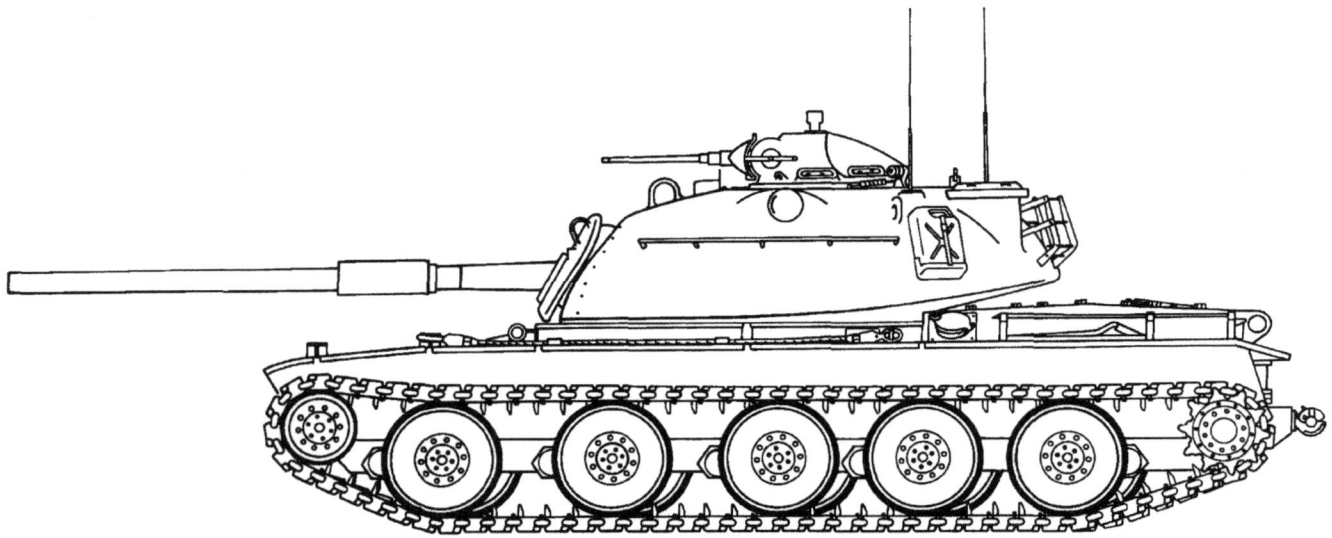

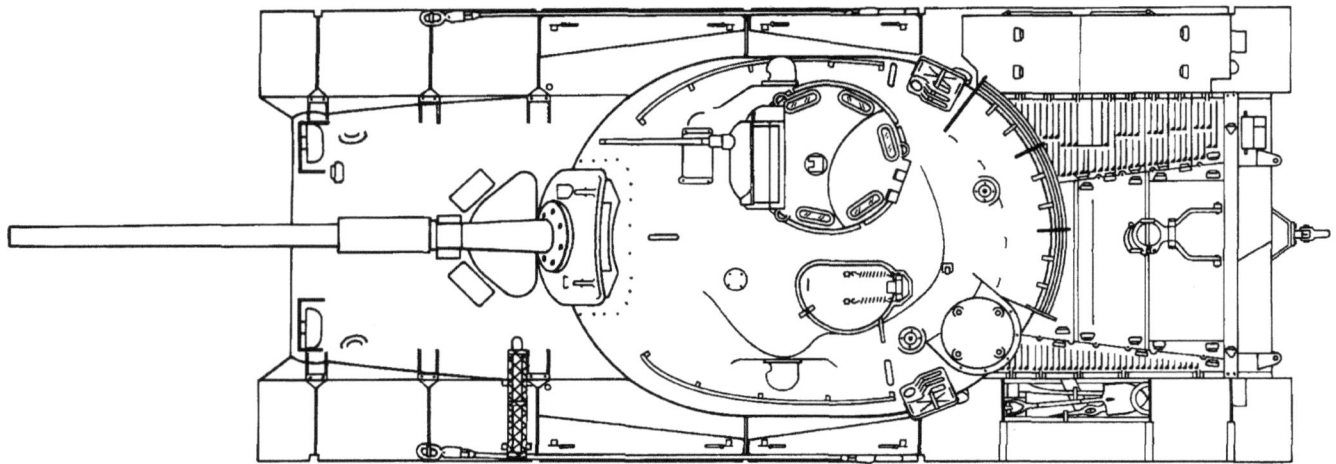

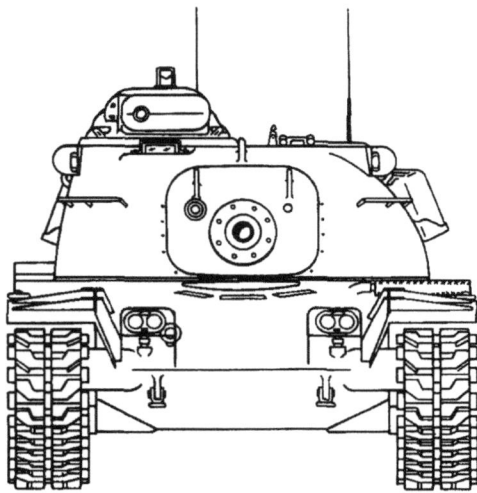

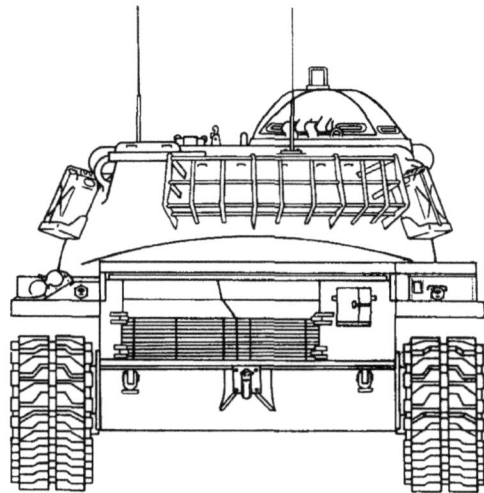

Scale 1:48

© D.P. Dyer

105mm Gun Tank T95E5

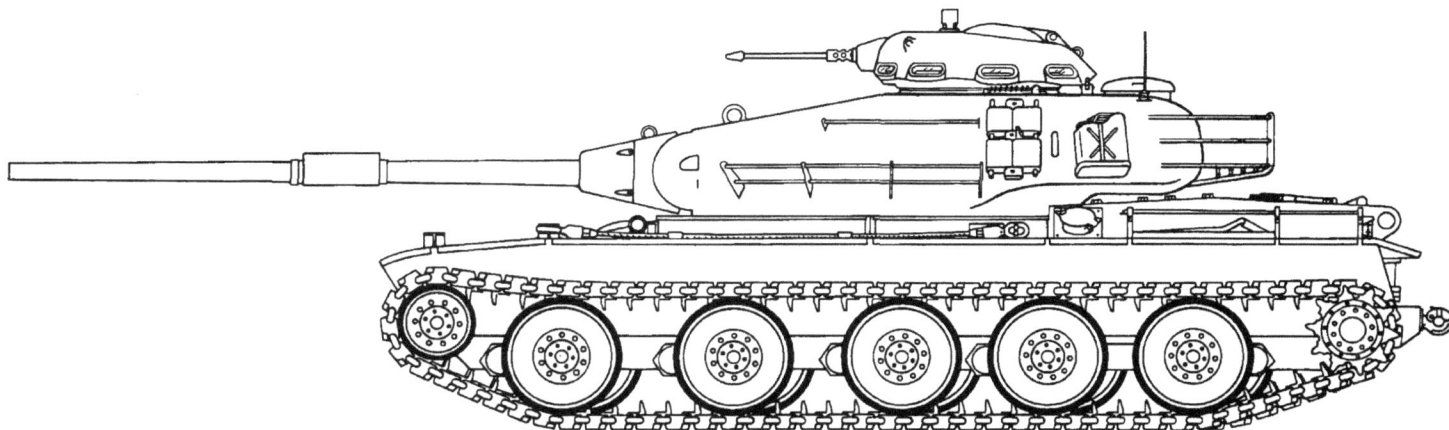

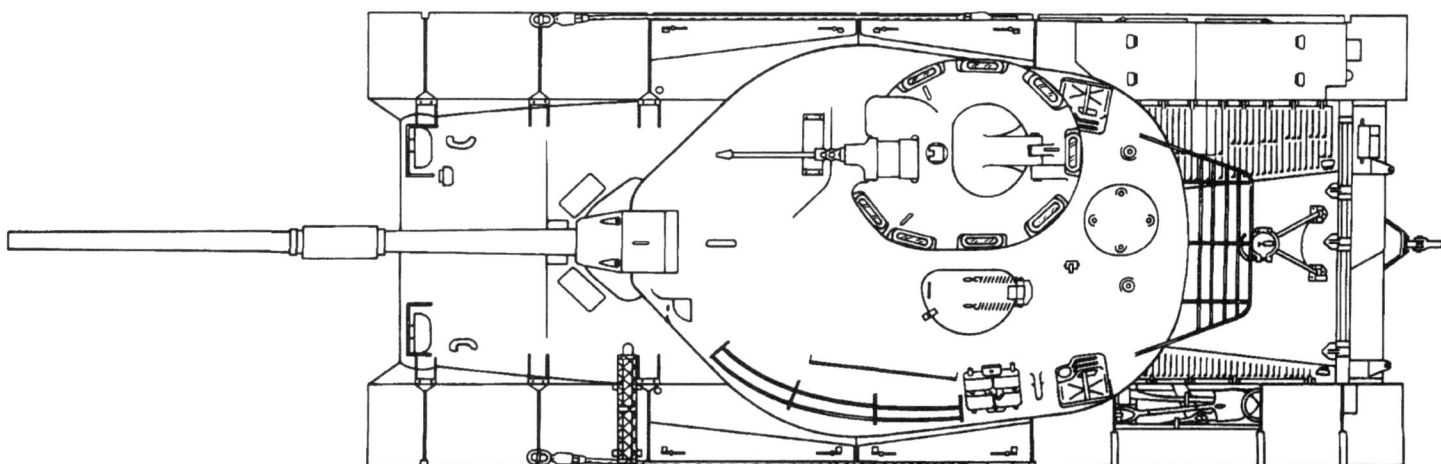

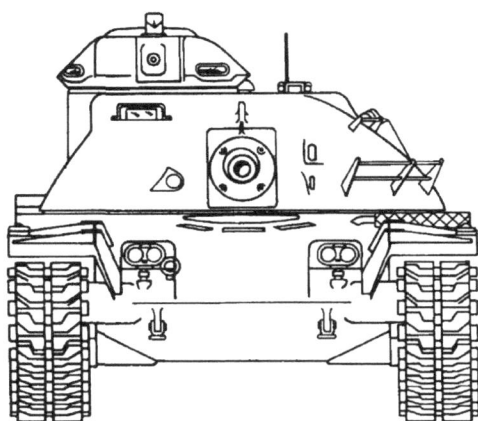

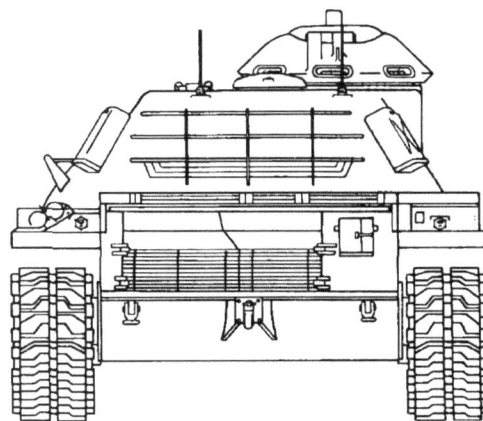

Scale 1:48

105mm Gun Tank T95E7

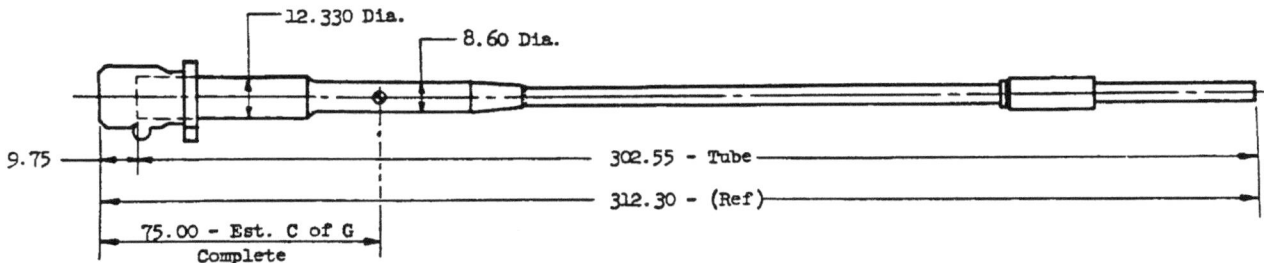

The dimensions of the 105mm smooth bore gun T210 are shown in the sketch above.

Although the T96 project was initiated in January 1955, it was not until July 1956 that accurate data were available on the size and configuration of the 105mm gun T210 and minor changes followed after that. The final dimensions of both the gun and its ammunition were larger than those originally supplied to the Ford Motor Company. The new gun dimensions required that the turret height be increased by two inches and the bustle be lengthened by five inches. The increased height in the modified turret provided adequate space for recoil and the space and weight saving advantages possible with the smaller gun and the rigid mount could no longer be realized. At this point, the combination gun mount was redesigned to include a recoil system. After the program was reoriented in October 1956 to use the T95 chassis for both the T95 and T96, this same turret was applied to the T95E4.

Because of the extreme length of the fixed ammunition (48 inches), the smooth bore 105mm gun T210E4 had to be mounted well forward in the turret to provide adequate space behind the breech for loading. With this arrangement, an equilibrator was required to compensate for the unbalanced condition of the gun. The concentric recoil mechanism in the combination mount was designed for a recoil distance of ten inches and the elevation range of the cannon extended from +20 to –9 degrees, except over the rear deck where no depression was possible. A coaxial .30 caliber machine gun M37 was installed on the left side of the combination mount and a T171E2 articulated telescope was located on the right. The type W fire control system was planned for installation as on the T95. A Cadillac Gage constant pressure hydraulic gun control system was provided for elevation and traverse. No stabilization was included. It was intended to install the larger T9 cupola for the tank commander. This cupola required a mounting ring 34 inches in diameter compared to a 29.75 inch diameter opening for the T6 cupola. A modified version of the T9 was later standardized as the M19 and installed on the M60 tank. The T9 was armed with the short receiver .50 caliber machine gun T175. Stowage was provided in the T95E4 for 40 rounds of 105mm ammunition with 12 rounds in the bustle, eight rounds in the turret basket, and 20 rounds in the hull.

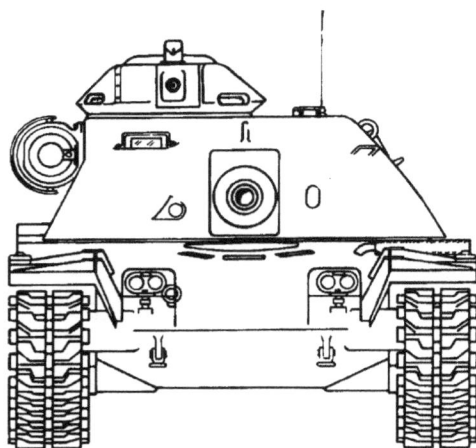

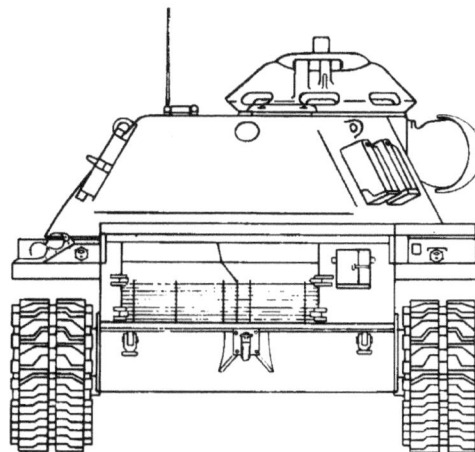

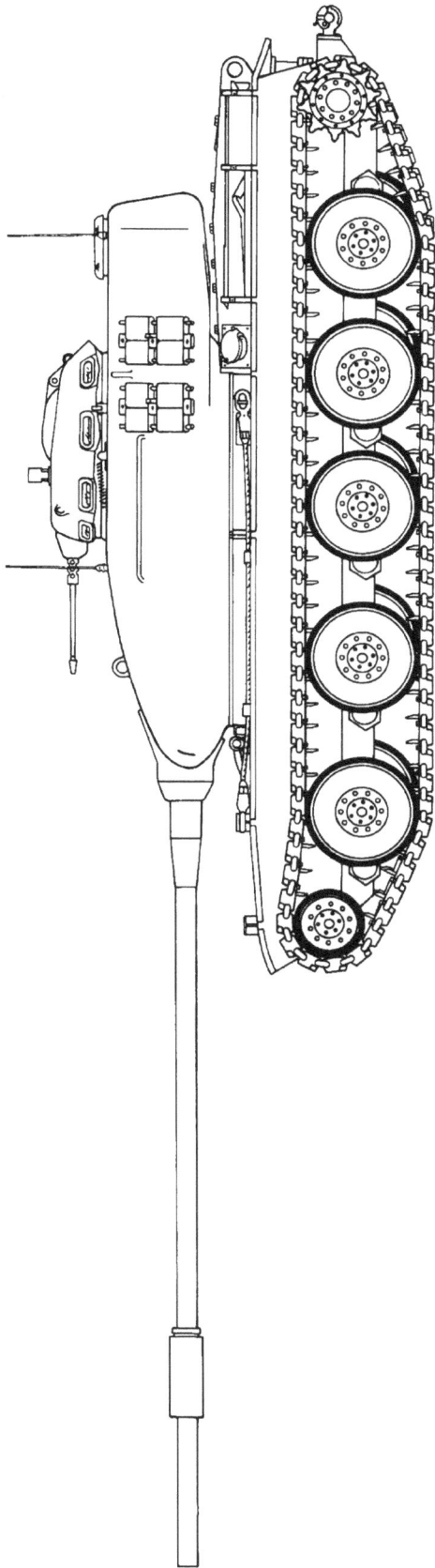

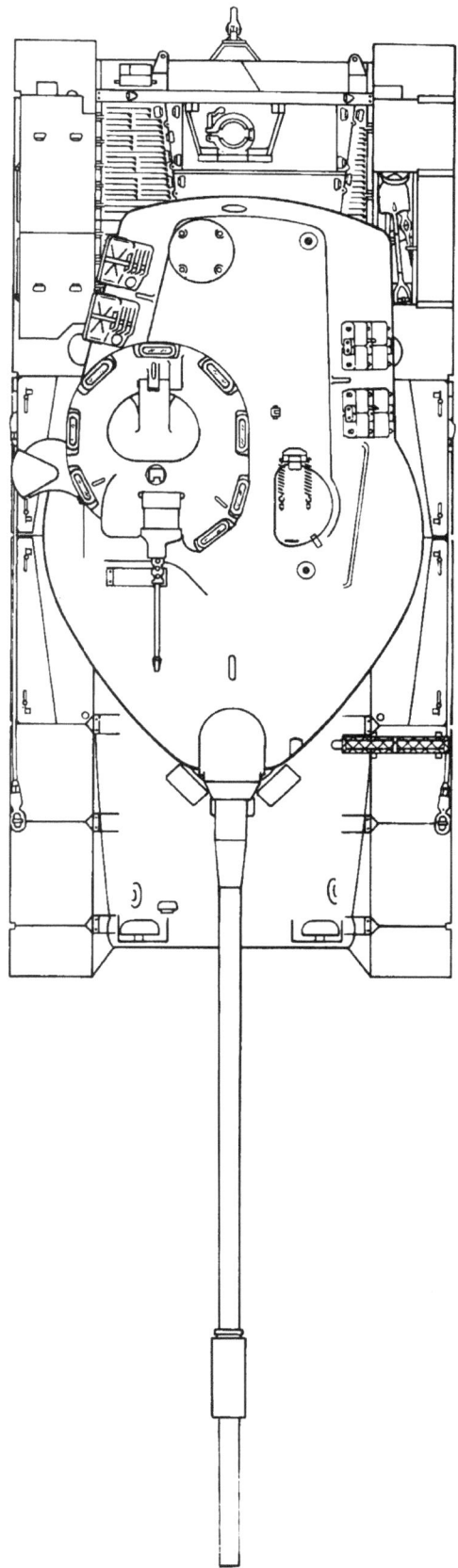

Scale 1:48

105mm Gun Tank T95E4

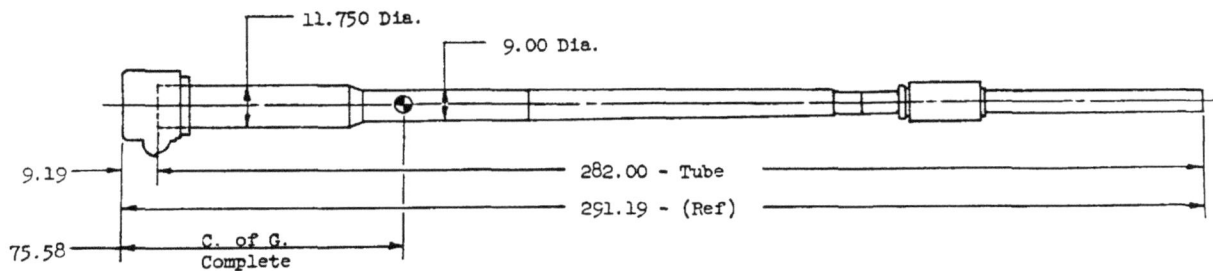

11.750 Dia.

9.00 Dia.

9.19

282.00 - Tube

291.19 - (Ref)

75.58

C. of G.
Complete

The sketch above shows the dimensions of the 120mm gun T123E6.

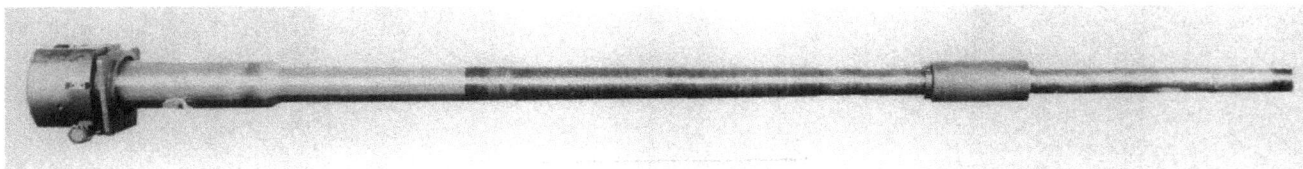

The 120mm gun T123E6 appears above and at the right.

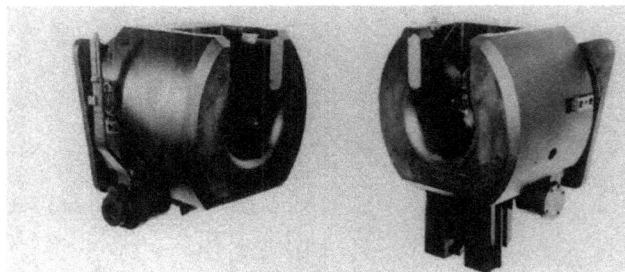

With the appearance of the lightweight 120mm gun T123E6, the contract with the Ford Motor Company was revised to replace the 105mm gun T210 in two of the four T95E4 tanks with the new weapon. These vehicles were redesignated as the 120mm gun tank T95E6. Because of difficulties in the smooth bore development program, interest now shifted to this vehicle for the new heavy gun tank and all work on the 105mm gun tank T95E4 was suspended in February 1958. The same mock-up was used for the T95E4 and the T95E6 with the appropriate modifications to the armament and stowage. Because of the shorter lengths of the two piece 120mm ammunition, less space was required for loading behind the gun breech. Thus the 120mm gun was moved to the rear in the turret compared to the 105mm weapon, allowing it to be balanced. With this installation, the mount could have been stabilized. The

T95E6 used the same Cadillac Gage gun control system as the T95E4. The concentric recoil mechanism in the combination gun mount was modified to reduce the recoil distance to five inches. A .30 caliber machine gun M37 was located on the left side of the mount with the T171E2 articulated telescope on the right. The remainder of the type W fire control system was the same as on the T95E4.

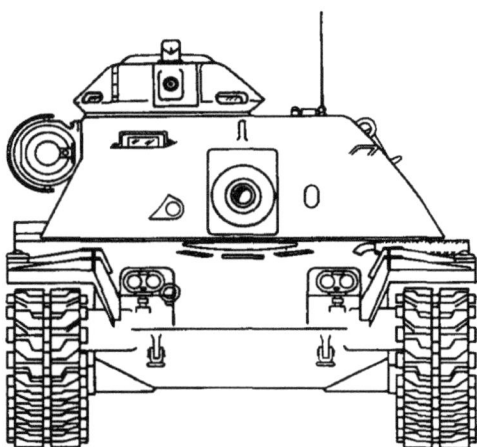

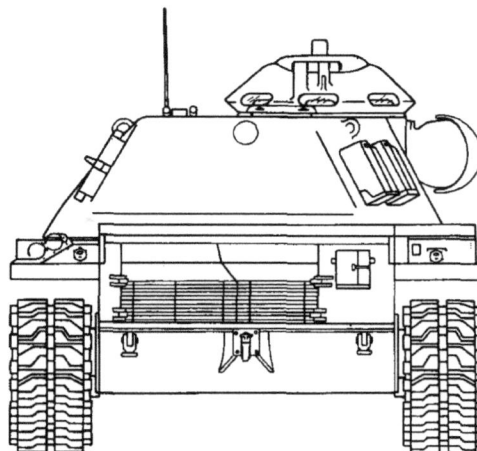

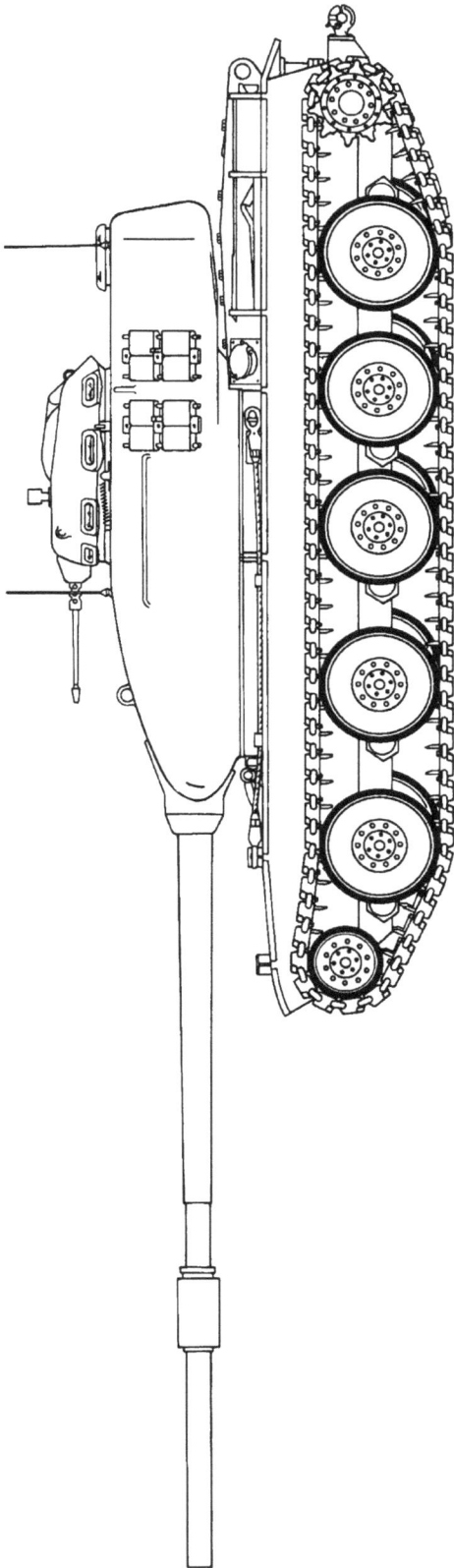

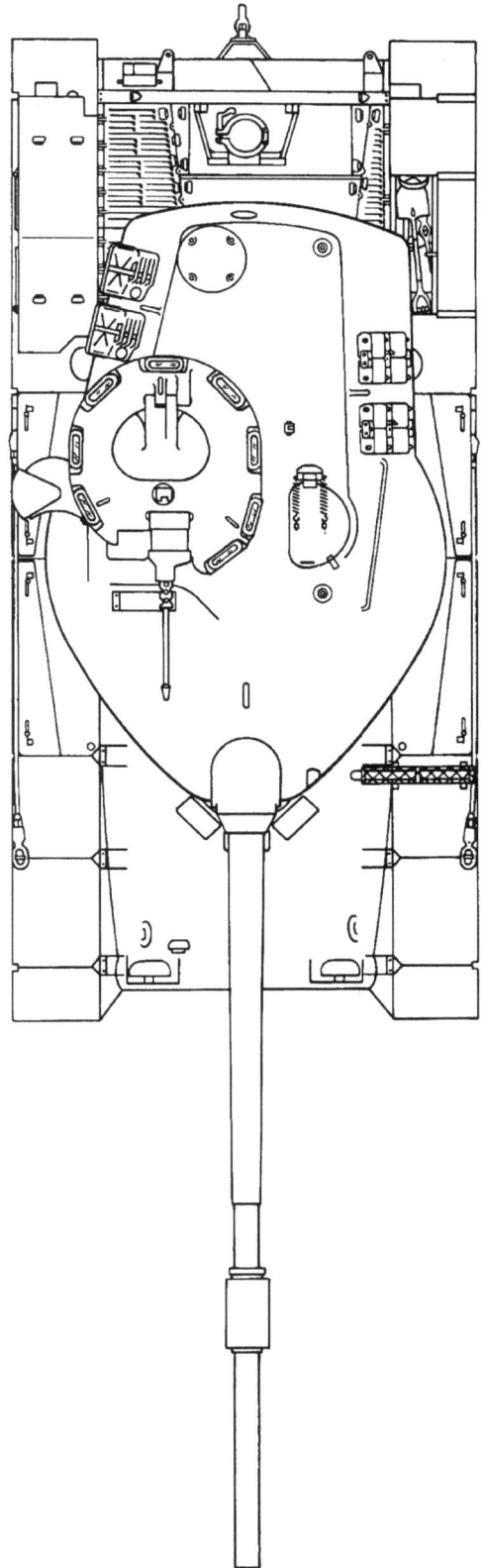

© D.P. Dyer

120mm Gun Tank T95E6

Above is an artist's concept of the 120mm gun tank T95E6. Three views of this vehicle are sketched at the left below.

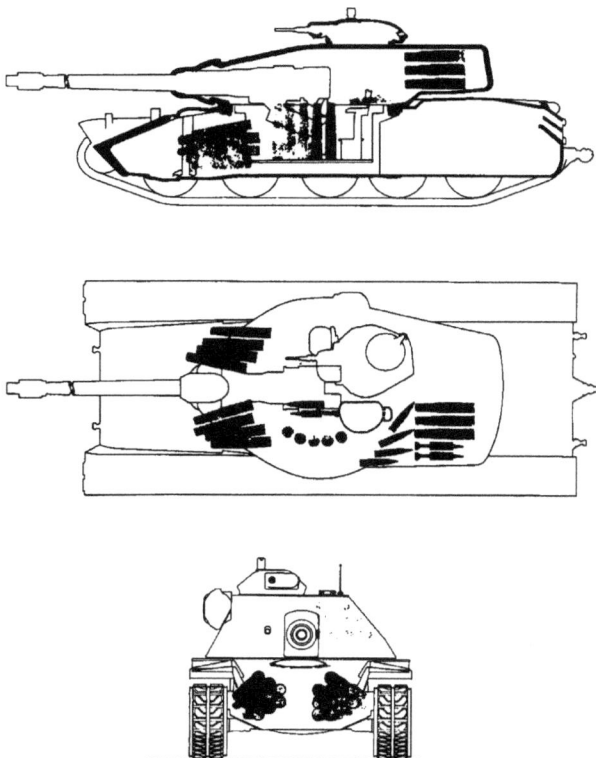

With the new armament, the stowage in the T95E6 was rearranged to accommodate 36 rounds of 120mm ammunition. Eight rounds were located in the turret bustle, five rounds in the turret basket, and 23 rounds in the front hull with 11 on the right side of the driver and 12 on his left. A power rammer, a motor driven hoist, and a simple ejection system for spent cartridge cases were proposed to assist the loader in handling the heavy 120mm ammunition. The hoist was used to lift the projectile and cartridge case into place in the gun breech and the loading tray. They were then chambered in the gun by manually releasing the locking mechanism of the spring loaded ramming hammer. When the gun was fired, the recoil compressed the spring and recocked the rammer for the next round. After firing, the loader placed the empty cartridge case on two guide rails suspended from the turret bustle roof. A spring loaded push rod was then rotated into contact with the base of the case. This automatically opened a small port in the rear wall of the turret bustle. Actuating a trip release then ejected the cartridge case from the turret. An electric motor then compressed the ejector spring to the cocked position and closed the bustle port.

The armor on the turret for the T95E4 and T95E6 offered greater protection than that on the turrets of the heavy tank M103 series or the original T96 proposal. From directly in front, it was invulnerable to the armor piercing

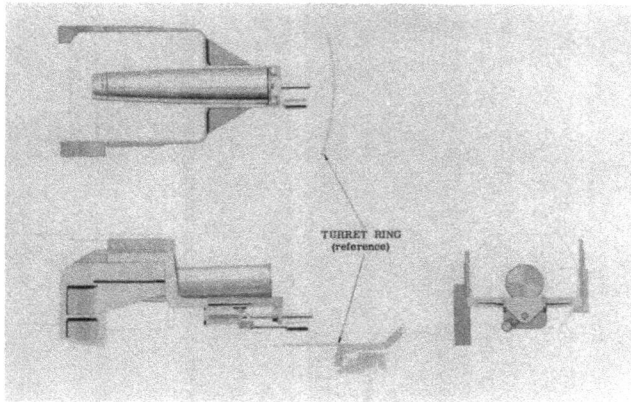

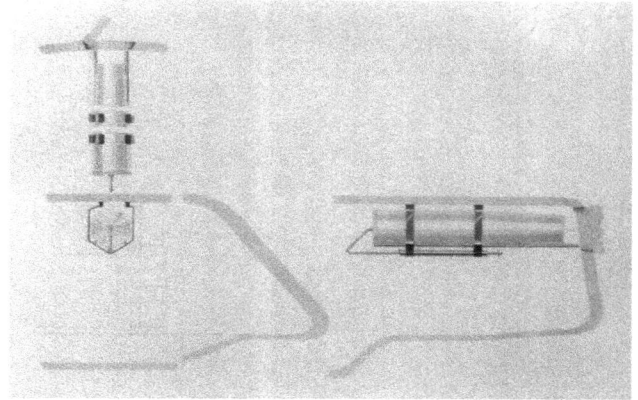

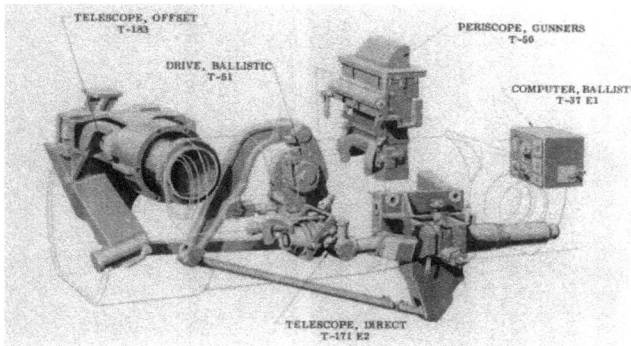

Above, the rammer and shell case ejector proposed for the T95E6 are shown at the left and right respectively. At the left is the T95E6 version of the W fire control system. Below is the T95E6 turret mock-up modified for application to the XM60 tank. The sketch at the bottom of the page shows the protection afforded by the T95E6 turret.

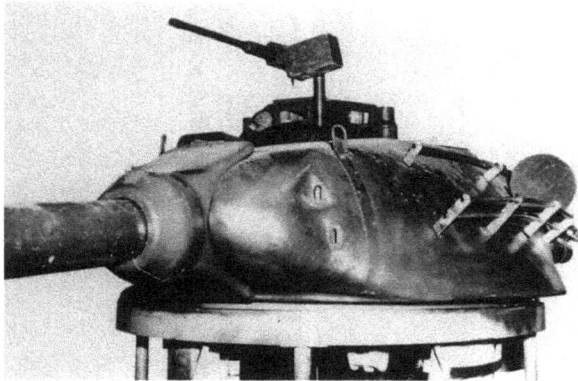

round of the Soviet 100mm tank gun at point-blank range. The M103 and the T96 turrets could be penetrated from the front by this round at ranges up to 2500 yards and 1500 yards respectively. Both the T95E4 and T95E6 turrets were out of balance and the condition was much worse in the case of the T95E4, even though both tanks used the same turret casting. This was due to the necessity of mounting the 105mm gun T210E4 farther forward in the turret to provide adequate loading space for the long fixed round of ammunition. Calculations indicated that to balance the turrets satisfactorily, counterweights of 547 pounds and 208 pounds would have to be added to the turret bustle on the T95E4 and T95E6 respectively. This brought the estimated combat weight to 89,241 pounds for the T95E4 and 90,319 pounds for the T95E6.

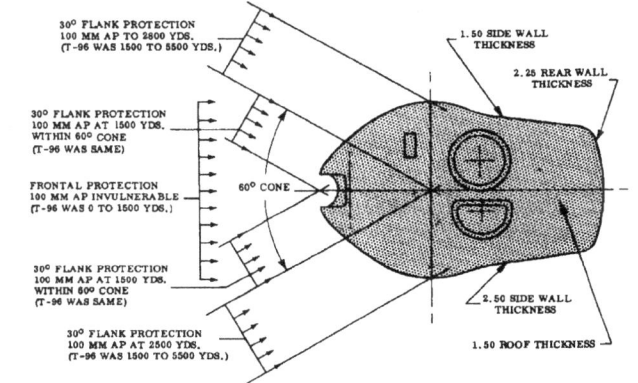

Above and at the left below is the full size mock-up of the 120mm gun tank T95E6. The mock-up of the T9 commander's cupola can be seen below at the right. A drawing of the integrated commander's station appears at the bottom of the page.

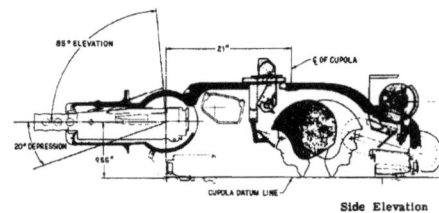

Side Elevation

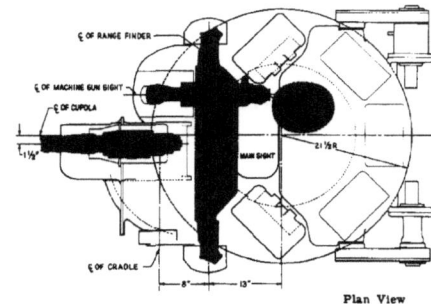

Plan View

Initially, it was intended to install the T9 cupola on the T95E6 as well as the T95E4. However, a new integrated commander's station was under development for future use. This program was planned in three phases. Phase I was proposed as only a maximum vision commander's station to be installed for test purposes on the T95E1. It consisted of an armor shell and hatch with six wide angle periscopes, three in the hatch and three in the cupola shell, to provide a 360 degree view for the tank commander. Also, a pintle mount was to be included for a .50 caliber machine gun. Phase II was referred to as a stripped down version of the integrated commander's station. It was intended to use this model to evaluate the type Y fire control system. The cupola was armed with the short receiver .50 caliber machine gun T175E1 supplied by a 400 round doughnut shaped ammunition box located above the cupola race ring. A T57 short base, coincidence type, optical range finder was installed in the cupola along with the commander's main sight. The latter consisted of an M20 head piece in the cupola roof which could accept a variety of electronic or optical vision devices. Data from the range finder were transmitted electrically to the ballistic computer in the turret. Five wide angle periscopes were mounted, three in the hatch and two in the cupola roof, to provide 360 degree vision. Armor on the cupola ranged in thickness from 2.5 inches in the

88

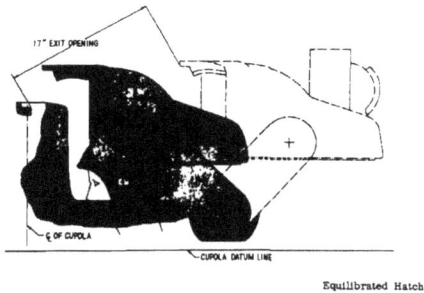

Equilibrated Hatch

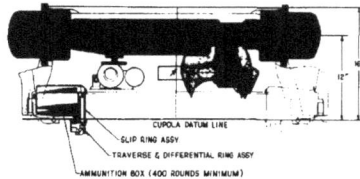

Range Finder Installation

The hatch and range finder arrangement in the integrated commander's station is sketched at the left. Above is the proposed installation of jettisonable fuel tanks on the T95E6.

front to 1.75 inches on the sides and rear, all at 30 degrees from the vertical. The cupola roof was one inch thick. All optical devices outside the armor were to be protected with ½ inch thick steel shutters when not in use. Phase III of the program was to include all the features of Phase II plus power traverse for the cupola as well as main gun and cupola target designation systems. This equipment allowed the commander to automatically rotate the main gun or the line of sight of his .50 caliber machine gun, range finder, and main sight on to the center line of any one of his periscopes. This was accomplished by pressing a button under the appropriate periscope. The cupola was designed with an equilibrated hatch cover which would require a maximum hand effort of 35 pounds to open. The vertical rise of the hatch cover on opening was about seven inches with a stop at four inches. This provided a direct vision slot for the tank commander over a sector of about 250 degrees with overhead protection from blast.

Despite the reduced fuel consumption of the AOI-1195-5 engine compared to the power plant in the M48 series tanks, the range of the T95 series was considered to be inadequate. To alleviate this problem, a jettison fuel tank kit was designed which increased the available fuel supply by approximately 50 per cent. The kit consisted of two standard 55 gallon drums carried on a rack at the top rear of the hull. The drums fed into the main fuel tanks through quick-disconnect fittings and they could be jettisoned in an emergency from inside the tank. The drums were mounted transverse to the vehicle center line with approximately 12 inches between them. This provided space for the main gun barrel when it was in the travel lock. A panel below the drums deflected the heat from the engine exhaust.

In April 1957, the Subcommittee on Automotive Equipment recommended to the Ordnance Technical Committee that siliceous cored armor be adapted to the T95 tank. Several siliceous cored armor castings already had been procured for ballistic tests. After the suspension of the T95E4 program, procurement was authorized for two additional T95 chassis to mount the T95E6 turrets. This brought the total number of T95 chassis under the program to eleven. The first pilot T95E6 (chassis number 10) was to have a cast homogeneous steel armor turret, but the hull was to be fabricated from siliceous cored armor. The second pilot T95E6 (chassis number 11) was to utilize siliceous cored armor in both the turret and the hull. The T95 program was terminated before either pilot was completed.

Although the jettison fuel tanks provided a stopgap method of extending the range of the T95 series vehicles, it was not a satisfactory solution to the problem. The development of the X type diesel engine originally proposed for the T95 was unsuccessful and by mid 1958 other more efficient power plants were under consideration. A survey of commercially available diesel engines indicated that a militarized version of the General Motors Corporation 12V71T might be suitable for tank installation. The 12V71T, built by the Detroit Diesel Division of General Motors, was a 12 cylinder, 60 degree V, liquid-cooled, supercharged, compression ignition engine operating on a two stroke cycle. With a displacement of 852 cubic inches, it then developed 570 gross horsepower at 2300 revolutions per minute. As converted for Ordnance use, it weighed about 3600 pounds and was governed to a maximum speed of 2400 revolutions per minute.

A program also had been initiated to develop a diesel engine specifically for tank use. Both air-cooled and liquid-cooled designs were considered for an 1100 cubic inch displacement, V type, supercharged, compression ignition engine intended for installation in the T95 chassis. A contract was awarded to Continental Motors Corporation for the development of the air-cooled engine designated as the AVDS-1100. A similar contract went to the Caterpillar Tractor Company to design and build the liquid-cooled LVDS-1100. At that time, the use of the 12V71T was considered to be an interim solution pending the development of the ultimate tank diesel. Since the adoption of the air-cooled or the liquid-cooled version depended upon future tests, the engine was referred to during the interim period as the VDS-1100. Both the AVDS-1100 and the LVDS-1100 were four stroke cycle engines and they were expected to develop about 550 gross horsepower.

Taking the new power plants into consideration, it was proposed in late 1958 that the entire T95 program be reoriented once again. The following eight types remained under the proposed new program.

1. 105mm gun tank T95E5 (105mm gun T254 in a M48A2 turret on the T95 chassis)
2. 120mm gun tank T95E6 (120mm gun T123E6 in a T95E4 type turret on the T95 chassis. Two were being fabricated by Ford with siliceous cored armor in both hulls and one turret)
3. 105mm gun tank T95E7 (105mm gun T254E2 in a T95E1 turret on the T95 chassis)
4. 90mm gun tank T95E8 (90mm gun M41 in a M48A2 turret on the T95 chassis with the 12V71T diesel engine)
5. 120mm gun tank T95E9 (T95E6 with the 12V71T diesel engine)
6. 90mm gun tank T95E10 (T95 with the ultimate VDS-1100 diesel engine)
7. 120mm gun tank T95E11 (T95E6 with the ultimate VDS-1100 diesel engine)
8. 120mm gun tank T95E12 (T95E6 type turret with a 2 meter base range finder, a full solution fire control system, and the ultimate VDS-1100 diesel engine)

Although these variants envisioned further development of the T95 series, time had run out for the project. The slow pace of the development discouraged many former supporters of the T95. By early 1958, many senior officers believed that the T95 would provide only a marginal advance and that the quickest and most cost effec-

tive way to obtain an improved tank would be to install a more powerful gun and a compression ignition engine in the M48A2. The Bureau of the Budget (BOB) also believed that the tank modernization program was moving much too slowly and recommended that the Army investigate all possible means of immediately replacing the M48A2 with an improved tank. This led to the action by BOB on 1 May 1958 prohibiting further procurement of the M48A2 after the fiscal year 1959. This decision was no surprise to the Army and confirmed the preference of many senior officers for the development of an interim main battle tank. The Ordnance Tank Automotive Command (OTAC) also had expected this move and a proposal dated 4 June 1958 presented a preliminary program to meet this requirement. The new tank was referred to as the XM60 weapon system and it was later standardized as the 105mm gun tank M60. It was essentially the M48A2 rearmed with the 105mm gun T254E2 and powered by the new Continental AVDS-1790 diesel engine. Although originally intended as only an interim main battle tank, it was destined to serve with numerous modifications for several decades.

The 105mm gun T254E2 with the British tube had been selected as the main armament for the XM60 after the comparative tests at Aberdeen Proving Ground in October 1958. The smooth bore 90mm gun still suffered from excessive dispersion. This resulted from thermal distortion of the long thin barrel as well as problems with the experimental ammunition. The lightweight 120mm gun had the disadvantage of two piece ammunition. This reduced the rate of fire when only one loader was available. On 28 January 1959, the Chief of Ordnance was directed to terminate all work on vehicles utilizing the smooth bore 90mm gun T208 and the 120mm gun T123E6. On 7 July 1960, OTCM 37478 officially terminated the T95 project, except for the evaluation of the new diesel engines which might be useful in future development programs. It directed that the field testing of these engines in the T95 chassis be continued to permit their evaluation against other existing vehicle-engine combinations. The OTCM also directed that work on the T95E7 turret with the 105mm gun T254E2 be continued in order to develop an improved turret for the M60 tank. This work resulted in the long nosed turret eventually adopted on the M60A1. Thus the T95 program came to an end and the pilot chassis were put to use as test beds to develop components for future tanks.

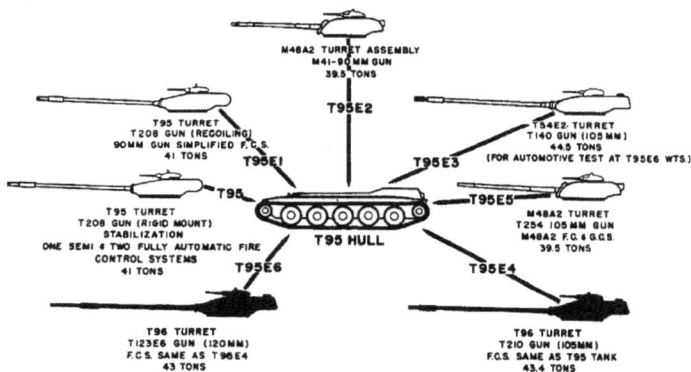

The proposed variants of the T95 tank series are outlined here.

PART II

THE ULTIMATE MAIN BATTLE TANK DESIGN

Above, the new rear engine compartment grill doors on the T95E8 appear at the left and at the right, a T95E8 is under test at the Yuma, Arizona Test Station during 1961. Note that the original T114 tracks have been replaced by the 24 inch wide T127 tracks.

THE REQUIREMENT FOR A NEW MAIN BATTLE TANK

After the standardization of the 105mm gun tank M60 and the termination of the T95 project, the main battle tank program reverted to the study of various design proposals and the development of components for future use. Four T95E8 tanks powered by the General Motors 12V71T diesel engine were utilized in this program. Rebuilt by the Ford Motor Company from four of the original pilot tanks, they were assigned the new registration numbers 9B1052, 9B1053, 9B1054, and 9B1055.

The first pilot T95E8 (registration number 9B1052) was tested at the Yuma Test Station from May 1959 to April 1960. The tank was subjected to full load cooling and drawbar pull tests to evaluate two engine cooling systems. Neither provided adequate cooling in the desert environment

and recommendations were made to increase the air flow through the radiator. During these tests, the tank was fitted with the original 21 inch wide T114 track. Later, all of the T95E8 pilots were equipped with the 24 inch wide T127 track. The fourth T95E8 pilot (registration number 9B1055) was used at Detroit Arsenal to evaluate air conditioning equipment for hot weather operations.

Four other T95 chassis were provided as test beds to evaluate the ultimate VDS-1100 diesel engines. Registration numbers 9B1051 and 9B1047 (chassis numbers 6 and 5) were assigned for the AVDS-1100 and registration numbers 9B1044 and 9B1043 (chassis numbers 7 and 4), in that order, had the LVDS-1100 installed for evaluation at the Yuma Test Station during 1962.

Below are left front (left) and right rear (right) views of the T95E8 power pack consisting of the General Motors 12V71T engine with the XTG-411-3 transmission.

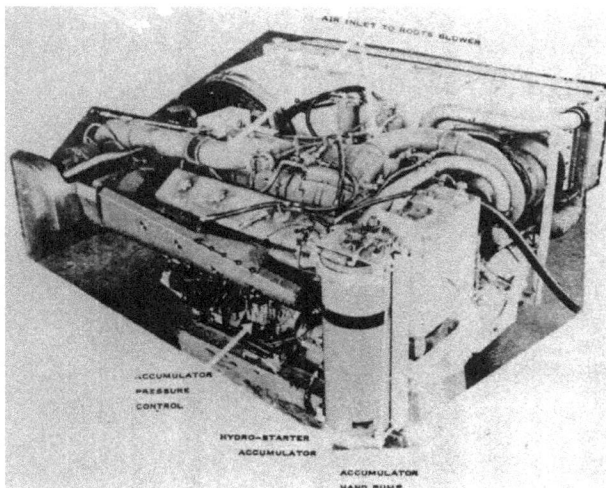

93

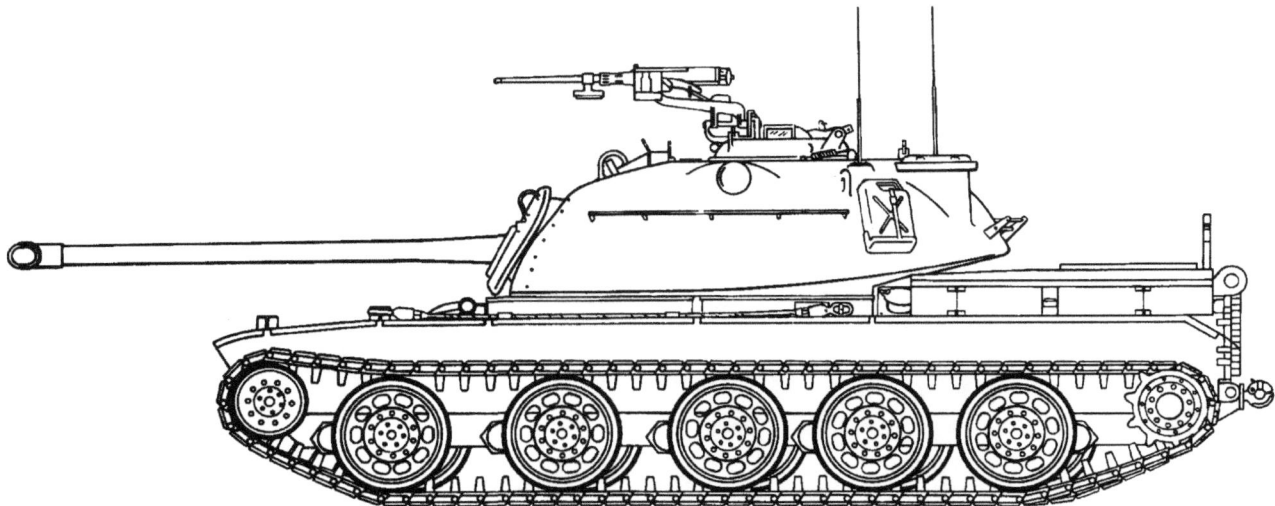

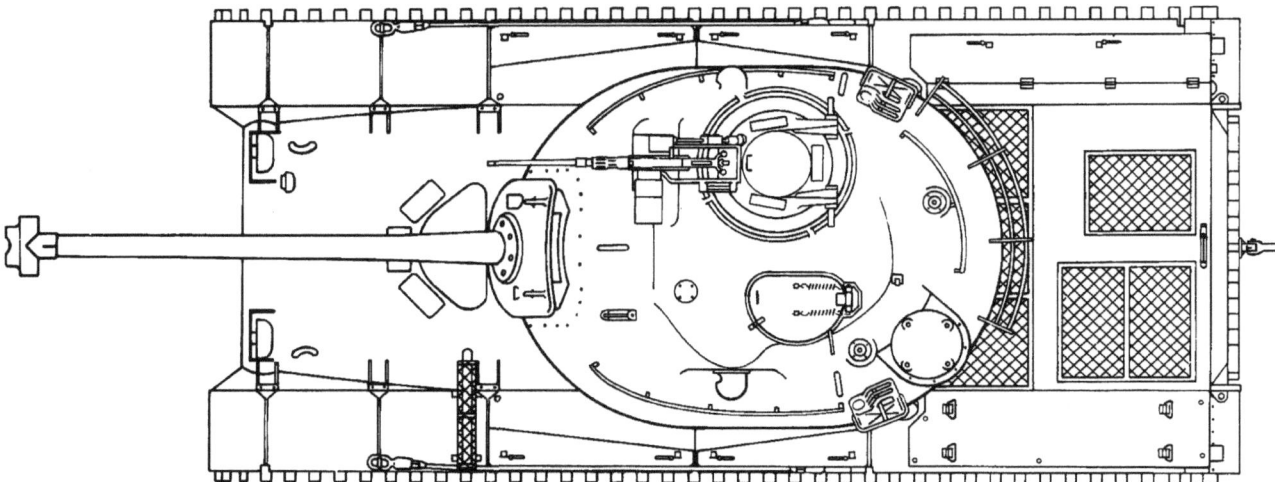

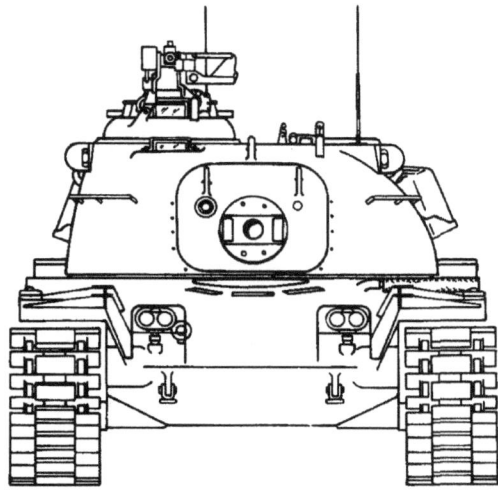

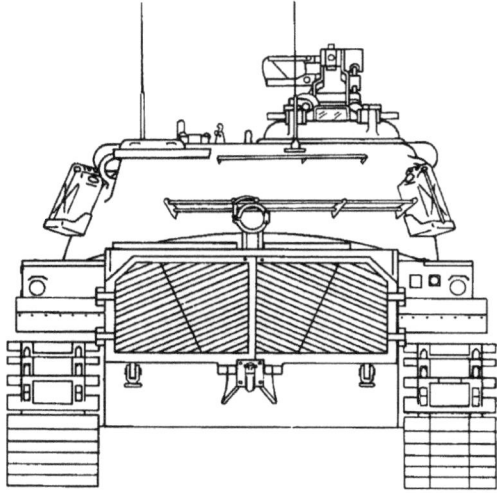

90mm Gun Tank T95E8

Above is the fourth pilot T95E8, registration number 9B1055. Below, the T95E8 is being used for an air conditioning study at Detroit. Note the M48 type turret with the low silhouette commander's hatch fitted on this tank.

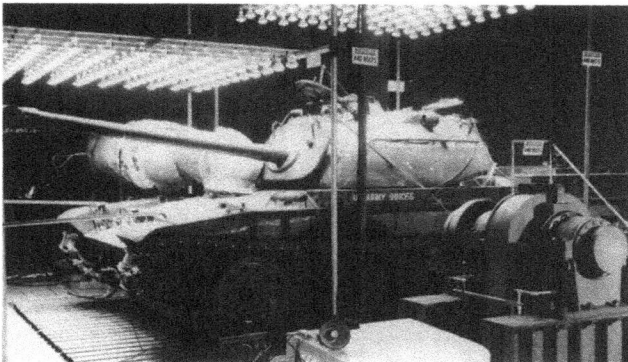

At the left, the Caterpillar LVDS-1100 engine is installed in T95 chassis, registration number 9B1044. Below, the LVDS-1100 engine is at the left and the Continental AVDS-1100 engine is at the right.

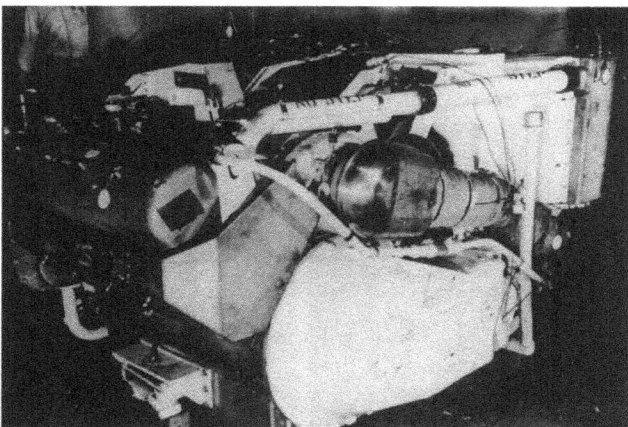

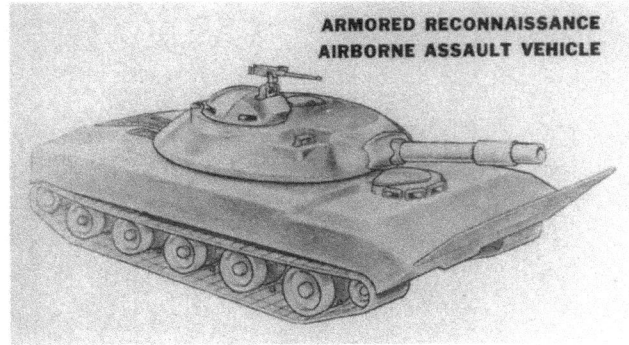

Above are artist's concepts of a missile firing main battle tank (left) and the proposed armored reconnaissance airborne assault vehicle (right).

As mentioned earlier, General Maxwell D. Taylor, the Army Chief of Staff, established the Ad Hoc Group on Armament for Future Tanks or Similar Combat Vehicles usually referred to as ARCOVE. This group was formed in February 1957 under the Technical Advisory Panel on Ordnance which provided administrative support and it included representatives from the panels on electronics and aeronautics. As previously described, ARCOVE reviewed the armament requirements for future combat vehicles and made its recommendations in May 1957, followed by a formal report in January 1958. A major recommendation was that a maximum effort be made to develop a small missile with line of sight command guidance for use in tanks by 1965. To fund this program within budget limitations, a sharp reduction was recommended in conventional tank weapon development. In particular, cancellation was recommended of the program for the smooth bore cannon with the hypervelocity kinetic energy penetrators. This, of course, contributed to the decision to cancel the T95 project. The effect of this action was to shift the emphasis of United States tank weapon development to the use of chemical energy warheads such as the shaped charge and the squash head projectile. It was to have a significant effect on future tank development.

The Fourth Tripartite Conference on Armor and Bridging was held at Quebec, Canada during October 1957. As with the earlier Tripartite Conferences, it was to have a strong influence in determining the characteristics of future combat vehicles. This conference established the desirability of replacing the light, medium, and heavy gun tanks with two new combat vehicles. A new concept of an Airborne Reconnaisance/Airborne Assault Vehicle (AR/AAV) was substituted for the earlier light tank and the medium and heavy vehicles were combined into a single, all purpose, main battle tank (MBT).

In line with the ARCOVE recommendations, work was intensified on component development for the new fighting vehicles. Primary interest centered on the new weapon system around which the vehicles were to be designed. OTCM 36753, entitled Combat Weapon System (Pentomic) and dated 13 February 1958, officially initiated the development program. The leading candidate that emerged from the various design studies was the combat vehicle weapon system (CVWS) Shillelagh. It consisted of the 152mm closed breech gun-launcher XM81 which was capable of launching the XM13 guided missile or firing conventional ammunition. The XM81 was approximately half the weight and length of the 105mm gun T254E2 now standardized as the M68 in the M60 tank. Thus it was extremely attractive for combat vehicle installation. Fitted with a separable chamber breech mechanism, it was designed to use conventional ammunition with a completely combustible cartridge case and primer. Such ammunition eliminated the problem of removing spent cartridge cases from the tank as well as the necessity of salvaging them for future use. Unfortunately, it also created some new problems that did not become obvious until later in the program. Three types of conventional ammunition were under development for the XM81. These were the XM409 HEAT-MP, the XM410 white phosphorus, and the XM411 training round. As its designation indicated, the XM409 was a multipurpose projectile combining the armor defeating characteristics of the shaped charge with the blast effect of a standard high explosive round. Such a combination reduced the types of ammunition required, simplifying the vehicle stowage. All of these rounds were spin stabilized by the rifling in the XM81. The XM409 had a muzzle velocity of 2260 feet per second and its spin compensated shaped charge warhead was estimated to penetrate seven inches of rolled homogeneous steel armor at 60 degrees obliquity.

At the right is the 152mm gun-launcher XM81.

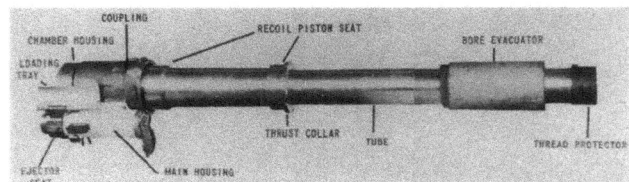

The operation of the Shillelagh missile is illustrated at the right and the missile itself can be seen at the lower right.

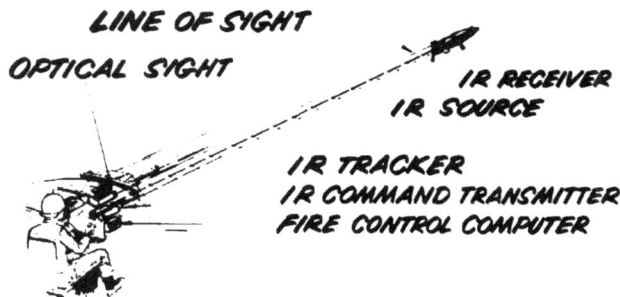

LINE OF SIGHT
OPTICAL SIGHT
IR RECEIVER
IR SOURCE
IR TRACKER
IR COMMAND TRANSMITTER
FIRE CONTROL COMPUTER

Developed under contract with the Aeroneutronics Division of the Ford Motor Company, the XM13 Shillelagh missile was ejected from the gun-launcher at approximately 260 feet per second. The solid rocket motor then boosted its speed to about 1060 feet per second. The missile flew a line of sight trajectory to the target under the control of an infrared (IR) tracking and command system. As the gunner held the sight on the target, the infrared tracker measured the displacement of the IR source in the missile tail from the line of sight. This error signal was processed by the electronic computer and a correction command was transmitted to the missile by a xenon arc lamp. In the gun-launcher, the missile rode on top of the rifling lands and engaged a keyway in the bottom of the tube to prevent rotation.

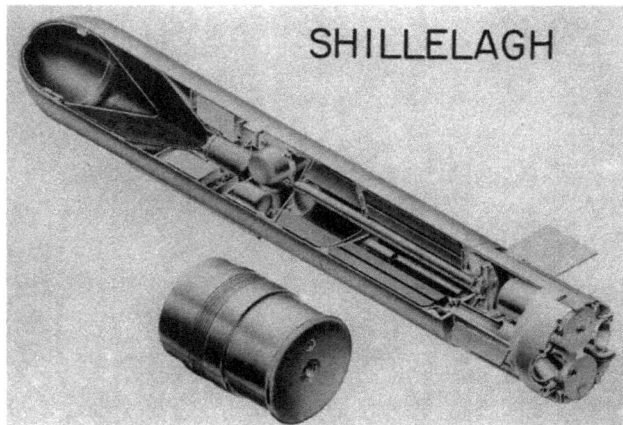

SHILLELAGH

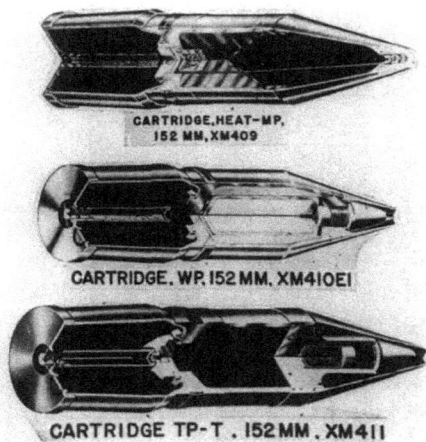

CARTRIDGE, HEAT-MP, 152 MM, XM409

CARTRIDGE, WP, 152 MM, XM410E1

CARTRIDGE TP-T, 152 MM, XM411

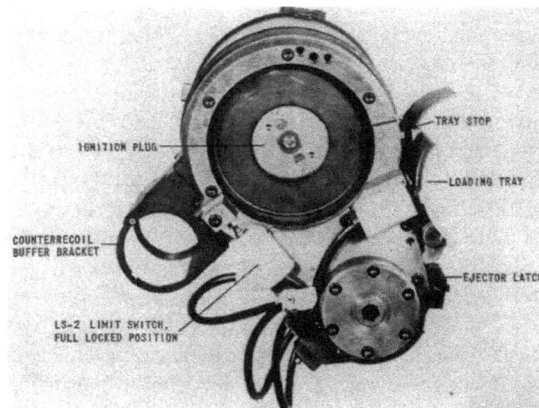

The combustible case conventional ammunition for the 152mm gun-launcher is above. At the upper right is a rear view of the XM81 gun-launcher with the breech closed. Below are two views of the open gun-launcher breech.

The proposal drawings above for the main battle tank (MR) powered by the AVDS-1100 engine are shown with (left) and without (right) radiological protection.

The tank development program approved in August 1957 by the Army Chief of Staff included a long range goal based largely on the ARCOVE recommendations which called for the development of the AR/AAV and a new MBT. The former was intended to replace both the 76mm gun tank M41 and the 90mm self-propelled gun M56 used by the airborne forces. Several designs were reviewed and a concept presented by the Cadillac Motor Car Division of General Motors Corporation was selected for further study. The evaluation of the preliminary design and the construction of a full scale vehicle mock-up were completed in May 1960. The contract with Cadillac was then revised to provide for detailed engineering design incorporating recommended changes and the construction of a final engineering mock-up. Thus by the time of the Research and Engineering Program Review at OTAC in July 1960, the project for the AR/AAV was well underway. This vehicle, armed with the Shillelagh weapon system, eventually appeared as the M551 Sheridan.

Unfortunately, the rapid progress in selecting a design for the AR/AAV did not extend to the program for the new main battle tank. Military characteristics of a new main battle tank for the mid-range time period were approved by the United States Continental Army Command (USCONARC) and submitted to the Ordnance Tank Automotive Command (OTAC) in March 1960. The mid-range time period indicated that the vehicle would be ready for production in about five years and it was referred to as the MBT(MR). As a result of a meeting with senior Armor officers in May 1960, OTAC was requested to prepare a number of concepts for the MBT(MR) which would fulfill the requirements outlined in the military characteristics and, at the same time, minimize complexity and provide a minor degree of protection from radiation. These concepts were presented during a meeting at OTAC on 6-7 July 1960. One proposal was selected for detailed study and the construction of a mock-up. This particular concept utilized the driver-in-turret design and was armed with the Shillelagh weapon system, although other types of armament also were under consideration. Three versions of the basic concept were presented at OTAC on 13 October together with a review of the mock-up.

Two versions of the proposed MBT(MR) were powered by the AVDS-1100 engine with the XTG-411 transmission. Of the two vehicles, one was provided with some radiological protection which consisted of a neutron attenuating layer based on a mixture of butyl rubber and polyethylene installed inside the turret. This rubbery material also was expected to be useful in cushioning personnel impacts with the turret walls. Layers of high density polyethylene and the engine fuel were to be used to provide similar protection inside the hull. In addition, a 1/16 inch thick coating was to be applied to the outside of the tank. Consisting of a mixture of epoxy resin and boron carbide, it was intended to prevent capture of thermal neutrons by the steel armor. This shielding provided an attenuation ratio of about 20:1 compared to 3-4:1 for the armor alone.

The third driver-in-turret concept was fitted with the LVDS-1100 engine and it did not include radiological protection. At that time, the study indicated that the height and length of the tank would have to be increased by four inches and two inches respectively to accommodate the liquid-cooled engine.

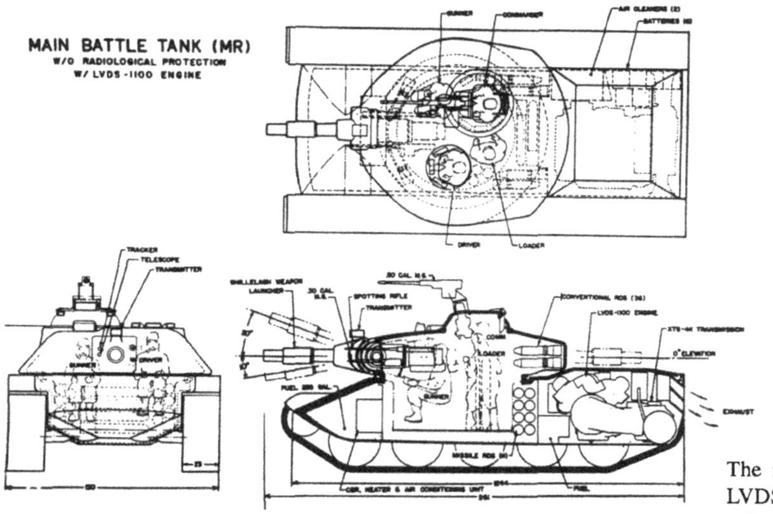

MAIN BATTLE TANK (MR)
W/O RADIOLOGICAL PROTECTION
W/ LVDS-1100 ENGINE

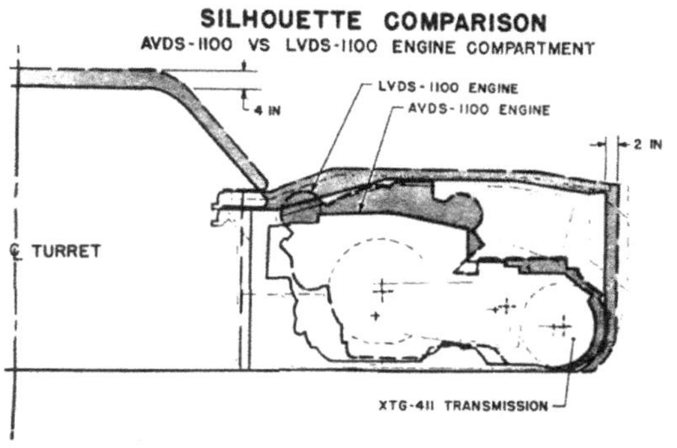

SILHOUETTE COMPARISON
AVDS-1100 VS LVDS-1100 ENGINE COMPARTMENT

The main battle tank (MR) concept at the left is powered by the LVDS-1100 engine and is without radiological protection. The larger volume required to house the liquid-cooled LVDS-1100 engine is illustrated above.

An integrated commander's station was proposed for all of the concepts with an electro-mechanical target designation system. This allowed the commander to align the main gun with his sight in both azimuth and elevation by simply pressing a button. As in the earlier Rex proposals, the driver's station was geared to the turret ring so that it counterrotated when the turret moved. Thus the driver faced forward at all times regardless of the turret position. The driver's controls were actuated electrically through a slip ring and duplicate controls were provided in the commander's station allowing him to drive the tank in an emergency.

Details of the proposed integrated universal commander's station can be seen in the sketch below.

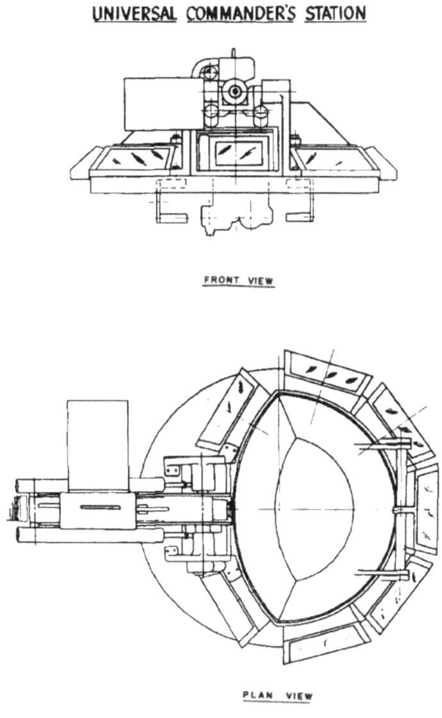

UNIVERSAL COMMANDER'S STATION

FRONT VIEW

PLAN VIEW

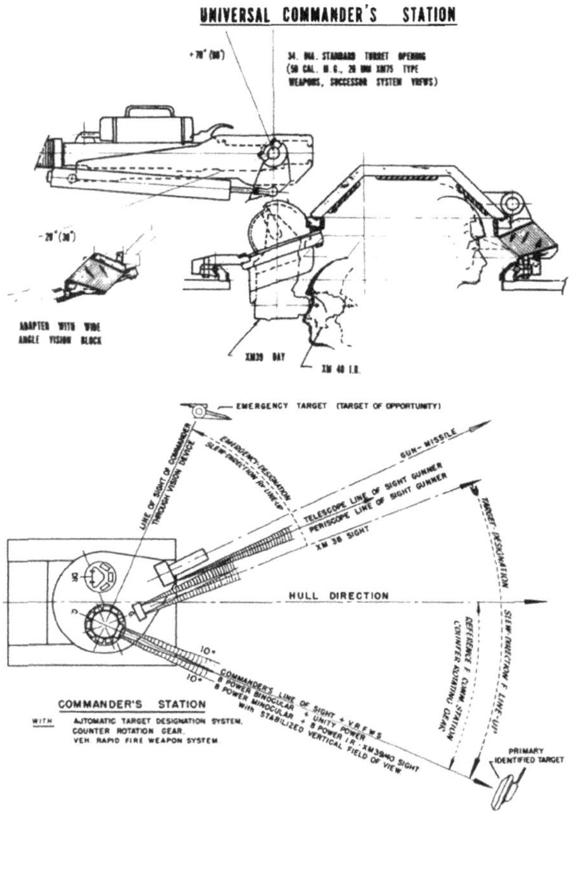

UNIVERSAL COMMANDER'S STATION

99

T95 MODIFIED
W/O RADIOLOGICAL PROTECTION
W/AVDS-1100 ENGINE

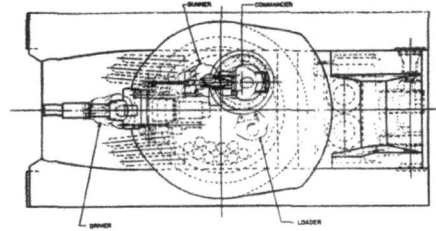

At the right is a drawing of the proposed modified T95 tank with the AVDS-1100 engine, but without radiological protection.

An additional concept also was presented showing a modified version of the T95 armed with the Shillelagh weapon system and powered by the AVDS-1100 engine. In this proposal, the driver was located in his usual position in the front hull and no radiation protection was included. The review on 13 October concluded that there was no significant cost advantage in utilizing the modified T95 to meet the MBT(MR) requirement and recommended the development of the driver-in-turret concept. At that time, the AVDS-1100 engine also was recommended based on the lower weight resulting from the smaller hull volume required.

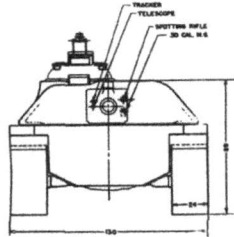

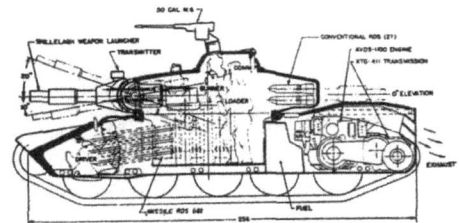

During this same time period another design was proposed to provide greatly increased radiological protection for the tank crew. This arrangement eliminated the manned turret and enclosed the crew in a protective pod constructed as part of the tank hull. This pod utilized layers of borated polyethylene, lead, and other materials, in addition to the steel armor, to provide protection against various types of radiation. Attenuation ratios of 200-400:1 were expected depending upon the shielding installed. Locating the crew in the hull also greatly reduced the vehicle silhouette making it a much more difficult target to hit, but it created some serious other problems. Separated from the crew compartment, the main armament had to be operated by remote control and required an automatic loader. The shielding and the low position of the crew in the vehicle necessitated the use of television for driving, observation, and fire control. Test rigs based on the M48A2 chassis were used to evaluate the various components such as the television equipment and other vision devices. During these tests, the driver operated the vehicle in both the normal and supine positions. After further analysis, it was concluded that the complex equipment required to operate such an arrangement was beyond the state of the art at that time for reliable service.

Above is the pod design with the externally mounted gun proposed for maximum radiological protection. Below, the Pod A test rig appears at the left and the Pod B-1 test rig is at the right.

100

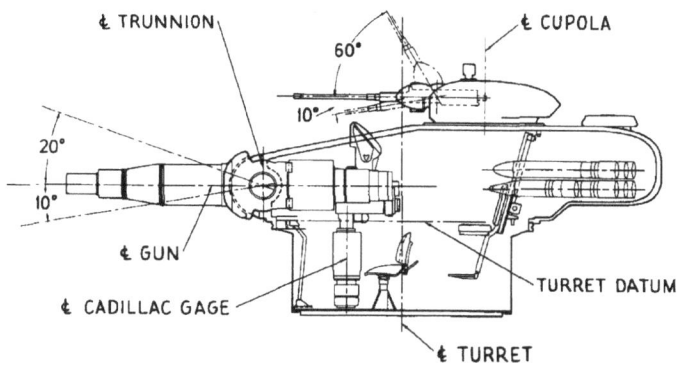 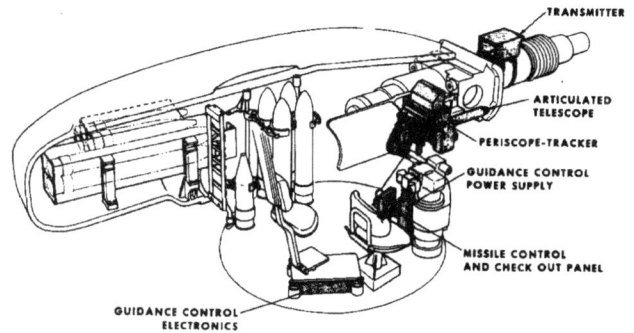

The sketches above show the Shillelagh weapon system and its fire control equipment installed in the T95 turret.

Three T95 turrets were modified to permit the installation of the 152mm XM81 gun-launcher. Mounted on M48 chassis, they were to provide test beds for the evaluation of the Shillelagh weapon system. Although this system was the preferred armament for the MBT(MR), by late 1961 problems with the XM13 missile required that the program be reorganized. The missile was reclassified as an applied research project and it was obvious that there would be some delay before it would be available for service. On 10 January 1962, representatives from various ordnance organizations met at OTAC to review armament systems that might be suitable replacements, if the Shillelagh missile could not be developed in a timely manner. Time was particularly critical for the AR/AAV (the XM551 Sheridan) which required a decision on the armament by April 1962. The possible delay was not as serious for the MBT(MR)

since the program was restricted to concept studies and component development. The requirements also differed for the tank because of its ability to carry a much heavier weapon system. Several backup weapons were considered and concept studies were prepared showing their application to the MBT(MR). The 152mm gun-launcher XM81 also was considered without the missile depending only on the combustible case conventional ammunition. It was expected that the Shillelagh or some other missile then could be introduced at a later date. The 105mm gun M68 as standardized for the M60 tank was considered as an alternate armament system. It had the advantage of being immediately available and its ammunition was already in production. Compared to the Shillelagh system, the use of the 105mm gun increased the overall tank weight by about 1700 pounds to between 43 and 44 tons.

Below is a proposal for a new main battle tank featuring an integrated fighting compartment (driver-in-turret) armed with the Shillelagh weapon system. Note that it includes both a spotting rifle and a laser range finder.

 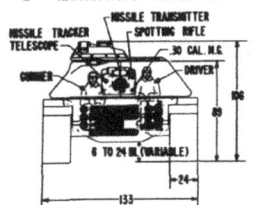 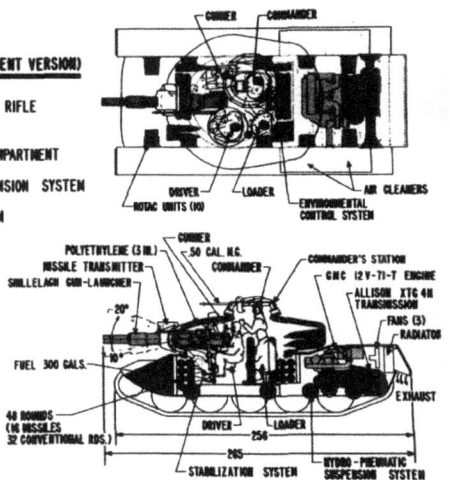

The test rig for the 120mm Delta gun shown here made use of the first pilot T95E8, registration number 9B1052.

Although the ARCOVE recommendations had shifted the emphasis to missiles for future tank armament, some work had continued on high velocity guns resulting in another candidate as a backup weapon. This was the 120mm Delta gun which was an outgrowth of the earlier work on the smooth bore 90mm gun T208 and the 105mm gun T210. In the 120mm weapon, the arrow type projectile was modified to a delta wing configuration in an attempt to minimize the problem with aerodynamic heating. The complete gun weighed 2856 pounds and it was designed with a separable chamber breech to fire combustible case ammunition. The APFSDS projectile weighed 17.6 pounds and had a muzzle velocity of 5300 feet per second. It was estimated that the Delta gun would increase the overall weight of the MBT(MR) by 2600 pounds resulting in a combat weight of a little over 44 tons. The vehicle width would increase from 133 inches to 136 inches. Although no pilot MBT(MR) was constructed, the Delta gun was installed in a test rig based on the chassis of the first pilot T95E8, registration number 9B1052, after it had been fitted with the 24 inch wide tracks.

Above is a drawing of the complete arrow type round for the 120mm Delta gun. The dimensions of the weapon itself are in the sketch below.

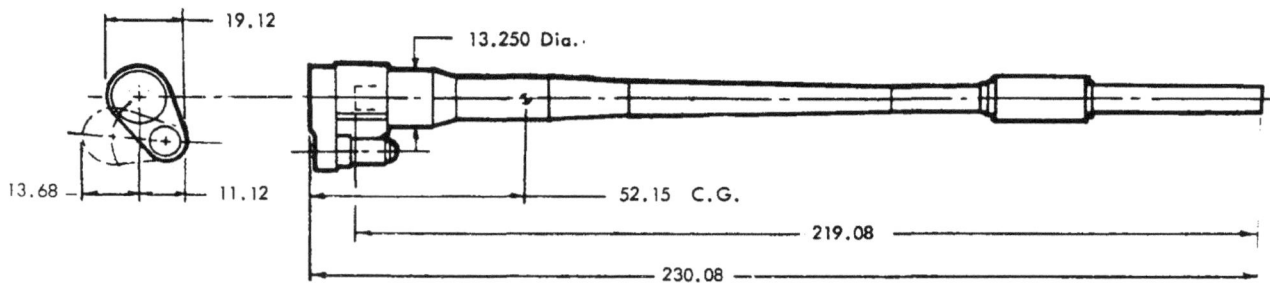

102

Additional photographs of the Delta gun test rig are above and exterior and interior views of the gun mount appear below.

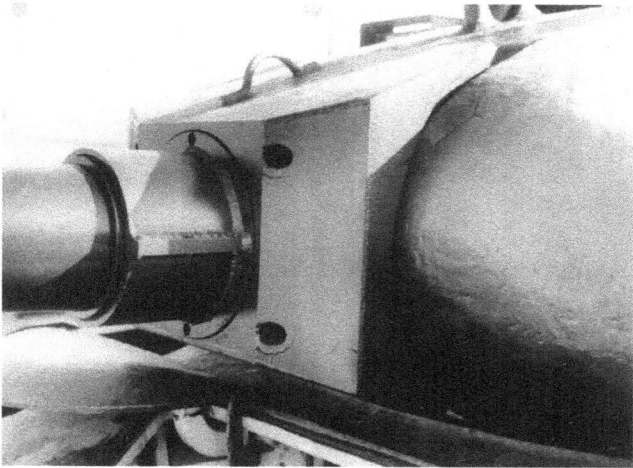

In the interest of standardization, the British X23 120mm gun also was considered as a possible weapon system for the MBT(MR). With a complete weight of 3926 pounds, this gun fired a 20 pound APDS projectile at a muzzle velocity of 4700 feet per second. The projectile was loaded separately with a bag charge propellant weighing 19 pounds. As with the Delta gun, the tank width was expanded to 136 inches when armed with this weapon and the combat weight increased by 3952 pounds to almost 45 tons.

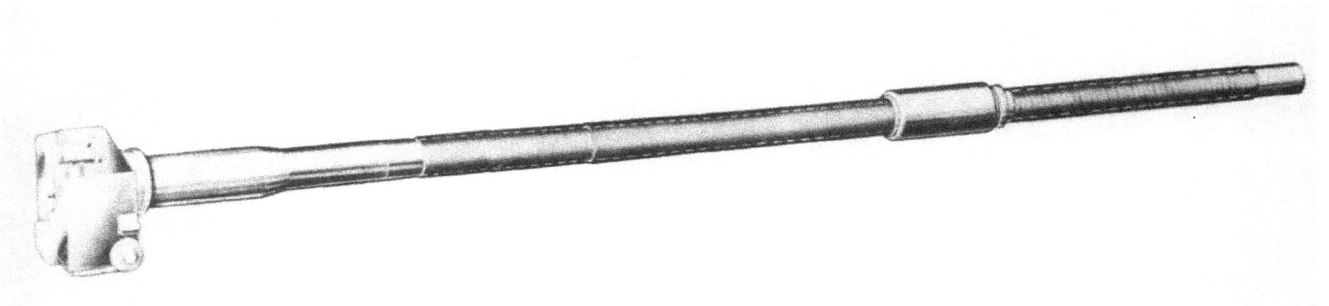

Above is the British XM23 120mm gun. The dimensions of this weapon are shown in the sketch below.

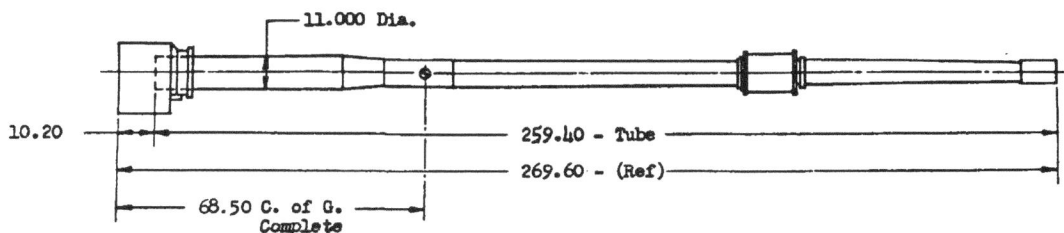

11.000 Dia.

10.20

259.40 - Tube

269.60 - (Ref)

68.50 C. of G.
Complete

At that time another promising weapon system was undergoing a feasibility study. Known as the POLCAT (post launch correction, antitank), this was a gun launched projectile equipped with a terminal guidance system. It used a detector in the projectile nose to home upon an illuminator signal reflected from the target. Although designed as an infantry weapon, the TOW (tube launched, optically tracked, wire guided) missile also was considered as possible future tank armament.

The operation of the POLCAT system is illustrated at the right. Below are drawings of a proposed new main battle tank armed with the 120mm Delta gun (left) or the British 120mm gun (right).

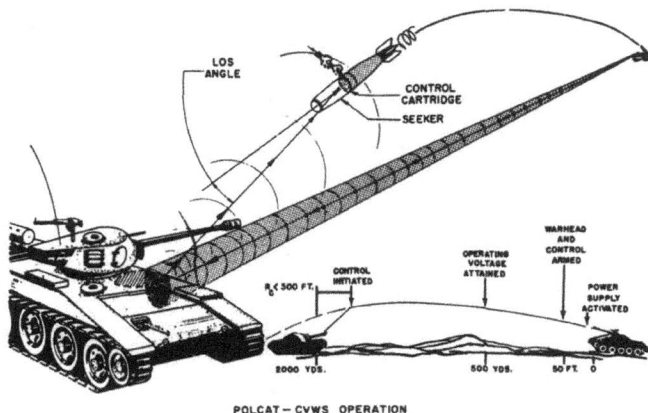

POLCAT - CVWS OPERATION

NEW MAIN BATTLE TANK
(INTEGRATED CREW FIGHTING COMPARTMENT VERSION)

● 120 MM DELTA ARMAMENT SYSTEM
● LASER RANGEFINDER

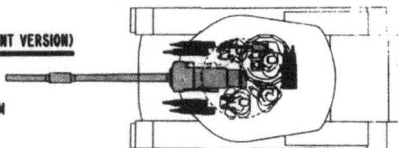

NEW MAIN BATTLE TANK
(INTEGRATED CREW FIGHTING COMPARTMENT VERSION)

● 120 MM U.K. ARMAMENT SYSTEM
● SPOTTING RIFLE & LASER RANGEFINDER

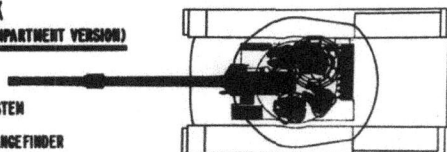

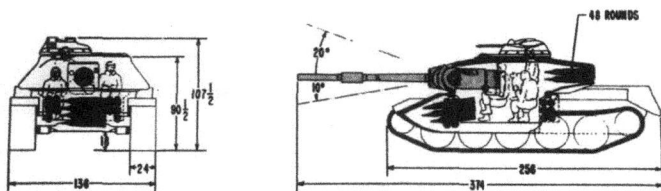

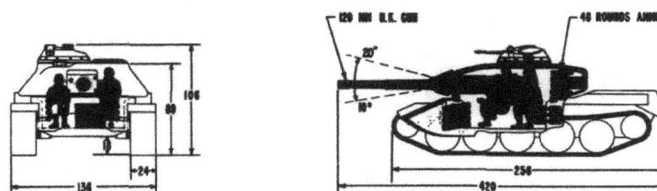

Although the driver-in-turret arrangement, also known as the integrated crew fighting compartment, was the recommended concept for the new tank, by 1962 another design approach required consideration. Proposed by Joseph Williams and Clifford Bradley, it was an outgrowth of the earlier pod concept which placed the entire crew in the hull and used a remote control gun. The new design sought to retain many of the advantages of the earlier

arrangement while eliminating its most serious drawbacks. The main armament was mounted in a fully rotating manned turret, but the gunner and the loader were seated low in the turret basket, one on each side of the main weapon. The hatches above their heads were only slightly above the top of the hull. The upper part of the turret was only a little wider than the gun mount and the tank commander was located directly behind the cannon with 360 degree vision above the weapon. The driver rode in the conventional position in the center front hull. Referred to as the compact turret, this design reduced the frontal area of the tank until it was comparable to a vehicle armed with a remote control weapon. However, it retained the use of proven fire control and vision equipment and the low frontal area turret weighed about half as much as the conventional design. The crew had direct access for serving the cannon and an automatic loader was not required. Armed with the Shillelagh weapon system, the compact turret configuration was eventually adopted for the 152mm gun-launcher tank M60A2.

NEW MAIN BATTLE TANK

● HIGH COMMANDER VISION
● SMALLER TARGET AREA
● LESS WEIGHT
● VARIABLE HEIGHT SUSPENSION
● SHILLELAGH WEAPON SYSTEM

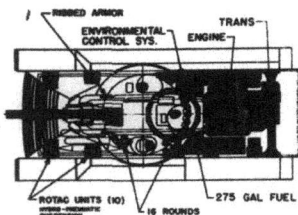

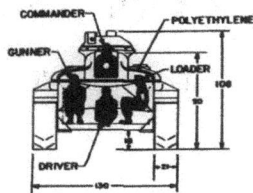

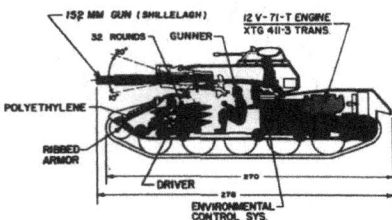

The compact turret design proposed for the new main battle tank is at the left. This version is armed with the Shillelagh weapon system.

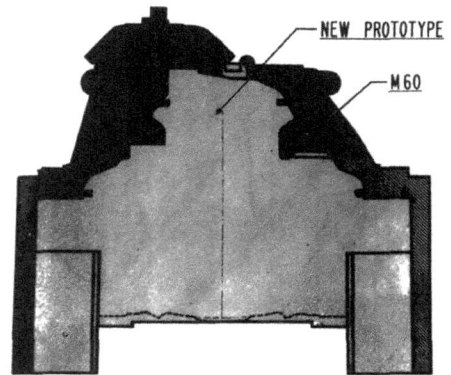

Above, the small frontal area of the MBT (MR) with the compact turret is obvious in the model photograph and in the comparison of its front silhouette with that of the M60 tank. Two versions of the compact turret mock-up appear below at the right.

During this period, there was considerable activity in the development of fire control equipment. In regard to range finders, the T53 Optar pulsed light range finder did not prove to be successful during its field tests. Despite the efforts to restrict the light pulse to a narrow beam, multiple reflections made it difficult to determine which represented the range to the intended target. Thus the tank commander had to visually estimate the range and select the most reasonable value. Although a multiple pulse unit with improved performance was developed by the Farrand Optical Company to replace the single pulse T53, a two meter base length optical range finder was specified for the T95E12. However, a new development now appeared that eventually was to make the pulsed light range finder a success. This was, of course, the laser. Its extremely narrow precise beam permitted easy ranging on a specific target. An experimental unit was assembled for test purposes by the Summer of 1963. In the meantime, other systems were being evaluated. One of these was the use of a matched spotter. This was a small caliber spotting rifle which matched the trajectory of the main weapon. The 15mm XM122 spotting rifle was under development for use with the 152mm gun-launcher when using conventional ammunition. Eventually, it was expected to closely match the trajectory of the 152mm round out to a range of 2000 meters. With this system, the spotting rifle was fired until a hit was obtained on the target. The main gun could then be fired with a high probability of a first round hit. Other experimental programs investigated the use of radar range finders on combat vehicles, but their immediate application was prevented by the limited resolution obtainable.

Below is the 15mm XM122 spotting rifle and the sketch at the right shows the installation of the early laser range finder.

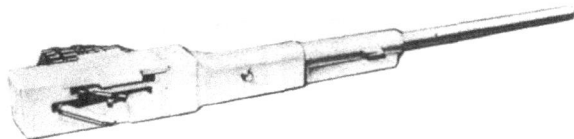

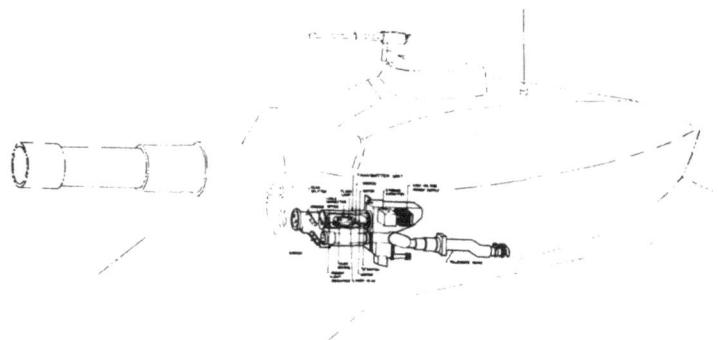

105

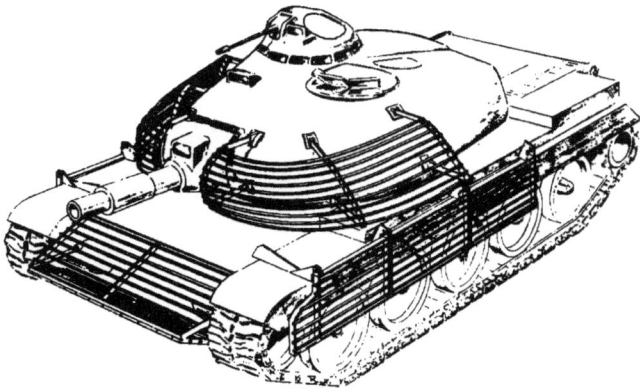

Above, a bar armor installation is sketched at the left and the model at the right features both ribbed and extendable bar armor. A cross section of the ribbed armor appears below at the right.

In addition to the siliceous cored armor previously described, several new approaches were investigated to increase the protection on the various MBT(MR) concepts. One of these was the use of ribbed armor. Ballistic tests indicated that greater protection could be obtained at the same weight or equivalent protection was possible at a lighter weight by using armor with this configuration. Ribbed armor actually was a form of spaced armor consisting of a series of ribs projecting from the main armor backup plate. Tests showed that weight savings of 15 to 30 per cent could be obtained compared to solid plate with equivalent protection against large caliber armor piercing projectiles. Unfortunately, it was not as effective against APDS rounds. Another innovation was the use of grill or bar armor to defeat shaped charge (HEAT) ammunition. This concept employed a screen of equally spaced steel bars to increase the standoff distance from the main armor plate. On some arrangements, the screen was moveable and could be extended or retracted when it was no longer required. Another concept being explored was the use of active or dynamic protection. One such project at Picatinny Arsenal was dubbed the Dash-Dot Device. With this equipment, sensors, such as infrared or doppler radar, detected the incoming projectile and fired linear shaped charges to destroy it. One limitation of this concept was the space required to mount and provide replacements for the linear shaped charges.

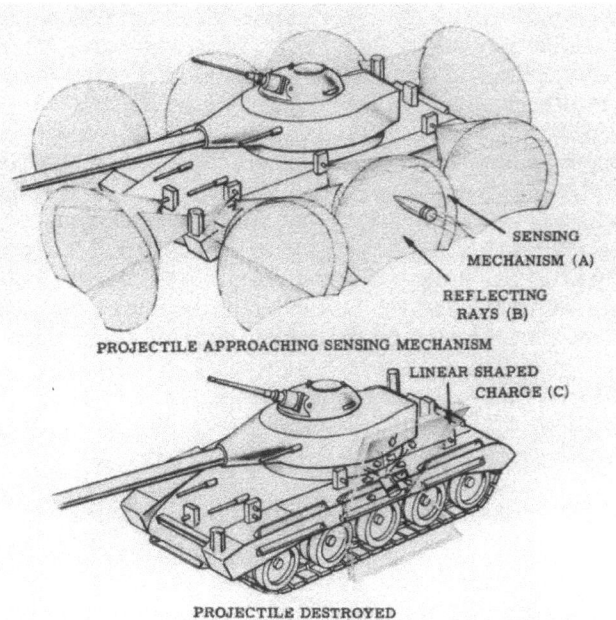

SENSING MECHANISM (A)

REFLECTING RAYS (B)

PROJECTILE APPROACHING SENSING MECHANISM

LINEAR SHAPED CHARGE (C)

The sketches at the right depict the proposed operation of the Dash-Dot Device.

PROJECTILE DESTROYED

Above are the left and right sides of the power pack using the General Motors 12V71T engine. The drawing below shows a proposed transverse installation of the AVDS-1100 variable compression ratio (VCR) engine in a main battle tank.

By March 1962, the AVDS-1100 and the LVDS-1100 were no longer being considered as power plants for the MBT(MR). Although their development had been relatively successful, they had, in effect, been squeezed out by the 12V71T engine which was immediately available without further development and the appearance of advanced technology engines for the future. Thus the MBT(MR) concepts from this period were designed around the 12V71T with provision for the installation of more advanced engines when they became available. The new technology covered the variable compression ratio (VCR) engine developed by Continental Motors and the very high output (VHO) engine under study at the Caterpillar Tractor Company.

At Continental Motors, the variable compression ratio pistons were installed in the AVDS-1100 engine and it was referred to during this period as the AVDS-1100 VCR. Initially, the installation of the new pistons raised the gross horsepower of the AVDS-1100 from 550 to 700 with a design objective of 850 for use in a new main battle tank. A later version of this engine, designated as the AVCR-1100 developed 1475 gross horsepower in the MBT70. The VCR pistons allowed the engine to develop high power with a relatively low structural weight. These pistons consisted of an outer shell fitted with the piston rings installed over an inner piston pin carrier. Oil chambers between the two parts were pressurized to extend the outer shell increasing the compression ratio to about 22:1 for good starting characteristics. During operation, the engine's cylinder pressure was limited to a maximum value by releasing oil and decreasing the compression ratio until it dropped to about 10:1. Thus the engine could develop high horsepower without exceeding the structural limitations of its lightweight construction.

The sketch at the right illustrates the operation of the variable compression ratio piston to limit cylinder pressure.

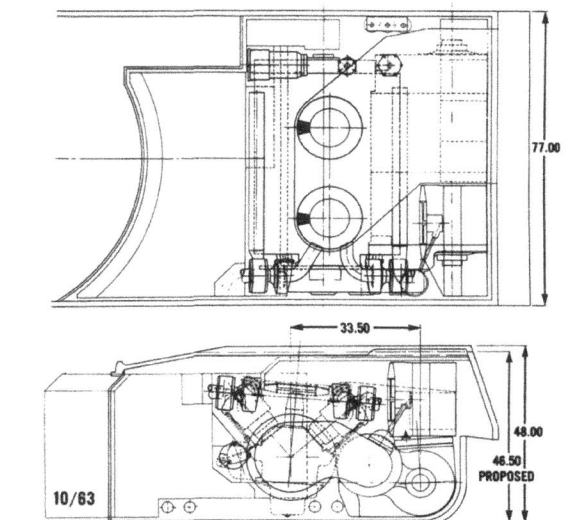

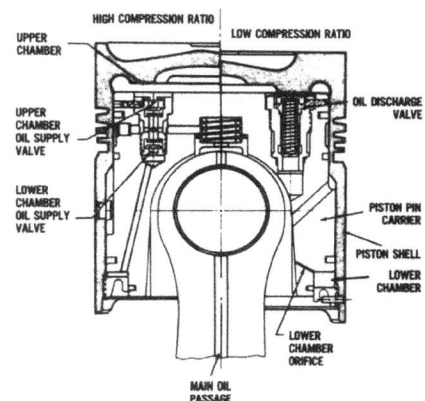

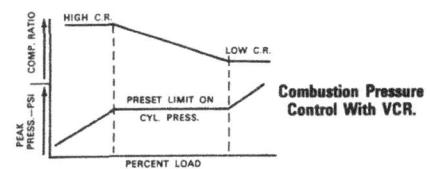

Combustion Pressure Control With VCR.

The configuration of the Caterpillar very high output engine can be seen in the photographs above.

The LVMS-1050 very high output (VHO) engine under development for the Army at Caterpillar was designed to develop 1000 gross horsepower at 2800 revolutions per minute. Both turbosuperchargers and aftercoolers were used to obtain this high performance in a relatively lightweight engine. Wide use of aluminum alloys held the weight of the first prototype to only 2483 pounds. With a bore and stroke of 4.5 and 5.5 inches respectively, the liquid-cooled V-12 had a displacement of about 1050 cubic inches. A prechamber type combustion system was featured with glow plugs to assure dependable starting at low temperatures.

Gas turbines also were being evaluated as tank power plants. In early 1961, the fourth pilot T95E8 was modified as a test bed for the Solar Saturn gas turbine and other types were under consideration. The XTG-411 transmission was used with the 12V71T engine in the various MBT(MR) concepts. More powerful engines would have required a new transmission. A new X-700 transmission

was planned to meet this requirement. It was intended to combine the best features of the XTG-411 and CD-850 designs and included pivot steer in neutral to increase the mobility in mud.

To further improve cross-country mobility, programs were initiated to develop new track and suspension systems and to enhance the performance of the components in existing systems. Foremost among these was the development of the friction hydropneumatic suspension. In this system, the torsion bars were eliminated and the wheels were individually sprung by means of hydraulic rotary actuators connected through damping valves to gas charged accumulators. Electronic controls enabled the driver to change the ground clearance of the vehicle by actuating solenoid valves which increased or decreased the volume of the fluid in the operating chambers of the actuators. The actuators rotated the road wheel arms to raise or lower the hull. This suspension provided a total wheel travel of 19

Below, the sketches show the friction hydropneumatic suspension (left) and the piston type hydropneumatic suspension (right).

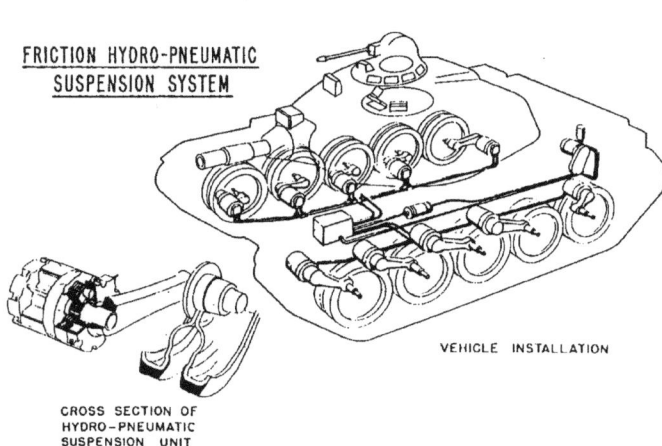
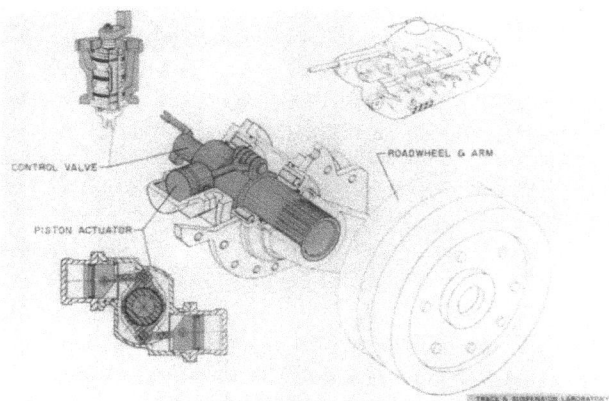

Above, the friction hydropneumatic suspension on test rig number three is shown raised (left) and fully lowered (right). Below, the suspension is lowered on the right side and raised on the left in one view. In the other, it is raised in front and lowered in the rear.

inches permitting a considerable increase in cross-country speed. The ability to increase the ground clearance also improved the mobility in deep mud.

Three T95 chassis were modified as test rigs to evaluate the friction hydropneumatic suspension system. The first completed 2700 miles of durability testing at the General Motors Proving Ground in 1962. The second test rig, registration number 9B2027, was fabricated incorporating a positive suspension lockout. The third rig, registration number 9B2030, was similar to number two and it was shipped to Aberdeen Proving Ground for test in November 1964. A piston type hydropneumatic suspension also was under development. This type replaced the hydraulic rotary actuators with piston type actuators. This suspension built by the National Water Lift Company also was evaluated on a modified T95 chassis.

Another method of achieving performance similar to that of the friction hydropneumatic suspension by using torsion bars also was under investigation. This arrangement used a torsion bar concentrically mounted inside a torsion tube and connected in series. This tube-over-bar suspension doubled the length of the torsion spring resulting in a wheel travel and spring rate comparable to that of the friction hydropneumatic suspension. An adjustable anchor for the torsion tube could be incorporated to permit changes in the ground clearance.

The components of the mechanical variable height suspension are sketched at the right.

Below, a T95 test rig is fitted with a later version of the hydropneumatic suspension which is shown lowered in front and raised in the rear.

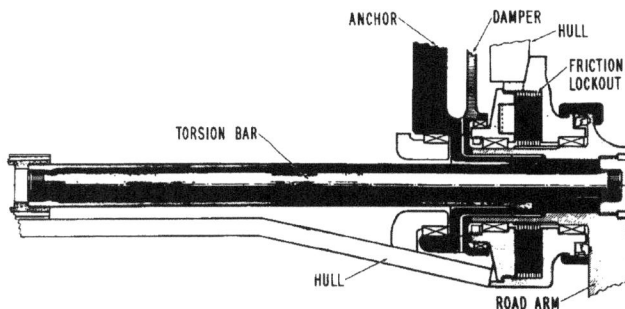

Above are exterior and interior views of the experimental driver-in-turret arrangement tested in an M48 tank at Fort Knox. The ring gear at the bottom of the right-hand photograph rotated the driver's capsule so that he faced forward at all times regardless of turret rotation.

Since no MBT(MR) pilots were constructed, several turret test rigs were assembled to determine the operating characteristics of the different turret configurations. The driver-in-turret arrangement already had been tested at Fort Knox in early 1955 using a modified M48 tank. The commander's cupola was removed and replaced by a new driver's station installed in a basket geared to the turret ring so that the driver faced forward regardless of the turret movement. Cables routed through the turret slip ring connected a modified tank control panel to the standard panel. A junction box and four servo mechanisms were used to connect the controls in the turret to the brakes, steering linkage, accelerator linkage, and the transmission. The test results concluded that the driver-in-turret position was superior to the conventional front hull location because of better vision, improved control of the tank, and relative freedom from dirt and water splash.

To further evaluate the driver-in-turret concept, a functional wooden mock-up was presented to the Armor Board at Fort Knox in November 1961 followed in December 1962 by the first hardware turret test rig. The turret of the latter, referred to as the integrated fighting compartment, was installed on a modified T95E8 chassis, registration number 9B1053. This vehicle retained the liquid-cooled General Motors 12V71T diesel engine with the XTG-411-4 transmission. Electrically actuated driving controls for the transmission, accelerator, brakes, and steering linkage were provided for both the driver and the tank commander. The

At the right is an early configuration of an integrated turret test rig on the second T95E8 chassis.

driver was seated in a counterrotating cupola and basket installed in the left front of the turret. A release on the rear of the driver's cupola ring allowed it to be locked in four positions in 90 degree increments relative to the hull. Armored mirrored covers for the driver's three periscopes could be positioned for optimium visibility. The gunner was in his usual position on the right side of the cannon in front of the tank commander. An XM38 periscope and an XM112 telescope were provided for his use. Initially, the test rig was fitted with a low silhouette commander's cupola with periscopes in the top and a .50 caliber M85 machine gun mounted on the top left side. Later, this was replaced with a mock-up of a new commander's station featuring a full ring of vision blocks giving a 360 degree view. The .50 caliber M85 machine gun was retained in approximately the same position. The loader rode in the left rear of the turret behind the driver. The manual driving controls were retained in the hull of the test rig for emergency use. The XM81E6 version of the 152mm gun-launcher was fitted as the main armament. Initially, wooden mock-ups were installed to represent the coaxial machine gun, a spotting rifle, the fire control equipment, and the missile control components. Ammunition racks were fitted with dummy ammunition.

110

Turret test rig number one, registration number 9B1053, can be seen in these photographs. Note the different cupola design compared to the vehicle at the bottom of the previous page.

Details of turret test rig number one are visible in the top view below. Also, note the surge tank added to the cooling system of the 12V71T engine in the T95E8 chassis.

Turret test rig number five installed on M60 tank chassis number 9B4470 is shown in these photographs. Note the 7.62mm machine gun in the commander's cupola mount.

The integrated fighting compartment type turret also was installed on the third and fifth turret test rigs. The former used the converted chassis of the first T95E8 pilot, registration number 9B1052. The fifth rig utilized an M60 chassis, registration number 9B4470. This was a modified M60 production vehicle powered by the Continental AVDS-1790 engine with the CD-850 transmission. The hull modifications included the installation of the electrical driving controls for use in the turret. However, the original mechanical controls were retained in the front hull for emergency use. The integrated fighting compartment turret installed on this tank differed in some details from that

Above at the left, the turret for test rig number five is shown before installation on the chassis without the 20mm gun. A closeup of the 20mm gun is at the top right. Below, the driver's station appears at the right with his seat folded up. Further details of the fifth turret test rig can be seen in the lower left and bottom photographs.

on the first test rig. Its general arrangement was the same, but a remote control 20mm Hispano-Suiza automatic cannon was mounted on the left side of the turret and a 7.62mm M73 machine gun replaced the .50 caliber M85 weapon on the commander's cupola. The 152mm XM81E6 gun-launcher was retained as the main armament with a coaxial 7.62mm M73 machine gun. A modified .50 caliber M85 machine gun was installed as a spotting rifle.

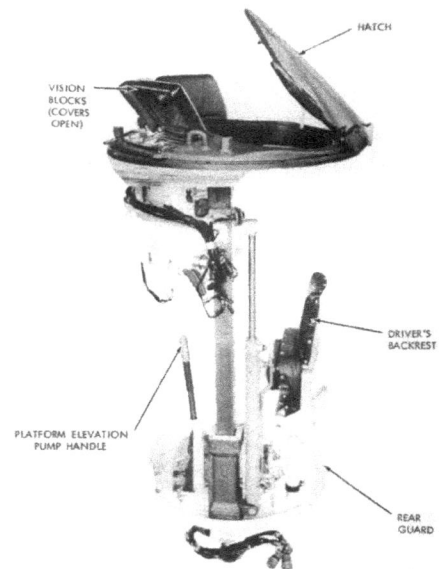

Above, the mock-up of the compact turret is installed on M60, registration number 9B4470. The commander's cupola is armed with the .50 caliber M85 machine gun.

Turret test rigs two, four, six, seven, and eight were intended to evaluate the new compact turret design. The second test rig mounted the compact turret on the modified T95 chassis, registration number 9B2027. Test rigs four, six, seven, and eight had the turret installed on the M60 chassis. An early mock-up of the compact turret also was mounted at one time on the hull of M60 9B4470, which also was used for test rig number five with the driver-in-turret arrangement.

In the compact turret, the gunner and the loader were located low in the turret basket on the right and left sides of the cannon respectively. The tank commander's station was directly behind the main gun. His cupola, like that on the fifth test rig, was armed with the 7.62mm M73 machine gun and it included a ring of vision blocks permitting an unobstructed 360 degree view above the cannon. The main armament was the 152mm XM81E6 gun-launcher with a coaxial 7.62mm M73 machine gun and a spotting rifle. As on the fifth test rig, a remote controlled 20mm Hispano-Suiza automatic cannon was installed on the side wall of the turret. This gun was located on the left side on test rig number four and on the right side for test rig number six. Both turrets four and six were mounted on M60 chassis 9B3488. Rig number six had a deflector bar installed on the front hull above the driver's hatch to limit the depression of the gun-launcher. No details are available on test rigs seven and eight.

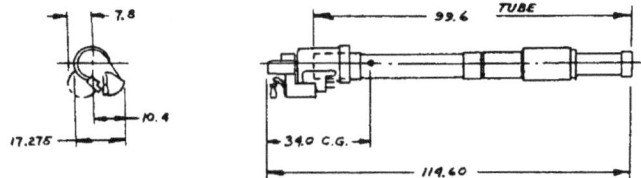

Turret test rig number four mounted on M60 chassis 9B3488 can be seen in these views. Note the 20mm gun installed on the left side of the turret. The dimensions of the XM81E6 152mm gun-launcher are given in the sketch above at the right.

Turret test rig number six appears in the photographs on this page. Like test rig number four, the turret is mounted on M60 tank chassis, registration number 9B3488.

Details of turret test rig number six can be seen in the top view below. Note that the 20mm gun is now installed on the right side of the turret.

Above is the model of the tank design concept by Robert W. Forsyth and John P. Forsyth which won the U.S. Armor Association tank design competition in late 1962.

Although no pilot model tanks were constructed during this period of experimentation, many concepts were studied and ideas for new approaches to main battle tank design were sought from numerous sources in the Armor community. In late 1962, the U.S. Armor Association sponsored a competition for a new main battle tank design. The winner, illustrated in the January-February 1963 issue of Armor Magazine, consisted of two tracked units connected and operated in tandem with a two piece gimbal ring type joint. This connection permitted the two units to roll and pitch independently about the axes within the joint. Designed by Robert W. Forsyth and John P. Forsyth, it illustrated the wide range of concepts under consideration at that time. The tank could be broken down into the two units thus meeting the weight requirements for transportation by the aircraft of that time. The vehicle was to be driven by a multifuel piston engine coupled to an electric generator in the rear unit. This provided power to four traction motors driving the sprockets, two located at the rear of the front unit and two at the front of the rear unit. The proposed main armament was carried on the front unit and consisted of a 155mm smooth bore, rocket boosted cannon and a 20mm Hispano-Suiza automatic gun. Provision also was made for the installation of a 7.62mm multibarrel machine gun on the rear unit. The vehicle was manned by a crew of seven with the commander/gunner, the gunner/loader, and the driver in the front unit and a four man tank support team in the rear. Weight was estimated to range from 24 to 32 tons depending upon the armor protection.

Although the wide range of component and concept studies had shown considerable progress, the time was rapidly approaching when a new tank would be required by the troops. Thus a design had to be selected and the effort concentrated on a development program to produce a new tank. However, political events were to have an overriding effect on this process. By late 1962, negotiations were in progress with the Federal Republic of Germany. These talks culminated in an agreement to jointly develop a new main battle tank for use by both countries. Signed on 1 August 1963 by Secretary of Defense Robert S. McNamara for the United States and Defense Minister Kai-Uwe von Hassel for the Federal Republic of Germany, this program was the primary tank development effort for the remainder of the decade.

The objective of the agreement between the United States and the Federal Republic of Germany was to provide a new main battle tank incorporating the latest design concepts which would be ready for production in both countries by 1970. Referred to as the MBT70, this tank was to be supportable by a common logistic system. The original agreement did not specify details of the new MBT, but it indicated that the first order of business was the preparation of a common list of military characteristics required by both countries. This was to be followed by the construction of a mock-up of the proposed vehicle. A two man Program Management Board (PMB) was established to implement the program. Initially the Board consisted of Brigadier General (later Major General) Welborn G. Dolvin as the U.S. Program Manager and Doktor Ingenieur Fritz Engelmann occupied the same position for the Federal Republic of Germany. Under the Program Management Board, a Joint Engineering Agency (JEA) and a Joint Design Team (JDT) were organized with equal representation from both countries. As can be seen from the organization chart, the JEA consisted of government civilian employees and military personnel while the JDT was manned by the industrial support contractors from each country. General Motors Corporation was awarded the support contract in the case of the United States. In the Federal Republic of Germany, an industrial consortium was

formed to provide similar service. Designated as the Deutsche Entwicklungsgesellschaft (DEG), it was incorporated in July 1964. Germany was selected as the location for the initial design phase of the program and the JEA and JDT were assembled at Augsburg in September 1964. The two groups were to relocate to Detroit, Michigan after the initial design phase was complete, but numerous delays prevented this until September 1966. Although they were supposed to be fully integrated, the groups were divided by differences in language, industrial practice, and engineering procedures making it difficult to resolve even simple problems. Thus the design was by committee with a compromise required for almost every decision or it had to be referred to a higher authority. For example, an agreement could not be reached in the JEA or the PMB on the use of the english or the metric system of measurement and the problem had to be settled at the Secretary and Minister of Defense level. The final decision permitted each country to use its own system within the components it manufactured, but all interface connections between components were to use the metric system. Since each country had an equal voice in the design, there were actually two chiefs in each group, although the senior representative from the visiting country was theoretically in charge in each location. All of these problems created havoc with the original development schedule and the time required to select the design concept increased from the estimated six months to two years. This, of course, also caused a similar delay in the date when the tank would be ready for production.

Below, the overall organization for the MBT70 program management is at the left and a more detailed chart for the Joint Engineering Agency is at the right.

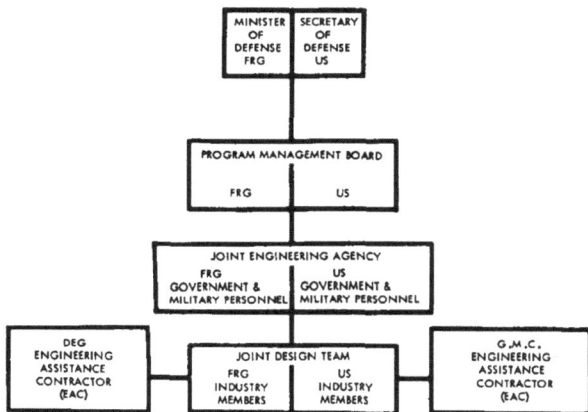

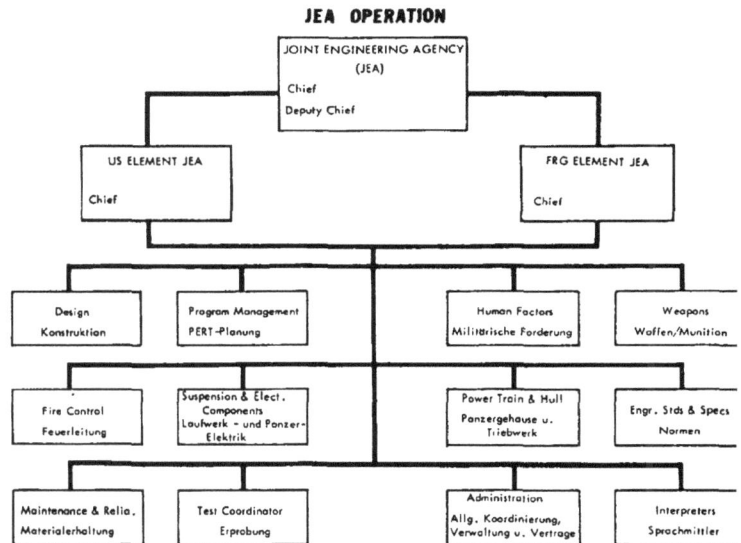

117

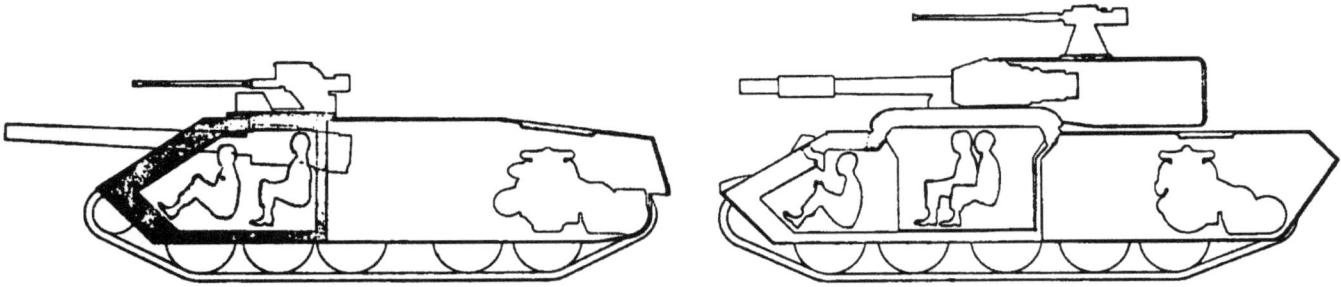

Two of the early configurations considered for the MBT70 program are sketched above. At the left is a casemate tank with limited traverse for the main weapon. At the right is a pod design with an externally mounted gun. Both arrangements provided the maximum radiological protection for the crew.

Preparation of the joint military characteristics for the new MBT required the resolution of several conflicting requirements. For example, the Germans were interested in a tank primarily for use in Central Europe while the United States required a vehicle suitable for operations worldwide. Both wanted a controlled crew environment with maximum armor protection and radiation shielding at the lowest possible weight. The Germans preferred a high velocity gun as the main armament, but the United States believed that a missile was essential to be effective at long ranges. Secondary armament was specified for use against both ground and air targets. To meet these joint military characteristics, a large number of design concepts were reviewed at Augsburg. They included just about every tank configuration previously studied at Detroit as well as a few new ones. The designs ranged from casemate tanks with limited traverse for the main weapon to pod concepts with externally mounted guns. Conventional arrangements were considered with the driver in the front hull and a variety of turret configurations. However, the driver-in-turret design was preferred by the United States team. One German concept mounted the entire crew in a ball type turret with three axis stabilization.

For the evaluation of the various proposals, a contract was awarded to the Lockheed Missiles and Space Company to perform a parametric design/cost effectiveness study. This study utilized a mathematical model of a tank which could be modified to represent any of the various design concepts. Computer analysis then evaluated the effectiveness of the proposed tank configurations with a wide variety of components against the expected threat under different combat conditions. By this means, the number of potential design candidates was reduced to five. Full scale mock-ups were fabricated of these five vehicles and they were subjected to further study, eventually resulting in two vehicle designs for final consideration. The final selection combined features from the two candidates. Key components were assigned to either the United States or the German team for design and fabrication. However, parallel development continued on some items to provide a backup if the prime candidate failed.

As mentioned previously, the American team preferred a new version of the 152mm gun-launcher for the main armament in the new tank. The original XM81 weapon was increased in length and development was started on a new kinetic energy armor piercing round. Designated as the XM150, the new gun-launcher also could fire all of the conventional 152mm rounds already available. Development continued on the Shillelagh missile extending its effective range and improving its reliability. Although Eugene W. Trapp, the senior civilian on the American side of the JEA, was successful in persuading the Germans to

Another design evaluated for application to the MBT70 appears below. This arrangement placed the entire crew, armament, and ammunition in a ball type turret with three axis stabilization.

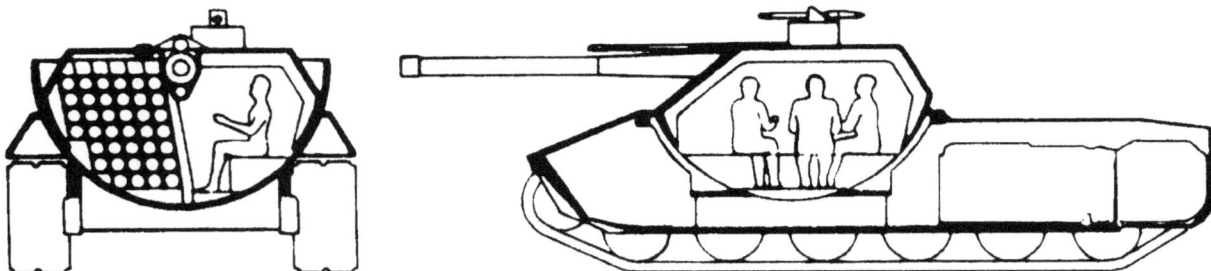

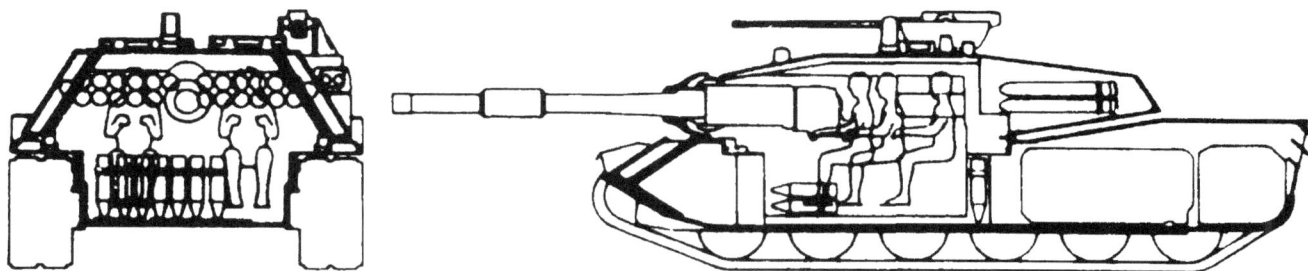

The basic design selected for the MBT70 can be seen in the sketch above. This is the integrated fighting compartment concept with the entire three man crew located in the turret. One of the automotive test rigs used in the MBT70 program appears in the photograph below at the right.

accept the driver-in-turret arrangement, they were less than enthusiastic about the 152mm gun-launcher. Early in the design study, the Delta-LASH system also had been considered as a main armament candidate. This consisted of the 120mm Delta gun previously described using the LASH (LAser Semi-active Homing) projectile in addition to the hypervelocity APFSDS round. Although this system was dropped early in the program, Germany continued the development of the 120mm high velocity gun and planned to arm at least some of their tanks with this weapon.

Below is a model of the final design selected for the MBT70. The tank commander's panoramic sight is visible on the turret roof as well as the television camera installation on the right front hull.

The first American pilot MBT70 appears above and below. Note the lack of any fire control equipment on this automotive pilot.

After selection of the design concept, the JEA and the JDT were relocated from Augsburg to Detroit, Michigan in July 1966. Initially, eight pilot tanks were authorized for construction by each country, however, this was later reduced to six. The first pilot was completed in the United States during July 1967 and it was displayed publicly for the first time in September, simultaneously with the first German prototype. Both were automotive pilots only and they were not fitted with any fire control equipment. The frontal area of both the hull and turret was protected by spaced armor and provision was made for the installation of radiation shielding to achieve an attenuation ratio of 20:1. The new tank was manned by a crew of three consisting of the commander, gunner, and driver. An automatic loader in the turret bustle eliminated the need for a fourth crew member. However, this also was not installed in the first pilots. The entire crew was located in the turret with the driver in a counterrotating capsule in the left front alongside the cannon. His controls were operated electrically through slip rings similar to those on the earlier MBT(MR) test rigs. The gunner was in the right front of the turret forward of the tank commander. With this arrangement, the turret could be controlled environmentally for protection against chemical and biological airborne contaminants and provided with heating and air conditioning for the comfort of the crew. The environmental control unit, manufactured in Germany, was installed in the left side of the hull. A ballistic grill on the upper left side of the hull covered

an air intake for this unit. An emergency escape hatch for the crew was located in the hull floor below the center of the turret. However, the slip ring had to be detached and lifted to one side before it could be used. To save weight, the escape hatch cover was made of cast titanium.

The travel lock for the 152mm gun-launcher can be seen in the rear view of the American MBT70 pilot below. Further details of the first American pilot are visible at the right.

Two views of the early German MBT70 automotive pilot above can be compared with a later German pilot tank below. The fire control equipment is installed on the latter vehicle.

Further details of the late German MBT70 pilot are shown in these photographs. Also, the 20mm gun is erected in the firing position. Note that the travel lock for the gun-launcher is located on the front hull of the late German pilot.

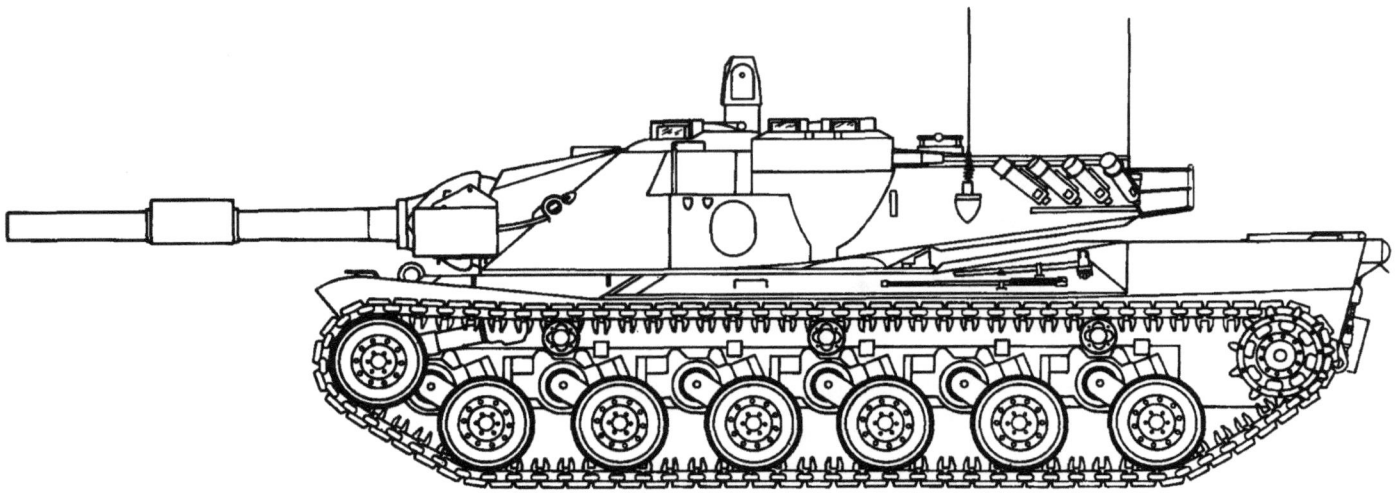

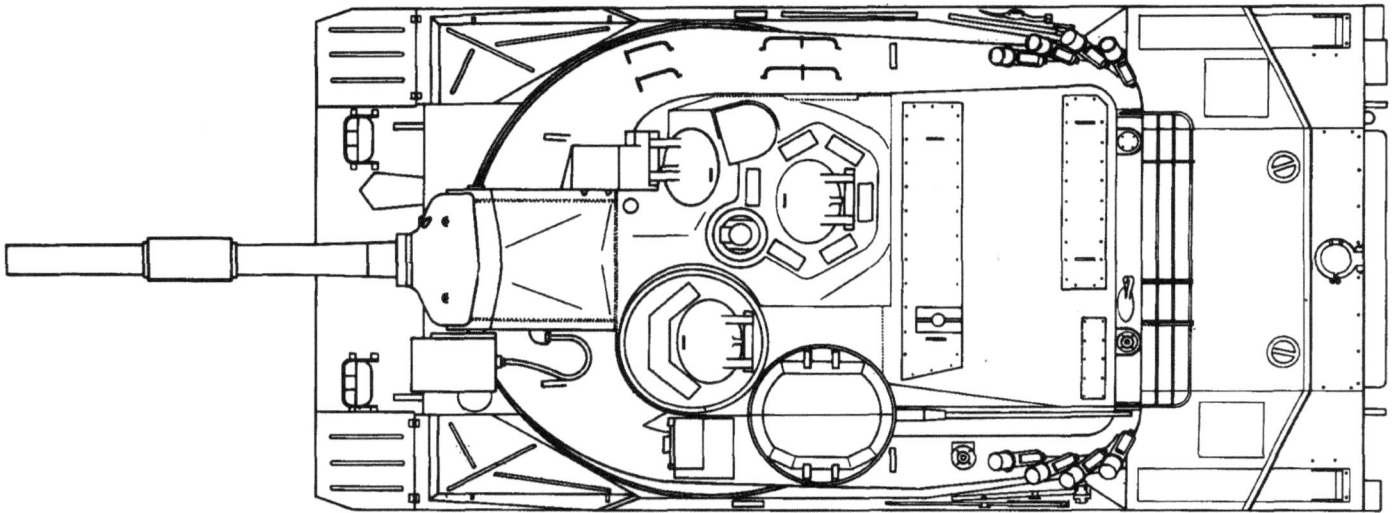

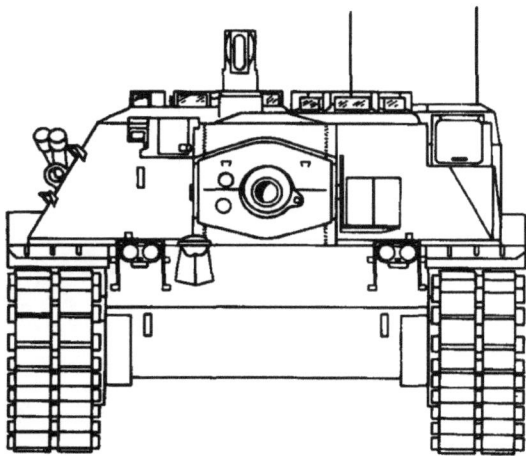

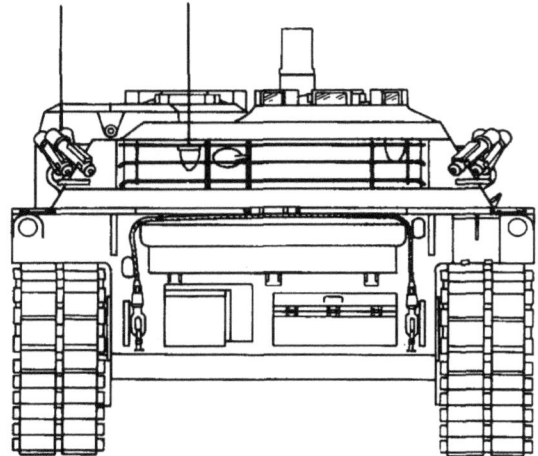

Scale 1:48

© D.P. Dyer

152mm Gun-Launcher Tank MBT70, United States Pilot

The second U.S. MBT70 prototype appears above with its fire control equipment and the driver's TV camera installed. Note the elimination of the bore evacuator on the gun-launcher. Below, a Shillelagh missile is launched from pilot number five during the test program.

The interior arrangement of the MBT70 is visible in the cutaway drawing above. Below is the American pilot number five. It is equipped with the missile transmitter and the commander's panoramic sight, but the searchlight has not been installed.

The American MBT70 pilot number two is shown in these photographs. The tank has the fire control equipment and the searchlight mount installed, but the light itself is missing.

Further details of the American pilot number two are visible in the view at the right. Note that the bore evacuator has been eliminated after the installation of the closed breech scavenging system. The American MBT70 pilot below has the searchlight mounted and the television camera installation can be seen on the right front hull.

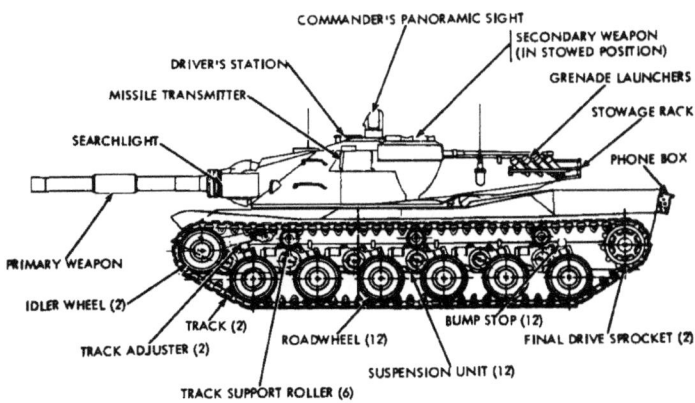

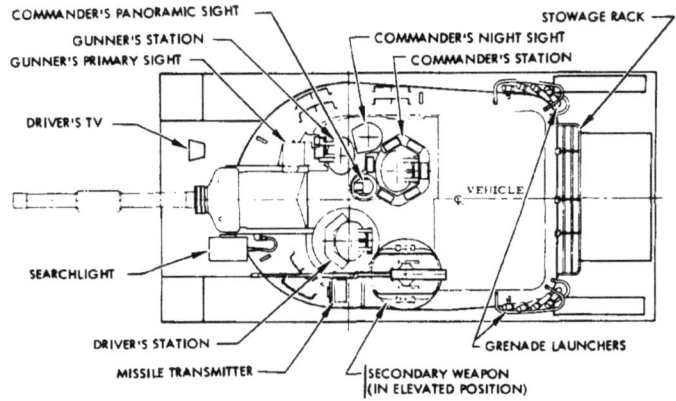

The various components of the MBT70 are identified in the sketches above.

The driver's station consisted of a capsule with a seat and platform suspended from a bearing in the turret roof. The low silhouette armored cupola above the seat contained a hatch and three periscope type vision blocks. The seat and platform were hydraulically adjustable in height to permit operation of the vehicle in a seated or semi-standing position. The latter was used with the driver's head exposed in the open hatch. The capsule counterrotated to keep the driver facing forward (0 degree or 12 o'clock position) regardless of the turret orientation. It also could be locked in the 50 degree position facing right or left as well as facing to the rear (180 degree or 6 o'clock position). The 50 degree left or right positions were to allow the driver to use one of his side vision blocks for driving in the event that the center block was damaged. The driver's main instrument and control panel was located inside the rotating capsule, but the auxiliary panel was fixed in the turret to the left of the driver when facing forward. In addition to the electrical controls, emergency mechanical controls were installed in the hull adjacent to the driver's capsule. However, both the capsule and the turret had to be locked in the forward position to permit their use.

Details of the MBT70 turret can be seen in the drawing below.

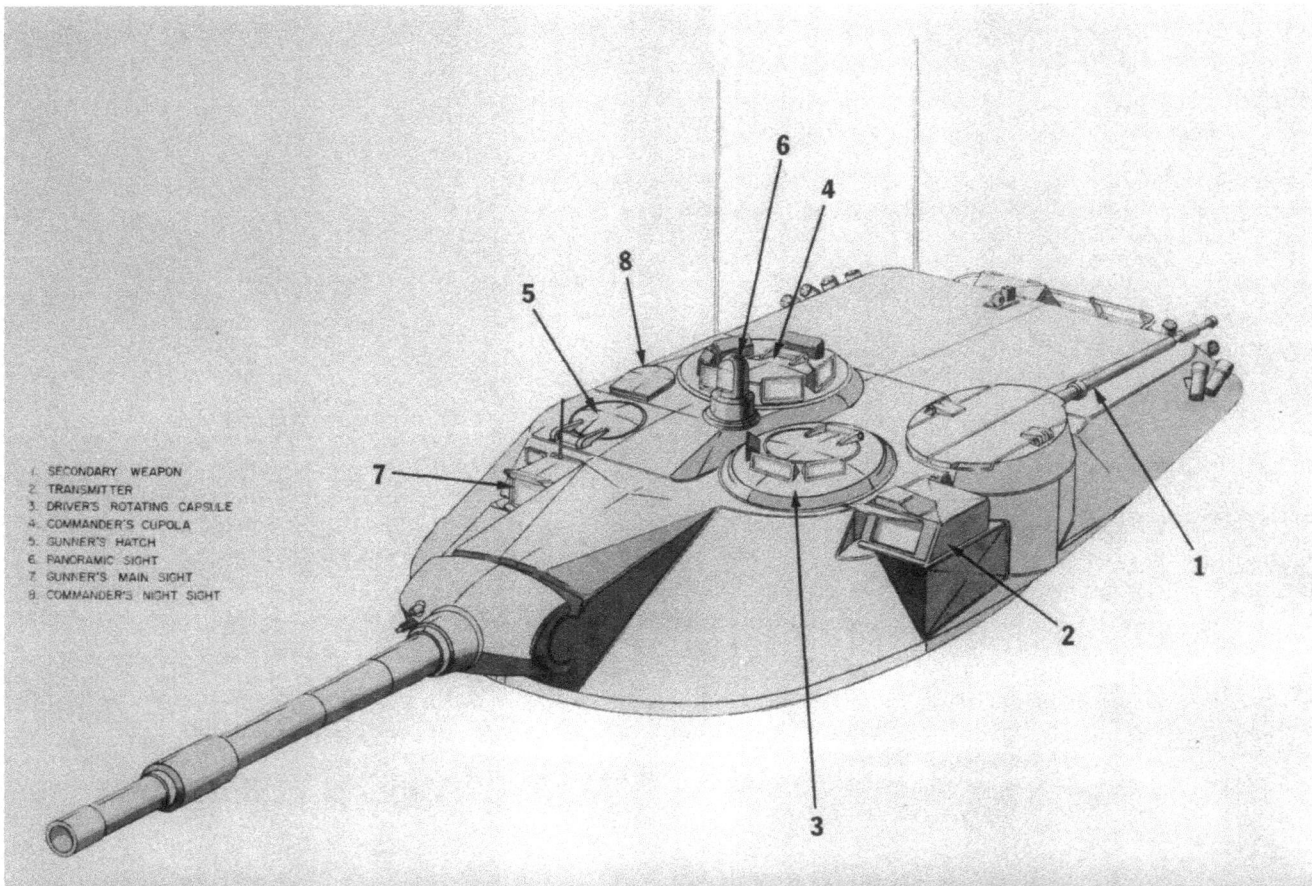

1. SECONDARY WEAPON
2. TRANSMITTER
3. DRIVER'S ROTATING CAPSULE
4. COMMANDER'S CUPOLA
5. GUNNER'S HATCH
6. PANORAMIC SIGHT
7. GUNNER'S MAIN SIGHT
8. COMMANDER'S NIGHT SIGHT

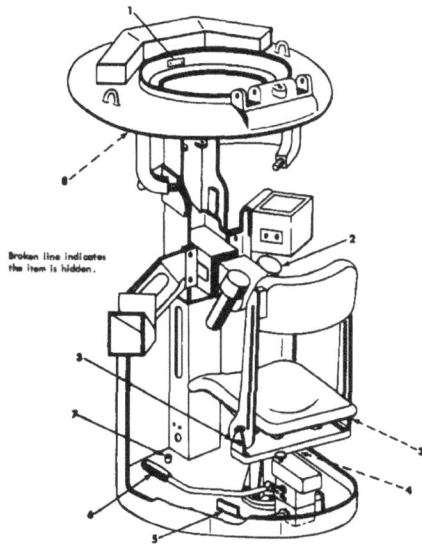

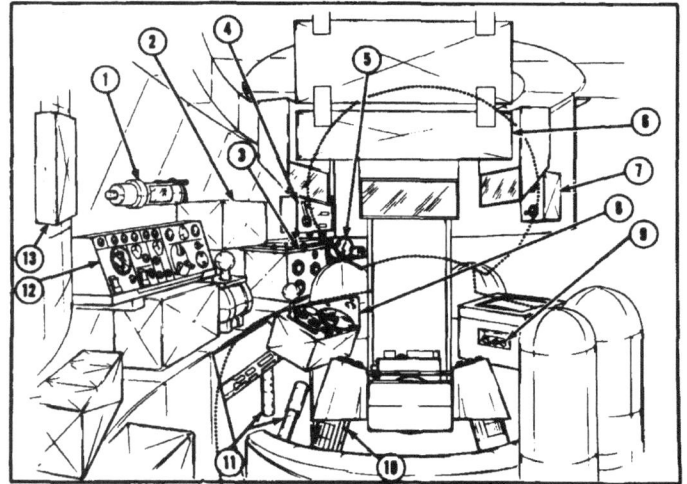

The numbers in these views of the driver's station indicate the following: Top left, 1. warning light, 2. control handle, 3 & 4. seat adjust, 5 & 6. platform elevation, 7. headlights, 8. capsule latch. Top right, 1.light, 2. F & H suspension controls, 3. power supply, 4. intercom, 5. transmission panel, 6. load panel, 7. periscopes, 8. instrument panel, 9. TV monitor, 10. control handles, 11. emergency controls, 12. auxiliary panel, 13. searchlight regulator. Below left, emergency controls, 1. brake, 2. transmission shift, 3, 4, 5. steering and throttle controls. Below right, driver's control handles.

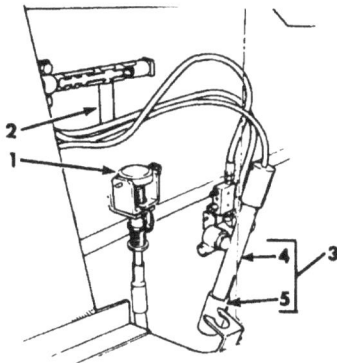

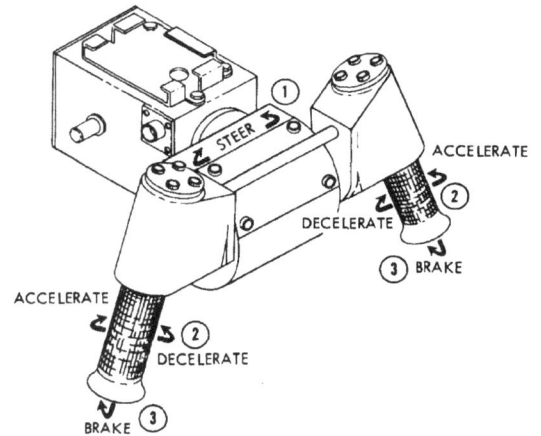

At the left is the left side of the driver's station in the German MBT70 pilot and below his controls can be seen through the turret hatch.

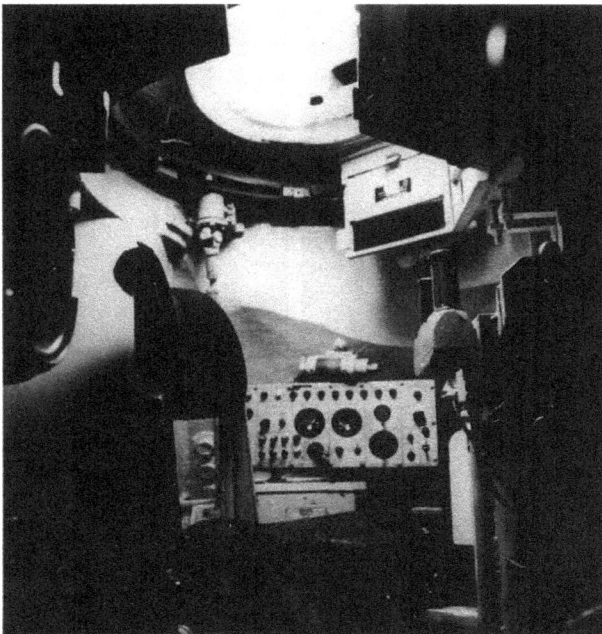

127

The gunner's station is at the left and his seat is seen above. Below, the gunner's control handles are sketched at the left and his weapon control panel is at the right. In the latter drawing, the numbers indicate the following: 1. & 2. secondary weapon controls, 3 & 4. weapon selector, 5. ammunition selector.

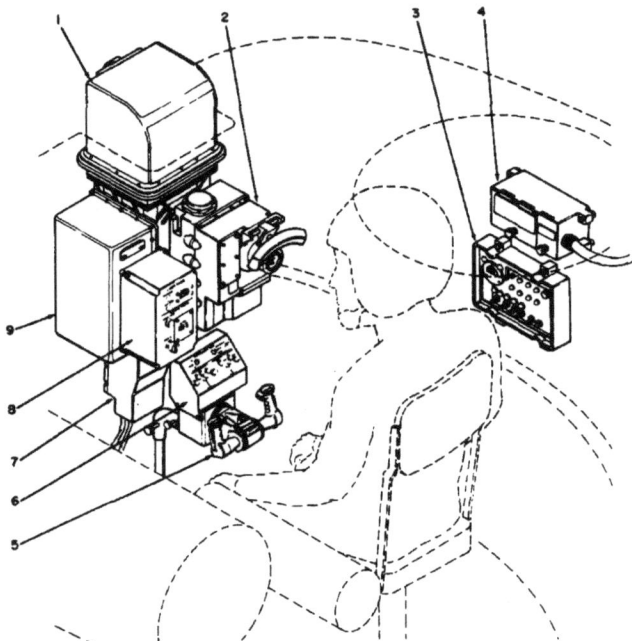

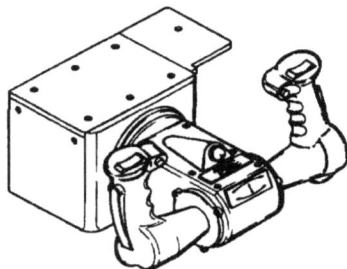

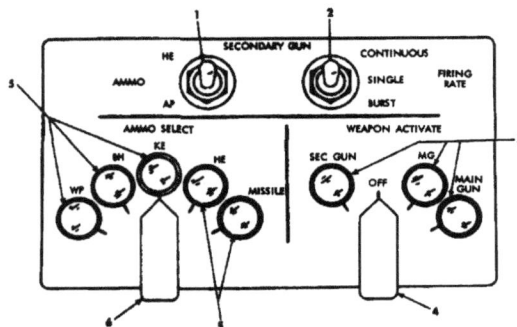

1. Gunner's periscope
2. Zoom telescope
3. Missile test checkout panel
4. Auxiliary control
5. Control handles
6. Weapon control
7. Missile tracker
8. Gunner's range control
9. Laser rangefinder

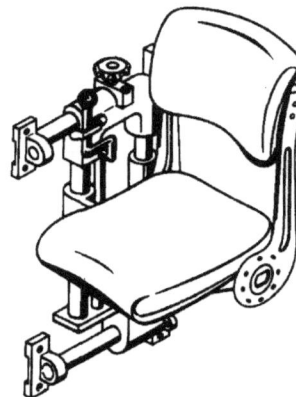

Below, the gunner's station on the German MBT70 pilot is shown looking forward (left) and to the right (right).

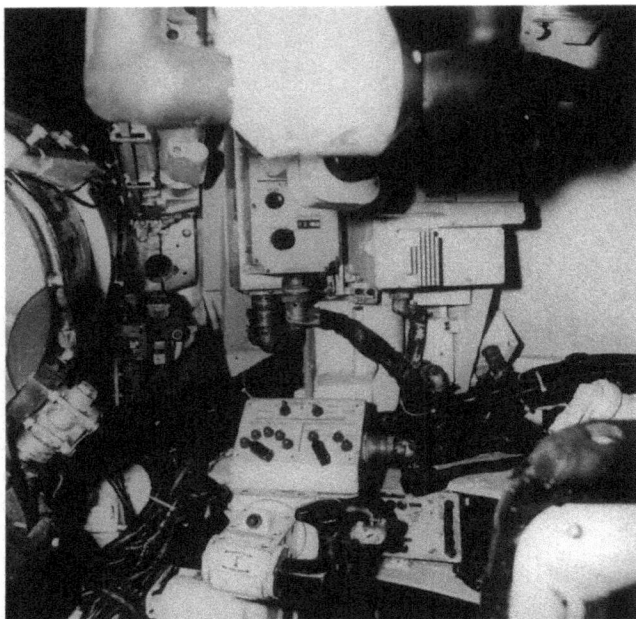

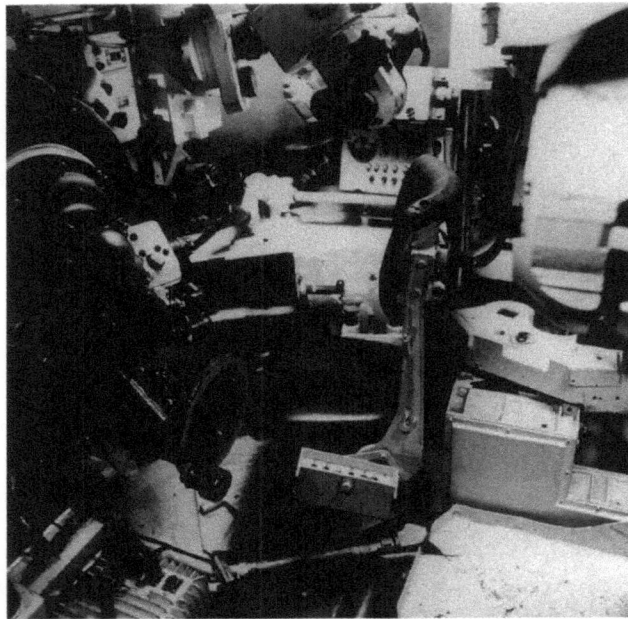

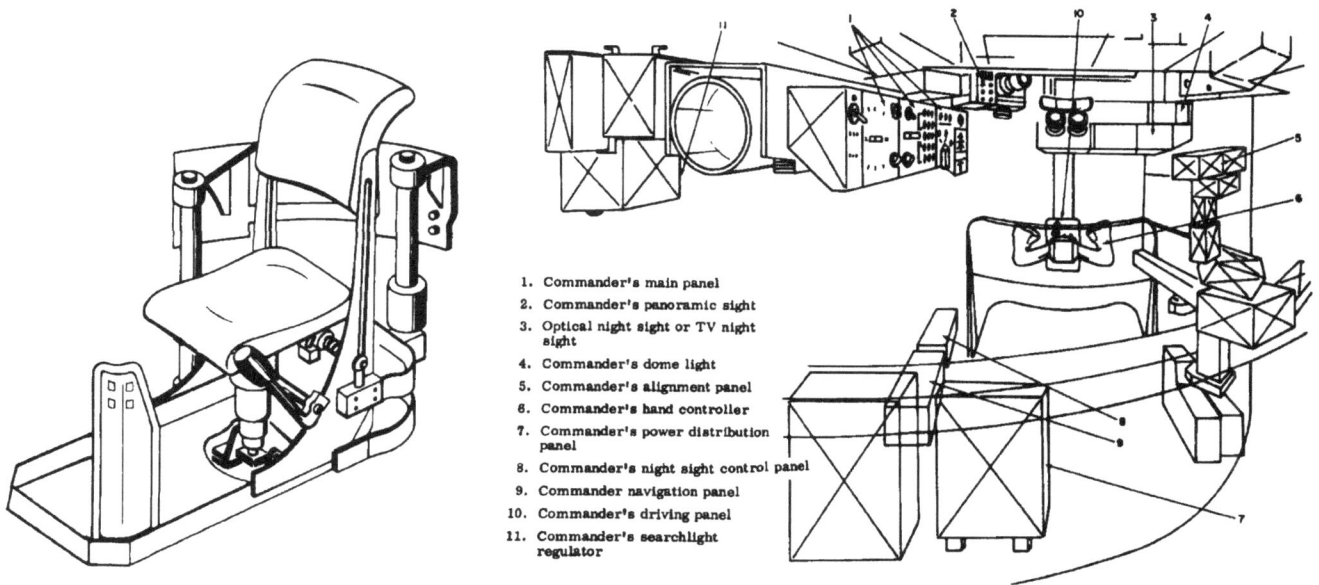

1. Commander's main panel
2. Commander's panoramic sight
3. Optical night sight or TV night sight
4. Commander's dome light
5. Commander's alignment panel
6. Commander's hand controller
7. Commander's power distribution panel
8. Commander's night sight control panel
9. Commander navigation panel
10. Commander's driving panel
11. Commander's searchlight regulator

Above, the tank commander's seat is at the left and an overall view of his station is at the right. The commander's hand controller is sketched below.

The gunner was provided with a hatch in the turret roof above his head. His seat was bolted to the side of the aluminum turret basket and it was mechanically adjustable. In addition to the gunner's primary sight, a periscope type vision block was provided to aid in acquiring targets and viewing the terrain. The gunner's auxiliary sight was an articulated telescope installed on the right side of the cannon above the coaxial machine gun. This sight included infrared viewing capability for night operations using infrared illumination from the searchlight.

The tank commander was located on the right side of the turret behind the gunner under a low silhouette fixed cupola with a hatch and six periscope type vision blocks. The latter provided a 360 degree overlapping view for surveillance close to the tank. The hatch covers for the commander and the driver were interchangeable. The commander's panoramic sight and night sight were mounted on the turret roof in front of the cupola to the left and right respectively. The commander's seat and platform were attached to the turret basket and they were hydraulically adjustable in height. Controls were provided to allow the commander to partially assume the duties of the gunner or the driver in an emergency and fire the armament or drive the tank. In the latter case, he could maneuver the vehicle, but he could not start or stop the engine.

Three auxiliary power subsystems driven by the main engine were required to generate electrical and hydraulic power. The electrical system consisted of a three phase alternator with an output of 20 kilowatts at 28 volts and eight 12 volt batteries connected in series parallel producing 24 volt direct current. Two hydraulic systems were

This photograph shows the tank commander's station in the German MBT70 pilot. The view is toward the rear of the turret.

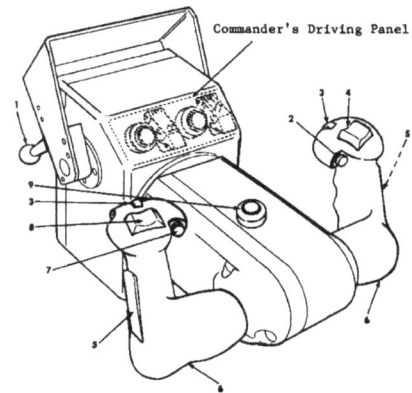

Commander's Driving Panel

129

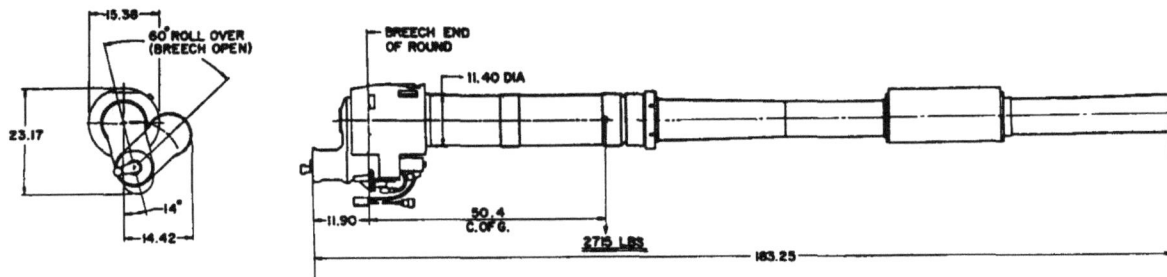

Dimensions of the 152mm gun-launcher XM150E5 are shown in the sketch above. The various components of this weapon are depicted in the drawing below.

provided. One operated at 1500 psi to power the turret drives and the automatic loader and the other developed 3000 psi for use with the hydropneumatic suspension.

The primary armament in the MBT70 prototypes was the 152mm gun-launcher XM150E5. This was a long barreled version of the gun-launcher used in the M551 Sheridan and the M60A2 tank. It launched the Shillelagh missile and fired all of the combustible case conventional ammunition provided for the short barreled weapons at a slightly higher muzzle velocity because of the longer tube. In addition, a kinetic energy armor piercing round was under development which was expected to equal or exceed the performance of the APDS round used in the 105mm gun M68. Designated as the XM578, it was an armor piercing, fin stabilized, discarding sabot (APFSDS) projectile with a long rod penetrator. Twentysix rounds of 152mm combustible case conventional ammunition or missiles were carried in the automatic loader in the turret bustle. This loader, designed by Rheinmetall was originally intended to be used in both the American and German prototypes of the MBT70. However, at a later date, an automatic loader developed by General Motors was installed in the American pilots. Access to the loader was through a hinged door in the bottom of the turret bustle when the turret was turned 90 degrees to the hull center line. The loader consisted of a continuous link carrier magazine and a rammer assembly. Any one of five types of ammunition could be selected. These were the high explosive, kinetic energy armor piercing, bee hive (flechette antipersonnel round), white phosphorus, or the Shillelagh missile. The magazine positioned the selected round for loading. At the same time, the gun-launcher was decoupled from the stabilization system, indexed to zero elevation, and locked in alignment with the loading tube. The round was then rammed from the magazine into the breech via the rail mounted loading tube. After loading, the weapon was automatically driven back to the stabilized line of sight. A firing rate of ten rounds per minute was expected with the Rheinmetall loader using

1. GUN-LAUNCHER TUBE
2. GUN-LAUNCHER BORE EVACUATOR CYLINDER
3. GUN SHIELD
4. GUN MOUNT TRUNNION YOKE
5. TRUNNION YOKE BEARING CAP (2)
6. RECOIL MECHANISM
7. SAFE-TO-FIRE INDICATOR
8. RECOIL MECHANISM RELIEF VALVE
9. GUN-LAUNCHER BREECH MECHANISM
10. BREECH MECHANISM HAND CRANK
11. GUN-LAUNCHER TORQUE KEY BRACKET
12. BUFFER STRIKE
13. COUNTERRECOIL BUFFER

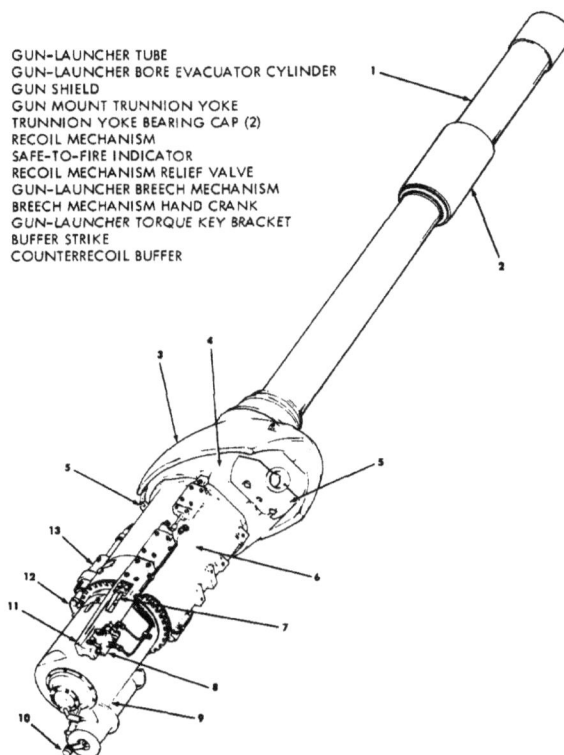

The breech of the XM150 gun-launcher appears at the right in the open position.

130

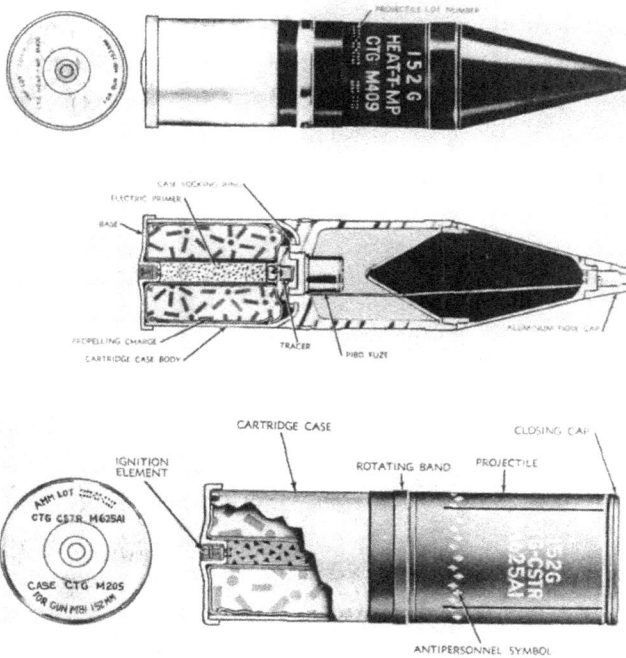

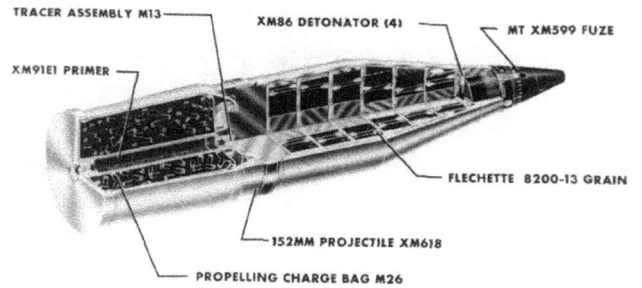

Above, the 152mm M409 HEAT-MP and the antipersonnel XM617 (flechette) rounds are shown at the left and right respectively.

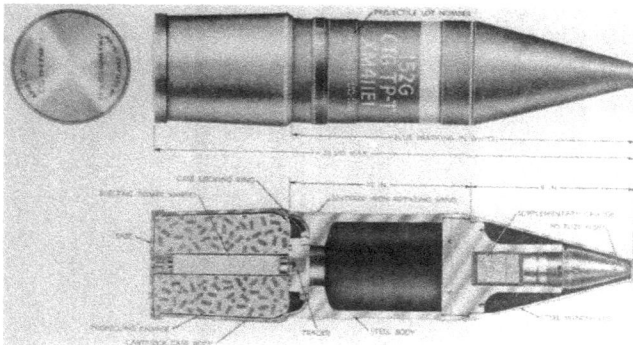

The 152mm M625A1 canister and the XM578E1 APFSDS armor piercing rounds appear above and at the right respectively. Below is the 152mm XM411E1 target practice round.

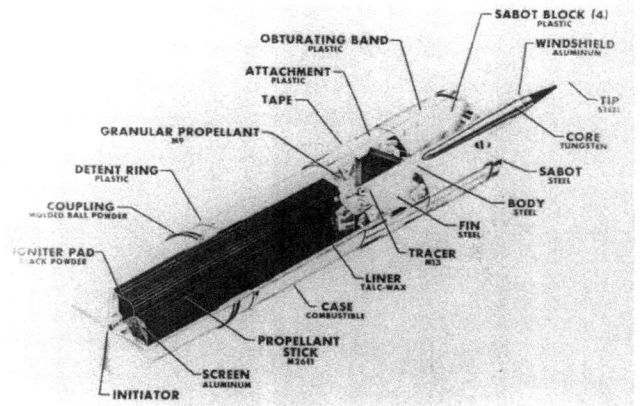

the conventional ammunition when the tank was not moving. The ammunition was normally inserted in the loader through a port in the rear wall of the turret bustle. However, the rounds stowed elsewhere inside the tank could be placed in the loading tube and retracted into the magazine. Eight 152mm rounds were stowed on the rear hull bulkhead outside of the turret basket. These could be reached through an opening in the basket wall at the driver's station. Twelve more 152mm rounds were located in the basket forward and to the rear of the driver's capsule, bringing the total carried to 46.

Ammunition stowage after the adoption of the 24 round General Motors automatic loader is reflected in the sketch at the right. The 24 rounds shown here brought the total 152mm ammunition stowage to 48. The numbers indicate the following: 1. HEAT-MP, 2. APFSDS, 3. APFSDS or missiles, 4. 7.62mm ammunition boxes.

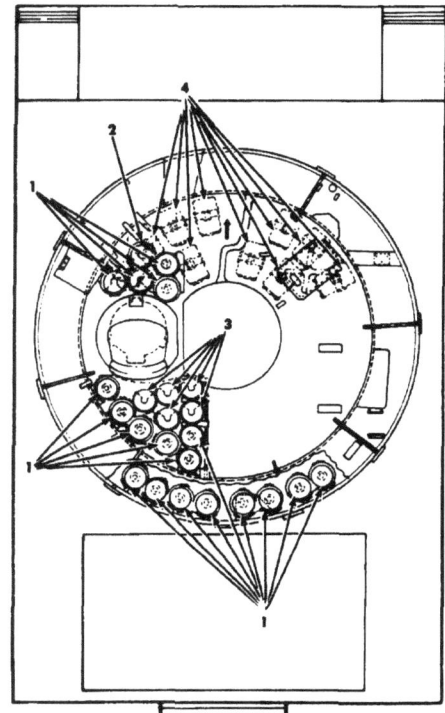

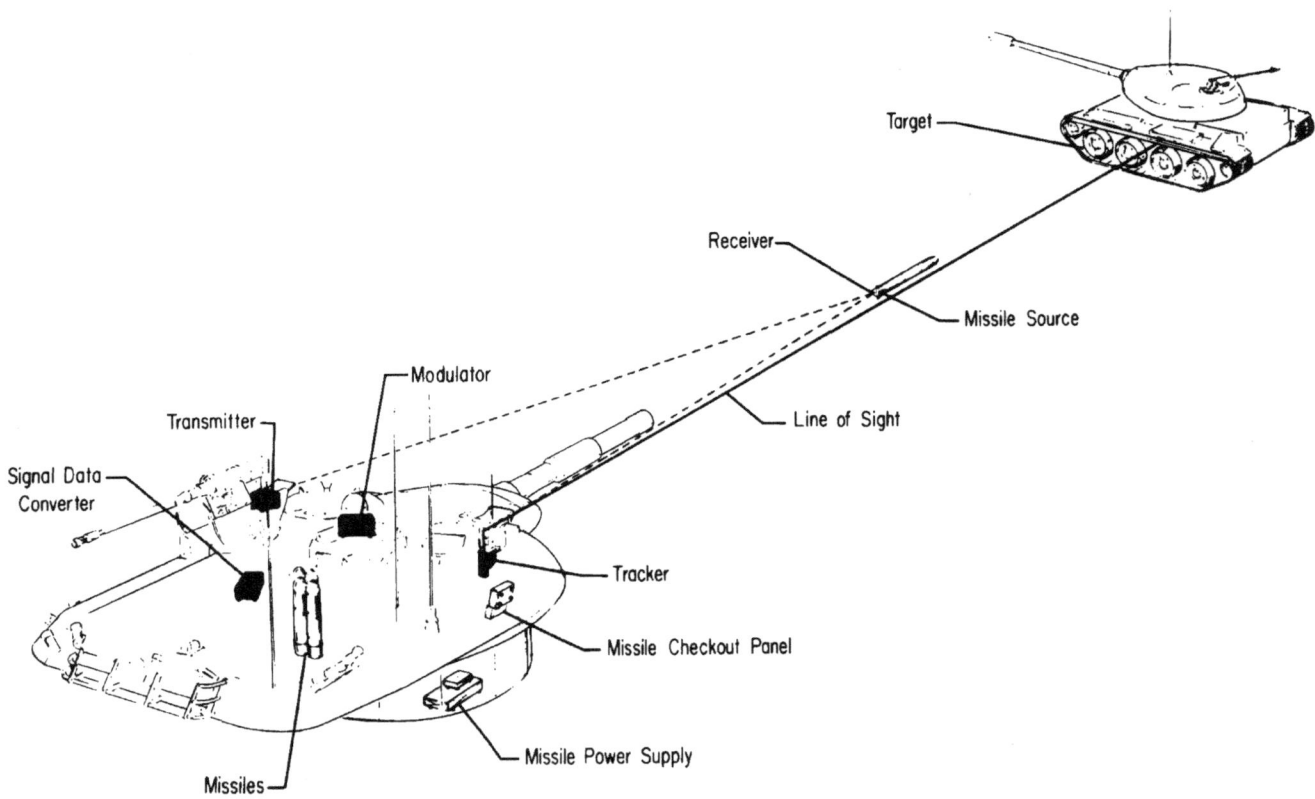

The operation of the Shillelagh missile system in the MBT70 is illustrated in the drawing above. Below, the missile components are sketched at the left and the markings for the surface attack (upper) and practice (lower) missiles appear at the right.

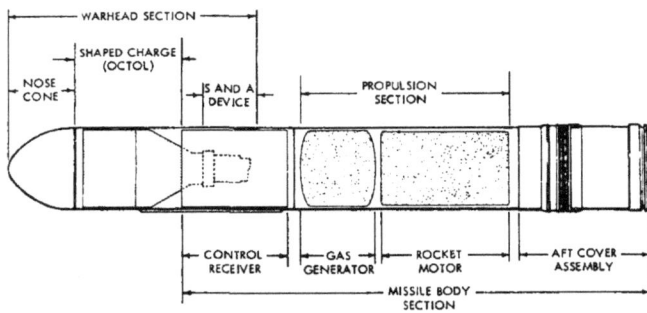

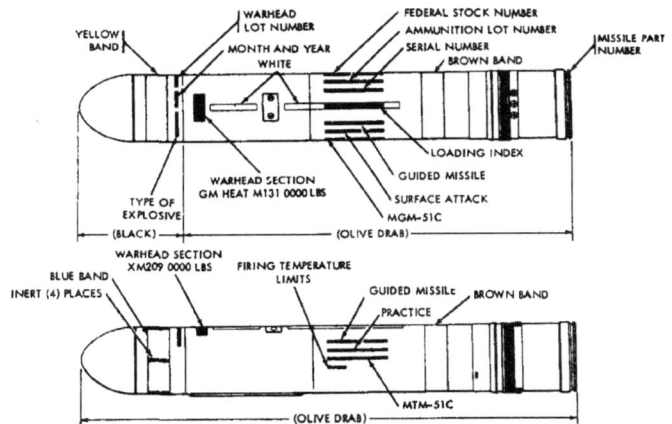

Below are two views of the Shillelagh missile being launched from the MBT70.

132

These photographs show the Rheinmetall 20mm automatic cannon erected in the firing position on the German MBT70 pilot. The commander's panoramic sight can be seen in the top left view. The components of the secondary weapon mount are sketched below at the right.

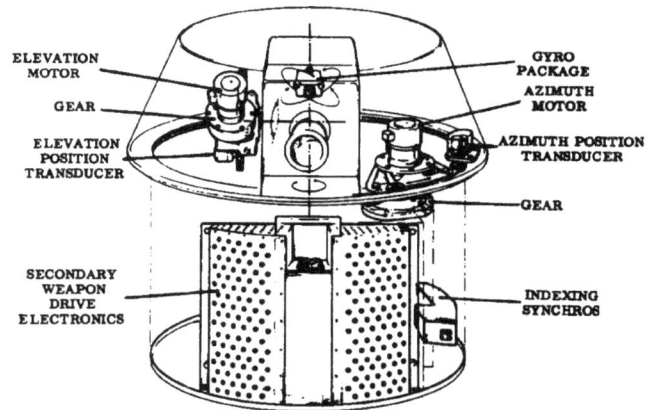

ELEVATION MOTOR
GEAR
ELEVATION POSITION TRANSDUCER
SECONDARY WEAPON DRIVE ELECTRONICS
GYRO PACKAGE
AZIMUTH MOTOR
AZIMUTH POSITION TRANSDUCER
GEAR
INDEXING SYNCHROS

Secondary armament on the MBT70 consisted of a remote controlled 20mm Rheinmetall RH202 automatic cannon installed in a watertight capsule at the left rear of the crew compartment in the turret. When not in use, this weapon was retracted under armored covers with the barrel extending to the rear along the top left side of the turret. When activated, the covers opened and the mount was elevated to provide the gun a 360 degree field of fire. It could be aimed and fired by either the gunner or the tank commander. Two belts containing a total of 750 rounds of 20mm ammunition were stowed within the gun mount. Either of two types of ammunition could be selected.

A 7.62mm M73 coaxial machine gun was installed on the right side of the gun-launcher below the gunner's auxiliary sight. Three thousand rounds of 7.62mm ammunition in belts were stowed on the forward turret basket wall and linked to the M73 machine gun. An additional 3000 rounds were stowed in 15 containers under the turret basket floor.

The 7.62mm coaxial machine gun and the blower to vent the powder gases appear in the drawing at the right.

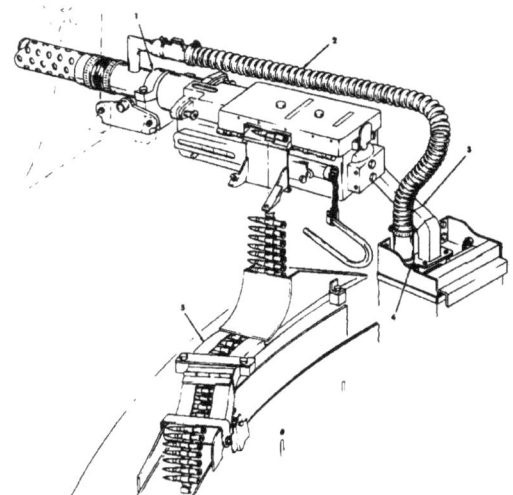

LEGEND

1. MOUNTING BRACKET
2. FLEXIBLE DUCT
3. SPENT BRASS COLLECTION SYSTEM
4. BLOWER AND VALVE
5. AMMUNITION FEED SYSTEM

133

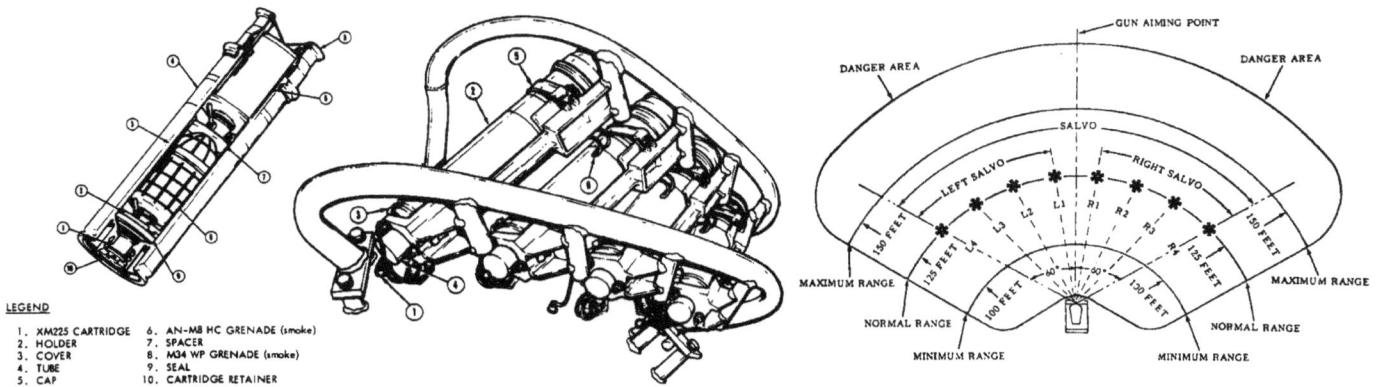

Above are the details of the smoke grenade launchers on the American MBT70. The components are as follows: 1.solenoid cap, 2. XM176 grenade launcher, 3. retaining clip, 4. solenoid. At the right are the grenade launchers installed on the German MBT70.

Eight XM176 grenade launchers were mounted on the turret bustle of the American pilots, four on each side. Each of these launchers contained two smoke grenades. Actuated from the commander's station, these launchers provided close-in protection and concealment for the vehicle.

The MBT70 was fitted with a highly sophisticated and complex fire control system. This allowed the tank commander to take over many of the gunner's functions in an emergency. The ballistic computer was a digital differential analyzer which received AC, DC, binary, and pulse signals from its own sensors, the gunner's control handles, accelerometers, the gunner's primary sight, the turret azimuth resolver, and the laser range finder as well as on-off selection signals from the gunner's weapon control, the gunner's range control, and the gunner's control handles. The various sensing devices included a powder temperature sensor, a gun bend sensor to predict barrel deflection due to temperature changes, and an air data sensor to measure air temperature, pressure, and cross wind velocity.

In addition to the six periscope type vision blocks, the panoramic sight and the night sight gave the tank commander 360 degree observation and were used by him to aim the gun-launcher and the coaxial M73 machine gun as well as the 20mm secondary weapon. The latter could be tracked through an elevation range of –10 to +65 degrees for use against ground targets or slow flying aircraft.

The control handles that the commander used to aim and fire the armament also were utilized to drive the tank, depending upon the operating mode selected. When the commander elected to operate a weapon or drive the tank, his controls overrode those of the gunner or the driver. The gunner's primary sight installed in the right front of the turret roof incorporated the laser range finder and the missile tracker. The commander's panoramic sight, the gunner's primary sight, the main weapon, the secondary weapon, and the infrared missile transmitter were stabilized in both azimuth and elevation. The missile transmitter was mounted on the exterior of the upper left turret wall protected by an armored cover. Needless to say, the front cover remained open during the flight of the missile. A roll gyro also provided information for cant correction to the ballistic computer. Two types of night sights were available for use by the tank commander. The first was a direct night sight utilizing a light intensifier tube observed directly through a binocular eyepiece. The second type incorporated a low light level television camera which transmitted the image to a television moniter. Moniters installed at the tank commander and gunner stations permitted both to use the system.

A German xenon searchlight was mounted on the left side of the gun shield. Similar to the unit installed on the German Leopard tank, it provided both white and infrared light. Provision was made for the installation of a land navigation system in command tanks which would automatically indicate the vehicle heading and position. A television camera could be mounted on the upper front hull to provide the driver close-in vision in front of the tank. A moniter like those used with the television night sight was installed in the driver's station. For deep fording, the camera was relocated to a base plate on one of the snorkel tubes. A passive night vision device could be installed from inside the tank to replace the driver's center vision block.

1. Laser rangefinder
2. Gunner's periscope assembly
3. Washer reservoir
4. Support structure (not visible)
5. Zoom telescope
6. Shutter substitute (not visible)
7. Window washer and wiper control
8. Turret gyro
9. Elevation rate gyro or roll rate gyro (typical)
10. Cable assembly set (representative drawing)
10A. Breech/loader control
11. Gunner's auxiliary control
12. Weapon control electronics
13. Gun and turret drive electronics
14. Turret azimuth resolver
15. Missile transmitter drive electronics
16. Power distribution panel
17. Air data sensor
18. Laser electronics
19. Vehicle velocity sensor
20. Powder temperature sensor
21. Vertical sensor
22. Gunner's primary sight electronics
23. Ballistic computer
24. 400 Hz power supply
25. Missile transmitter drive
26. Azimuth rate gyro
27. Gun trunnion synchro
28. Gun bend sensor
29. Gunner's control handles
30. Gunner's weapon control
31. Gunner's range control
32. Azimuth angular accelerometer
33. Elevation angular accelerometer

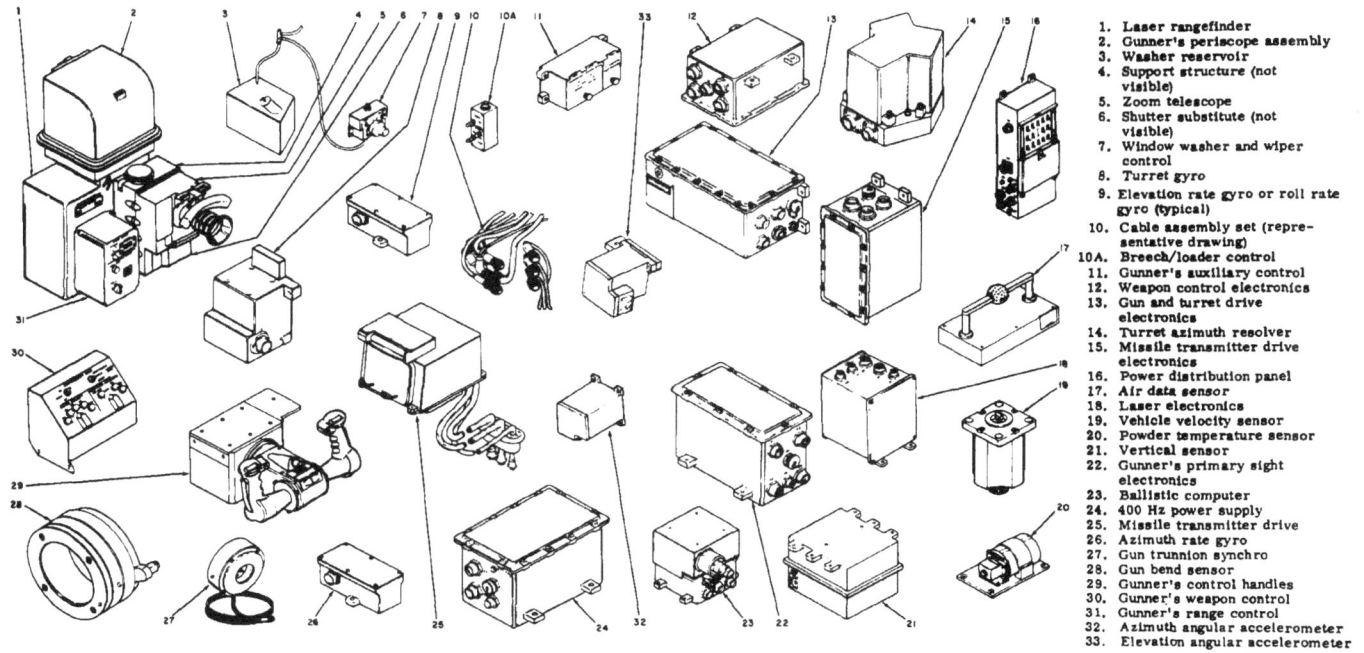

Various components of the MBT70 fire control system are illustrated above. Below at the left is the azimuth indicator. The numbers 1 and 2 refer to the scale and indicator light respectively. At the center below is a sketch of the xenon searchlight.

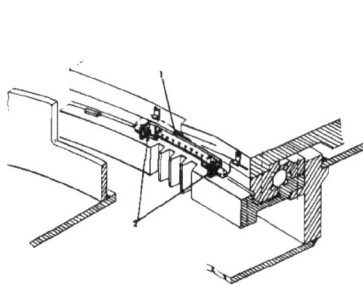

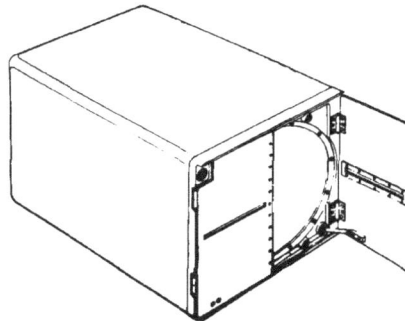

Above at the right is a view of the commander's night sight extended for operation. Below, from left to right, are the gunner's primary sight, the missile transmitter, and the commander's panoramic sight.

Above are left front and right rear photographs of the AVCR-1100 engine. The numbers refer to a component list no longer available. At the right are sketches of the cylinders for the AVCR-1100-2 (upper) and the AVCR-1100-3 (lower) engines. The numbers indicate the following: 1. rocker covers, 2. cylinder assembly, 3. cylinder hold down nut, 4. O-ring seal.

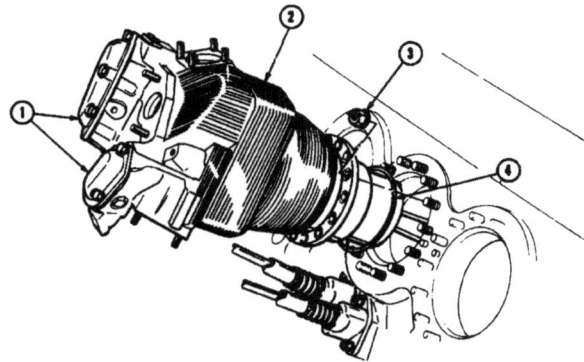

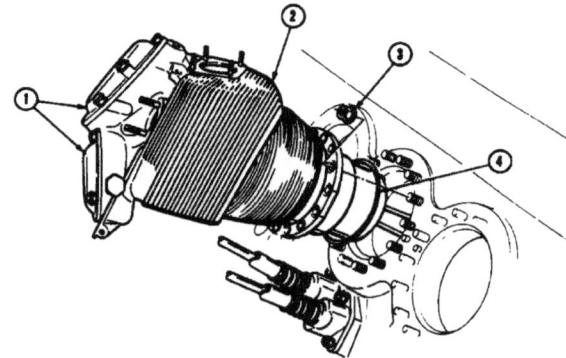

The American MBT70 pilot tanks were powered by the Continental AVCR-1100-2 or AVCR-1100-3 air-cooled diesel engine. The main difference between the two models was the displacement. The 4 7/8 inch bore of the AVCR-1100-2 was enlarged to 5 3/8 inches in the AVCR-1100-3. This increased the displacement from 1120 cubic inches to about 1360 cubic inches using the same five inch stroke. Both of these variable compression ratio engines were rated at 1475 gross horsepower at 2800 rpm. The engine was still under development and the displacement was increased to improve the reliability when operating at full power. As mentioned previously, the AVCR engines achieved their high output through the use of variable compression ratio pistons. Actuated by hydraulic

Below, the Daimler-Benz engine installed in the German MBT70 pilot is at the left and the Renk transmission is at the right.

136

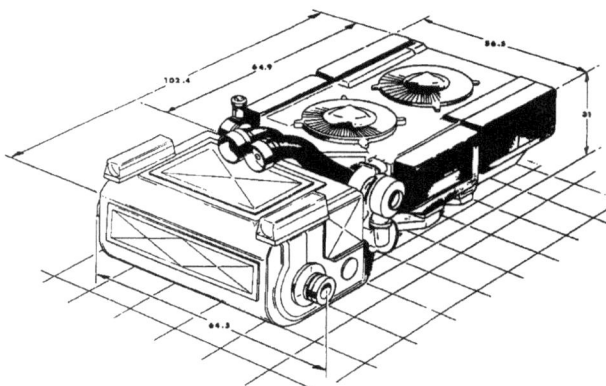

The power pack for the American MBT70 above and at the top right consisted of the Continental AVCR-1100-3 engine and the Renk HSWL 354 transmission. At the right is an exploded view of the Diehl 170 replaceable pad track shoe.

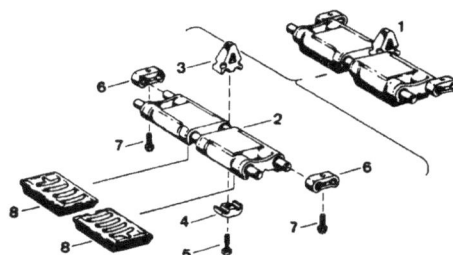

pressure, these pistons permitted operation at a high compression ratio (22:1) for starting, but limited the maximum cylinder pressure by reducing the compression ratio (to about 10:1) as the load on the engine increased. To avoid confusion in the contracting situation, the engine with the increased displacement was designated as the AVCR-1100-3, although this did not follow the usual nomenclature. At a later date, an improved version of this same engine received the proper designation of AVCR-1360.

The German design team had reservations about the lightweight AVCR engine and a somewhat heavier liquid-cooled diesel was developed by Daimler-Benz. This engine, which developed 1500 gross horsepower at 2600 rpm, was installed in the German pilots. Both the Continental and the Daimler-Benz engines utilized the German Renk HSWL 354 Powershift transmission. This unit featured four speeds forward, four speeds in reverse, automatic and manual shift, hydrostatic steering, and hydrodynamic braking. The transmission drove the Diehl 170 double pin tracks through planetary gear final drives. These tracks were 25.0 inches (635mm) wide with replaceable rubber pads.

Piston type hydropneumatic suspension systems were developed by the National Water Lift Company in the United States and by Frieseke and Hoepfner GmBH in Germany. The National Water Lift version was specified for installation on the first and second American pilots and the Frieseke and Hoepfner model on the remainder. Both types were fitted with twelve road wheels, six per track. The 3000 psi hydraulic system separately controlled four groups of suspension units, each consisting of three road wheel stations. These were located on the right front, left front, right rear, and left rear of the tank. Thus by setting the height of the suspension units in each quadrant, the tank could be raised, lowered, or tilted in any direction. A four lever manual control for the National Water Lift system was installed in the hull adjacent to the gunner's position. Each lever controlled the suspension units in one

Below, an inboard view of the National Water Lift suspension and track is at the left. In the view at the right, the National Water Lift suspension (left) is compared with the Frieseke and Hoepfner model (right).

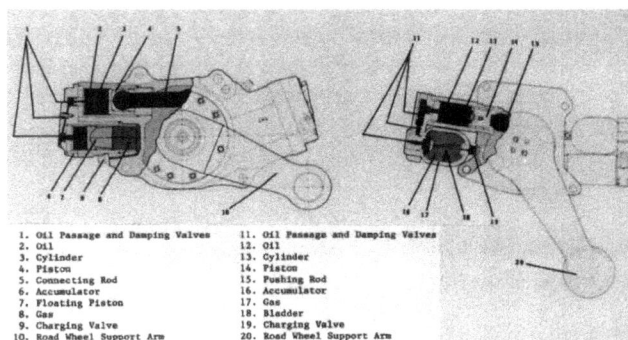

1. Oil Passage and Damping Valves	11. Oil Passage and Damping Valves
2. Oil	12. Oil
3. Cylinder	13. Cylinder
4. Piston	14. Piston
5. Connecting Rod	15. Pushing Rod
6. Accumulator	16. Accumulator
7. Floating Piston	17. Gas
8. Gas	18. Bladder
9. Charging Valve	19. Charging Valve
10. Road Wheel Support Arm	20. Road Wheel Support Arm

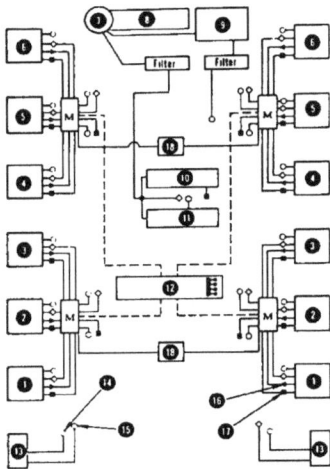

(M) Elevation Control Manifold consisting of:
  a. Elevation Control Valve
  b. Isolation Valve with Reset
  c. Low Supply Pressure Failure Valve
1-6. Hydropneumatic Suspension Unit
  7. PC Pump
  8. Pressure Compensator
  9. Reservoir
 10. Lockout Solenoid Valve
 11. Automatic Isolation
      Reset Valve
 12. Mechanical Height
      Control Actuators
 13. Compensating Idler
 14. Return
 15. Supply Pressure
 16. Fluid Transfer Between
      Units
 17. Lockout
 18. Quadrant Balance Valve

LEGEND:

◇ - Supply Pressure
○ - Return
■ - Lockout
○ - Isolation Reset
▶ - Fluid Transfer Between Units

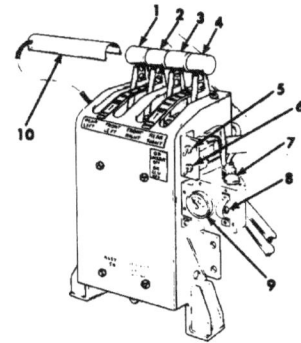

Above is a schematic diagram (left) and a drawing of the control console (right) for the National Water Lift suspension system. The numbers in the latter indicate the following: 1, 2, 3, & 4. the four quadrant operating levers, 5. lockout switch, 6. ECU switch, 7. light, 8. hydraulic pump switch, 9. fluid level, 10. lever lock. Below at the right are the automatic and manual controls for the Frieseke and Hoepfner suspension. The components on the automatic panel are indicated as follows: 1, 3, 7, & 9. the four elevation controls, 2. lockout light, 4. lockout switch, 5 & 6. switches, 8. warning light.

quadrant of the vehicle. They were operated by the gunner at the direction of the tank commander. The normal ground clearance of 17¾ inches was obtained by setting the console levers on the first graduation line below the maximum position. The minimum ground clearance was limited to about 5½ inches when the road wheel arms contacted the bump stops at the top of their travel range. The maximum ground clearance was approximately 29 inches. A toggle switch was provided to lock out the four corner suspension units (numbers 1 and 6 on each side) when firing the main weapon. Another toggle switch turned off the environmental control unit when raising the vehicle or checking the response time.

The Frieseke and Hoepfner hydropneumatic suspension featured either automatic or manual height adjustment selected by a toggle switch on a control panel installed to the left of the driver. Four rotary knobs were set to select the height for each quadrant of the tank. These knobs had detents at clearance heights of 10 inches, 17.5 inches, and 24 inches. A lockout switch also was provided for the corner suspension units. Track tension was adjusted hydraulically by shifting the idler at the front of each track. This idler was identical in size to the road wheels. Normally the track tension was automatically adjusted during lowering or raising. However, it also could be adjusted separately. When manual operation was selected, the suspension height was controlled by four toggle switches on the electrical switch box.

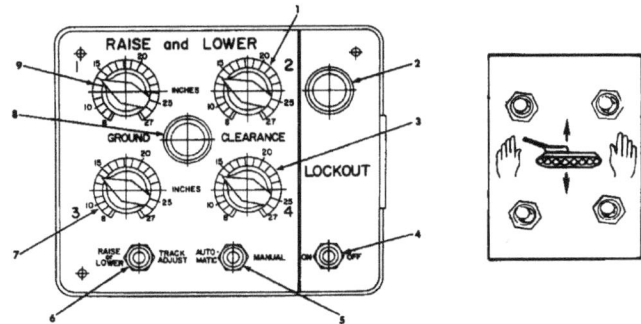

The Frieseke and Hoepfner suspension can be seen at the right in the photograph of the German MBT70 pilot.

138

The second American MBT70 pilot is shown above while raising (left) and lowering (right) the suspension. The ability to separately raise and lower the front and rear of the tank is illustrated below. Note the heavy black exhaust from the diesel engine during this operation.

Below, the tank is lowered to its minimum height at the left and, at the right, the front of the suspension is raised to assist in climbing the vertical wall. The weights added to the front hull and the turret were to simulate the weight of a fully stowed vehicle. The photographs on this page were taken at Aberdeen Proving Ground and were dated 27 March 1969.

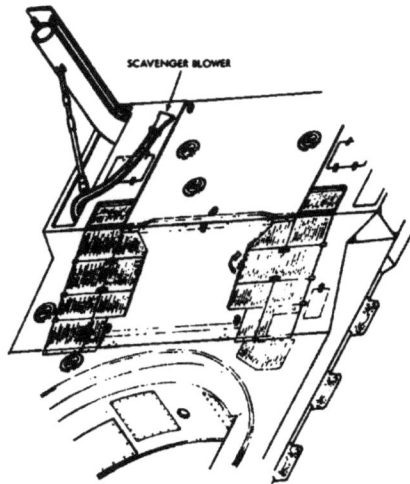

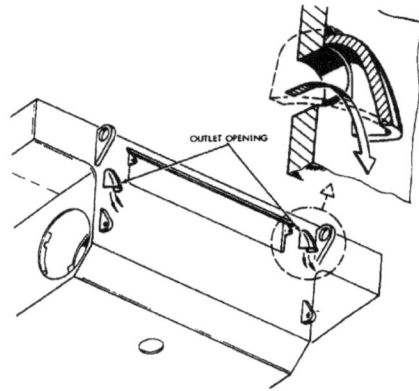

The top deck and rear hull of the American MBT70 are sketched above. These views were intended to illustrate the cleaning of the engine air intake filters using the scavenger blower. Note the fording snorkel in the erected position. Below is a top view of the German MBT70 pilot powered by the Daimler-Benz engine.

Two ballistic grills on top of the engine compartment provided intake air for the Continental engine in the American pilots. Portions of these grills were hinged to permit access to the engine and transmission for maintenance. The engine cooling air and exhaust gases were mixed and passed out of the vehicle through a rectangular armored grill extending across the upper portion of the rear hull plate. With the liquid-cooled Daimler-Benz engine in the German pilots, two large circular cooling air intakes were located on top of the engine compartment behind the turret. After passing through the radiators, this air and the engine exhaust exited through a rectangular grill on the rear of the tank.

Below, the fuel tank arrangement in the engine compartment of the American MBT70 is at the left and the vehicle fire extinguisher system is illustrated in the schematic diagram at the right.

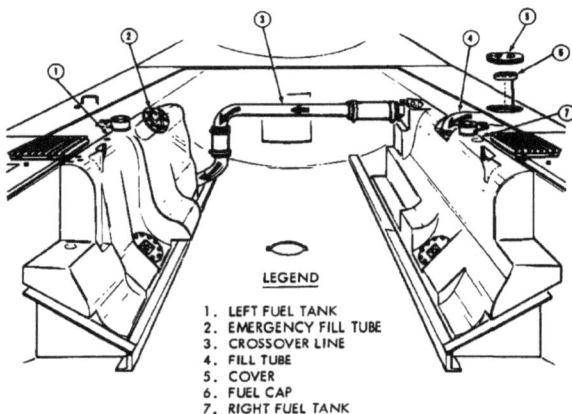

LEGEND

1. LEFT FUEL TANK
2. EMERGENCY FILL TUBE
3. CROSSOVER LINE
4. FILL TUBE
5. COVER
6. FUEL CAP
7. RIGHT FUEL TANK

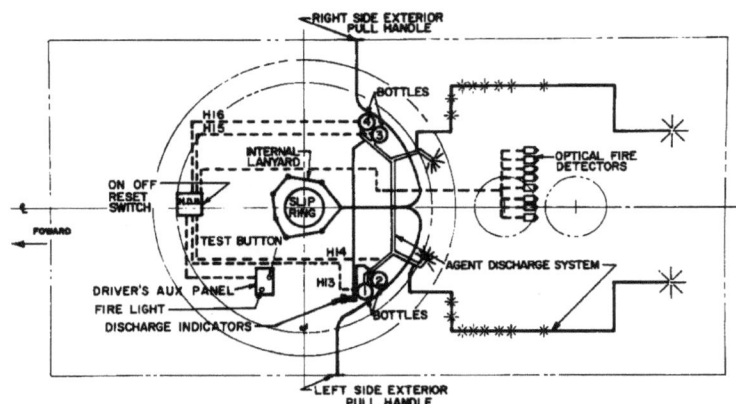

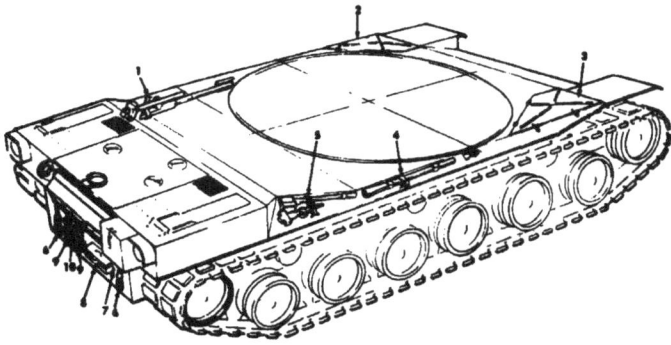

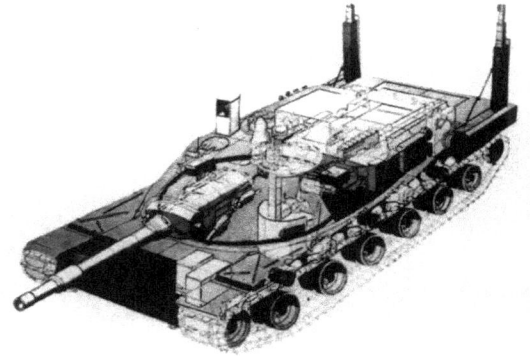

The hull arrangement of the American MBT70 can be seen in the sketch at the top left. The numbers indicate inspection points for the stowage after vehicle operation. The cutaway view at the top right shows the fording snorkels erected and reveals some features of the tank's interior. The snorkel in its stowed position is visible in the sketch below along with the air cleaner assembly.

The fuel system consisted of two aluminum tanks, one on each side of the engine compartment, and a multi-ply rubber tank installed in the front hull. A booster pump in each tank transferred fuel to a sump tank from which the engine fuel pump delivered it through the filters to the engine. Two watertight armored transverse bulkheads separated the crew in the center from the fuel in the front compartment and the engine compartment in the rear. To save weight, aluminum was used for the engine compartment floor and for access doors on the engine deck which were underneath the turret bustle when the gun was in the forward position.

The tank could ford water up to the turret top without special preparation. This was done by closing ports, inflating various seals, and erecting the engine air intake snorkels stowed at the rear corners of the tank. With the use of a deep submergence kit, the vehicle could operate in water 18 feet deep. The engine compartment was flooded when fording with the American pilots powered by the air-cooled Continental engine. However, use of the Daimler-Benz engine in the German prototypes required that the engine compartment be sealed and kept dry.

1. VACUUM HOSE ASSEMBLY
2. SNORKEL
3. AIR CLEANER
4. AIR CLEANER ELEMENT (3)
5. BLOWER
6. SELECTOR VALVE TO BLOWER DUCT
7. TURBOCHARGER
8. SELECTOR VALVE
   (Normal Position)
9. SELECTOR VALVE
   (Fording or Vacuum Position)

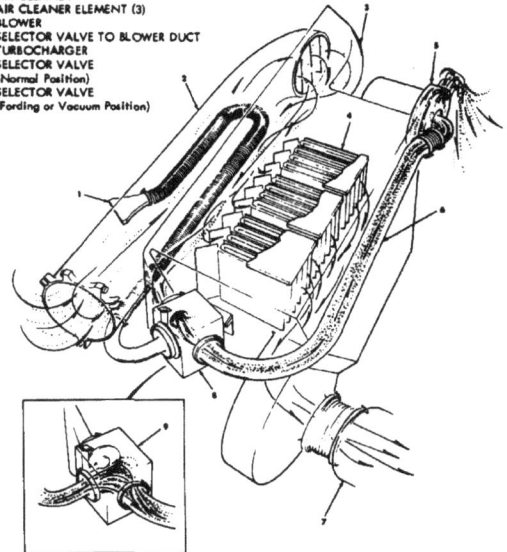

Below, the fourth American MBT70 pilot is climbing aboard its transport trailer. Additional weight also has been added to this tank to simulate a fully stowed vehicle.

Above, the fifth American MBT70 pilot is fitted with test instrumentation during its evaluation at Aberdeen Proving Ground. The commander's panoramic sight is mounted in front of his hatch and the television camera is installed on the front hull.

As new components became available, tests of the prototype tanks continued both in Germany and the United States. The first complete American pilot arrived at Aberdeen Proving Ground in late 1969. Needless to say, many development problems had to be worked out with the highly complex systems in the new tank. In Germany, there were difficulties with various fire control components and with the Rheinmetall automatic loader. The latter problem resulted in the General Motors automatic loader being installed in the American pilots. As mentioned earlier, the Continental AVCR-1100 engine was modified and its displacement increased, but questions still remained regarding its reliability. Also, the American manufactured turret drive did not perform satisfactorily without modification. Despite these problems, considerable progress was made and the pilot vehicles demonstrated exceptional performance as far as speed and mobility were concerned. The 152mm gun-launcher also operated satisfactorily once the difficulties with the missile and the combustible case ammunition were resolved. However, the different viewpoints of the American and the German representatives regarding their respective requirements once again became apparent. For example, the tank was now about four tons above the original weight limit. The Americans did not consider this excessive in view of the level of protection provided. However, to the Germans, this was a serious handicap and they requested design changes to reduce the weight. It was noted that some weight reduction could be achieved by eliminating the radiation shielding, but this

approach raised new objections. One of the reasons for the selection of the driver-in-turret arrangement was the ease with which such shielding could be provided for the entire crew located in such close quarters. If this protection were no longer a requirement, perhaps a complete redesign should be considered.

All of the development problems and proposed changes resulted in delays and rapidly escalating costs. In September 1969, Deputy Secretary of Defense David Packard requested a complete review of the program. The report of this review estimated that the cost to completion for the development and production engineering program would be 544 million dollars and the tanks could cost as much as 1.2 million dollars each. In view of these problems, it was not surprising that a Department of Defense news release on 20 January 1970 signaled the end of the joint program. In the future, each country would continue development to produce a tank best suited to their own requirements. For Germany, this resulted in the program that was to eventually produce the Leopard II. In the United States, the MBT70 was simplified to reduce costs and modified to use only American manufactured components.

The high power to weight ratio of the MBT70 is obvious in this photograph of the American pilot taken during the tests at Aberdeen Proving Ground.

142

AUSTERE MBT-70

The drawing above shows the austere version of the MBT70 after replacement of the 20mm cannon by the .50 caliber machine gun and the relocation of the missile transmitter to the main gun mount.

152mm GUN-LAUNCHER TANK XM803

After the termination of the joint program with the Federal Republic of Germany, the American team devoted its efforts to the design of a second generation MBT to meet the requirements of the United States Army. Following the general configuration of the first generation MBT70, the new tank was simplified to improve reliability and to reduce its cost. Also, as mentioned previously, all foreign components in the original tank were replaced by those manufactured in the United States. Initially referred to as the austere version of the MBT70, it later became the MBT70/XM803 and finally was designated as the full tracked combat tank, 152mm gun-launcher XM803.

The early sketches below show the transition from the MBT70 to the XM803 configuration. At the left, the 20mm cannon has been eliminated and not replaced. An early concept at the right mounted a .50 caliber machine gun on the side of the commander's panoramic sight and used the space for the former 20mm gun mount for ammunition stowage.

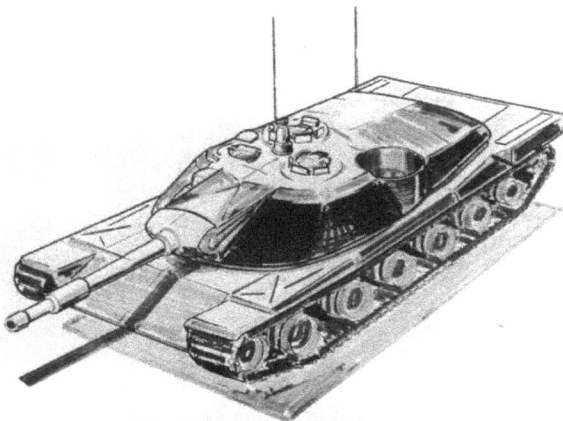

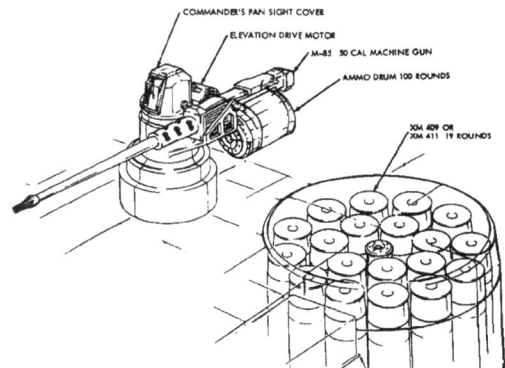

143

A model of the final XM803 design is shown in these photographs. The secondary weapon is now a .50 caliber M85 machine gun mounted on top of the commander's day/night sight. The missile transmitter is located on top of the main gun mount eliminating the need for a separate stabilizer system. Side skirts have been added to cover the upper part of the suspension.

The side skirts have been modified on the model of the XM803 in these views. Also, note the changes in the guard around the smoke grenade launchers.

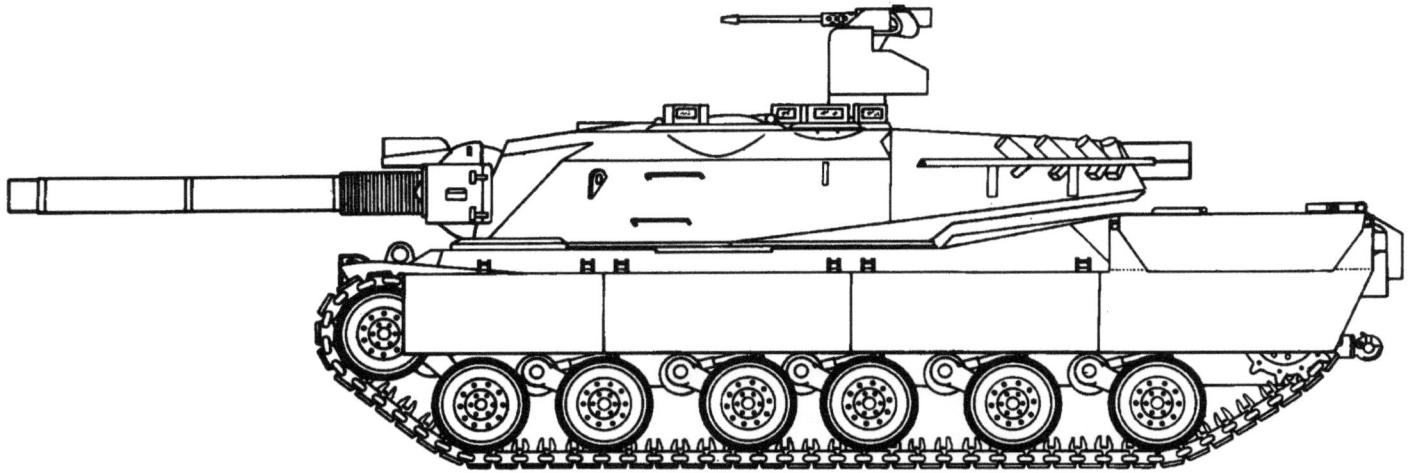

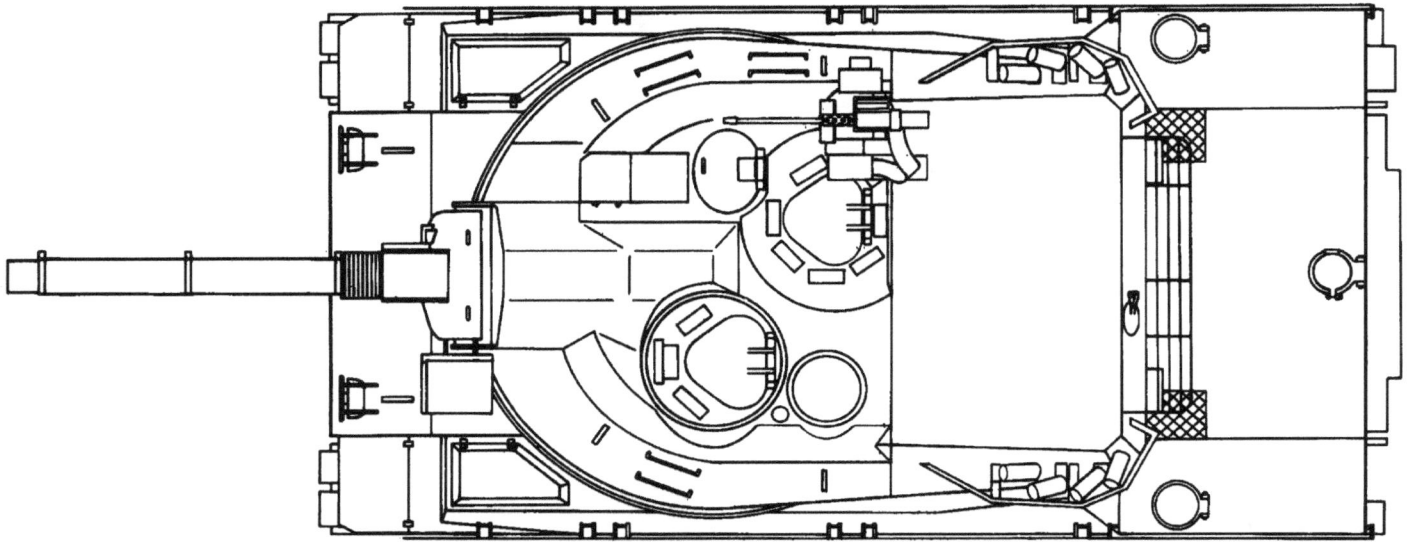

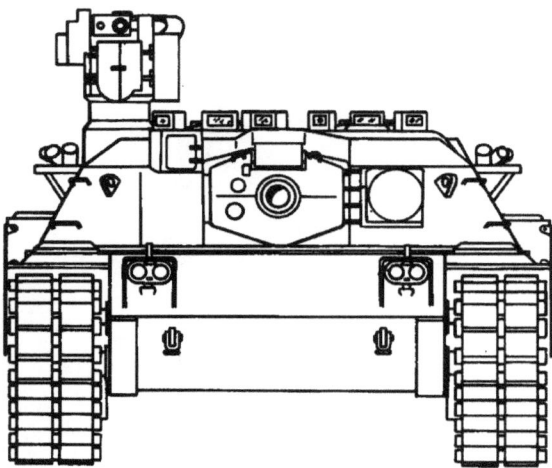

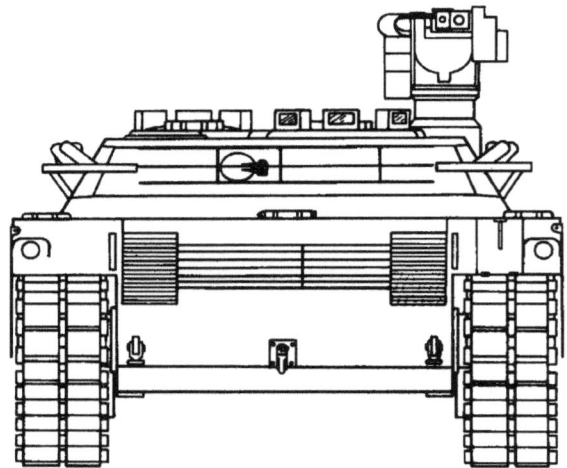

Scale 1:48

152mm Gun-Launcher Tank XM803

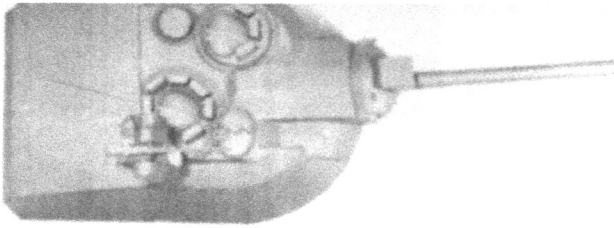

Above are two views of an XM803 turret model used during the design phase of the new tank. Below at the right is the full size turret mock-up of the XM803.

The turret on the XM803 was a welded assembly combining homogeneous cast steel armor, rolled homogeneous armor, high hardness steel plate, and alloy steel plate in a spaced armor configuration. It followed the same general arrangement as the prototype MBT70 turrets with the three man crew in the forward compartment separated from the automatic loader in the turret bustle by a one inch thick bolt-in aluminum alloy bulkhead. The bottom of the turret bustle structure contained a rectangular opening approximately 111 inches by 52 inches for the hinged baseplate carrying the automatic loader. A cylindrical shaped aluminum alloy turret basket was attached to the turret structure. The upper seven inches of the basket was approximately 100 inches in diameter. Below this point, the basket was reduced to a diameter of about 78 inches extending down for approximately 25 inches to about six inches above the hull floor. The turret was carried on a lightweight wire-race bearing with an inside ring diameter of 101 inches. This type of bearing was made from a high strength aluminum alloy with hardened steel raceway inserts.

As mentioned earlier, the MBT70 prototypes were fitted with a floor escape hatch beneath the center of the turret. However, the necessity of detaching the slip ring and tilting it out of the way made this hatch extremely difficult to use. Preliminary trials required about 30 minutes to open the hatch. Subsequent tests indicated that a trained crew equipped with wire cutters would still need about four minutes to clear the hatch for use. After review by General Creighton Abrams, the floor escape hatch was deleted on the XM803.

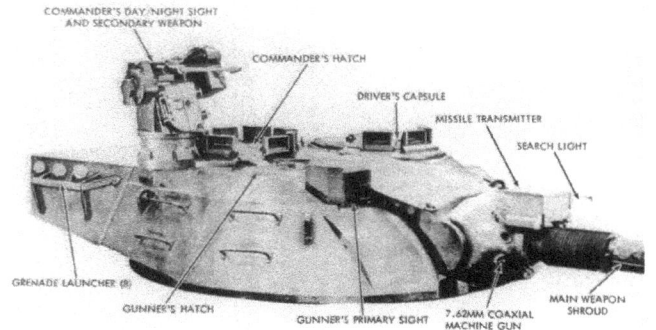

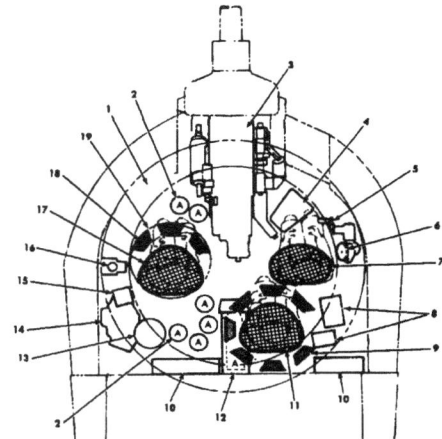

1. SHELF
2. MAIN WEAPON AMMUNITION (A)
3. MAIN WEAPON
4. BASKET FLOOR DOOR
5. MANUAL TRAVERSE MECHANISM
6. AZIMUTH INDICATOR
7. GUNNER'S HATCH COVER
8. COMMUNICATION EQUIPMENT
9. VISION BLOCK (7)
10. ELECTRONIC RACK ASSEMBLY
11. COMMANDER'S HATCH COVER
12. LOADING TUBE
13. CREW COMPARTMENT VENTILATOR
14. TRAVERSE GEAR BOX
15. TRAVERSE MOTOR VALVE
16. TURRET TRAVERSE LOCK
17. DRIVER'S HATCH COVER
18. VISION BLOCK (3)
19. DRIVER'S CAPSULE

The layout of the XM803 turret appears at the upper right and the turret armor assembly is depicted at the lower right. Below is a drawing of the turret basket.

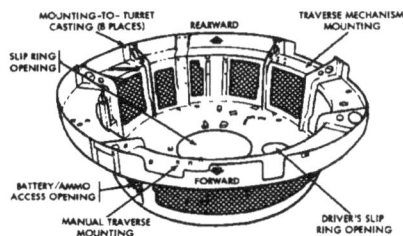

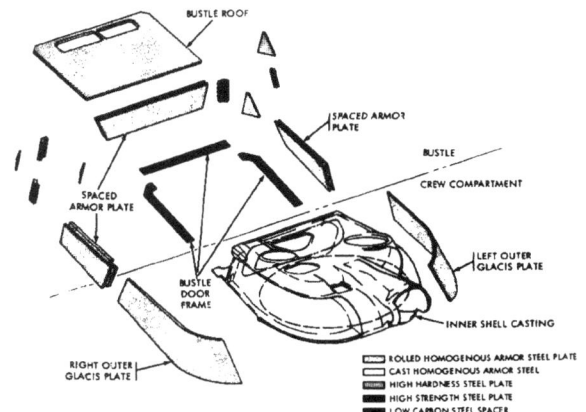

147

COMMANDER'S SIGHT POWER SUPPLY - DC

SECONDARY WEAPON (GFE)

COMMANDER'S SIGHT STABILIZED HEAD AND WEAPON MOUNT

COMMANDER'S SIGHT TELESCOPE

COMMANDER'S SIGHT AZIMUTH AMPLIFIER

COMMANDER'S SIGHT PRISM ELEVATION AMPLIFIER

SECONDARY WEAPON ELEVATION AMPLIFIER

60 DEG. ELEVATION

CAL. .50 MACHINE GUN

READY ROUND STOWAGE

15 DEG. DEPRESSION

COMMANDER'S DAY/NIGHT SIGHTS

360 DEG. CONTINUOUS ROTATION

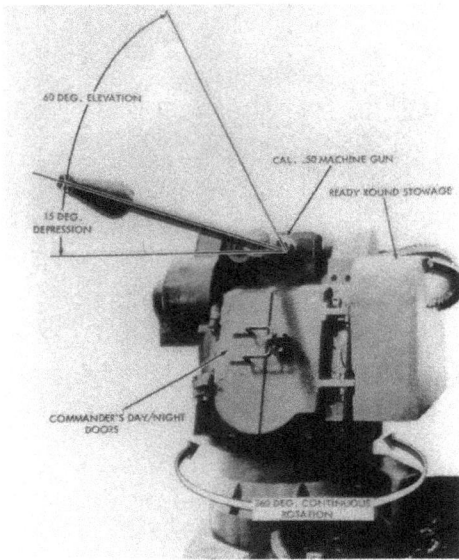

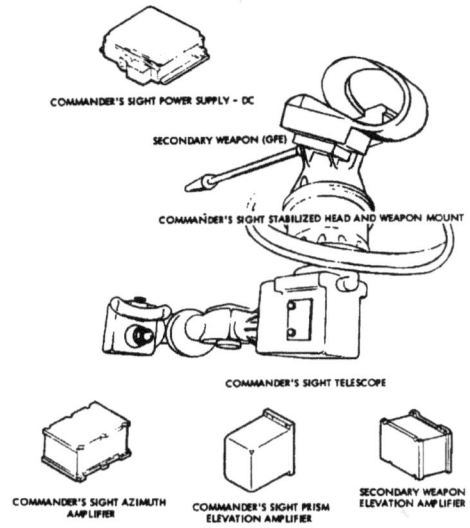

Above, the mock-up of the .50 caliber machine gun mounted on the commander's day/night sight is at the left and the sighting arrangement for this weapon is sketched at the right. Below at the right is a drawing of the tank commander's seat.

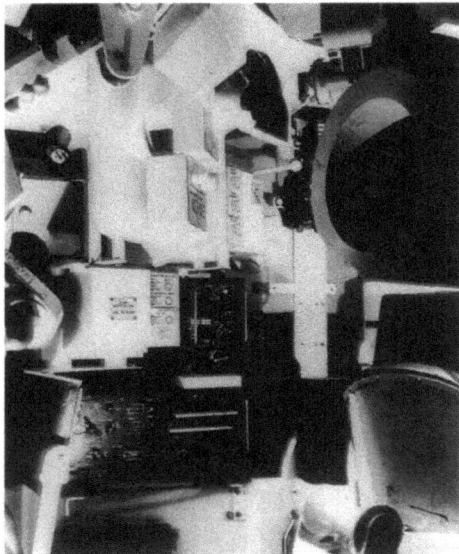

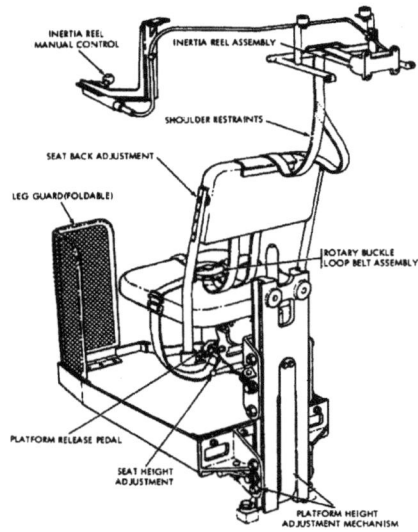

INERTIA REEL MANUAL CONTROL

INERTIA REEL ASSEMBLY

SHOULDER RESTRAINTS

SEAT BACK ADJUSTMENT

LEG GUARD (FOLDABLE)

ROTARY BUCKLE LOOP BELT ASSEMBLY

PLATFORM RELEASE PEDAL

SEAT HEIGHT ADJUSTMENT

PLATFORM HEIGHT ADJUSTMENT MECHANISM

Above is a view of the tank commander's station in the mock-up looking toward the right side of the turret. Below, the commander's station is seen through the turret hatch (left) and looking toward the rear of the turret (right).

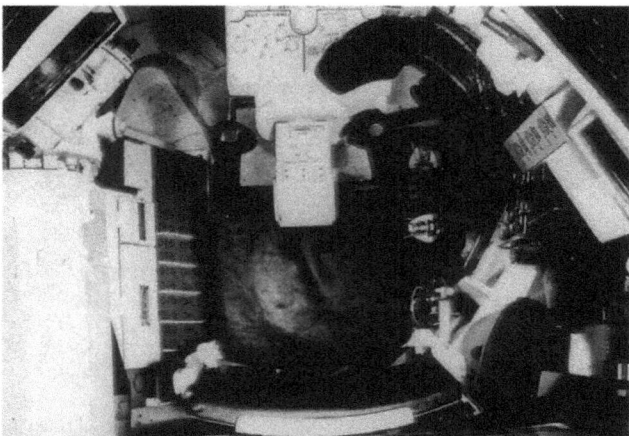

DAY/NIGHT SIGHT CONTROL AND ALIGNMENT PANEL (FIG. 4-8)

ELECTRONIC EQUIPMENT RACK

LOADING TUB

RADIO SET CONTROL (FIG. 4-19)

RADIO RECEIVER (FIG. 4-16)

COMMANDER'S SEAT

MAIN WEAPON ROUND (HEAT)

RADIO RECEIVER/TRANSMITTER (FIG. 4-17)

The location of the various components in the turret can be seen in the drawing at the right.

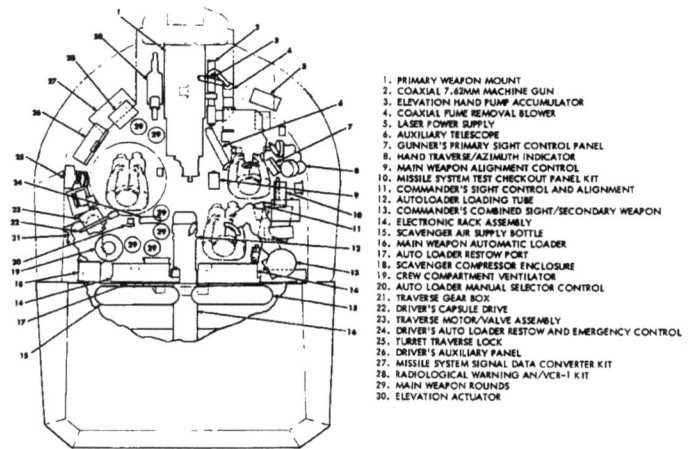

1. PRIMARY WEAPON MOUNT
2. COAXIAL 7.62MM MACHINE GUN
3. ELEVATION PUMP ACCUMULATOR
4. COAXIAL FUME REMOVAL BLOWER
5. LASER POWER SUPPLY
6. AUXILIARY TELESCOPE
7. GUNNER'S PRIMARY SIGHT CONTROL PANEL
8. HAND TRAVERSE/AZIMUTH INDICATOR
9. MAIN WEAPON ALIGNMENT CONTROL
10. MISSILE SYSTEM TEST CHECKOUT PANEL KIT
11. COMMANDER'S SIGHT CONTROL AND ALIGNMENT
12. AUTOLOADER LOADING TUBE
13. COMMANDER'S COMBINED SIGHT/SECONDARY WEAPON
14. ELECTRONIC RACK ASSEMBLY
15. SCAVENGER AIR SUPPLY BOTTLE
16. MAIN WEAPON AUTOMATIC LOADER
17. AUTO LOADER RESTOW PORT
18. SCAVENGER COMPRESSOR ENCLOSURE
19. CREW COMPARTMENT VENTILATOR
20. AUTO LOADER MANUAL SELECTOR CONTROL
21. TRAVERSE GEAR BOX
22. DRIVER'S CAPSULE DRIVE
23. TRAVERSE MOTOR/VALVE ASSEMBLY
24. DRIVER'S AUTO LOADER RESTOW AND EMERGENCY CONTROL
25. TURRET TRAVERSE LOCK
26. DRIVER'S AUXILIARY PANEL
27. MISSILE SYSTEM SIGNAL DATA CONVERTER KIT
28. RADIOLOGICAL WARNING AN/VCR-1 KIT
29. MAIN WEAPON ROUNDS
30. ELEVATION ACTUATOR

Another obvious change on the new tank was the elimination of the pop-up 20mm secondary weapon on the left side of the turret. This gun was replaced by a .50 caliber M85 machine gun mounted on top of the commander's stabilized day/night sight. In the XM803, this secondary weapon and the grenade launchers on the turret bustle could be operated only by the tank commander. The new day/night sight combined the functions of the separate day and night sights provided for the tank commander in the prototype MBT70. However, all of the television systems, including that for the driver, were eliminated from the XM803. In the new tank, the commander retained the ability to aim and fire the main weapon and the coaxial machine gun overriding the gunner. However, he could not take over any of the driver's functions. The new commander's station was equipped with seven periscope type vision blocks compared to six on the MBT70 prototypes.

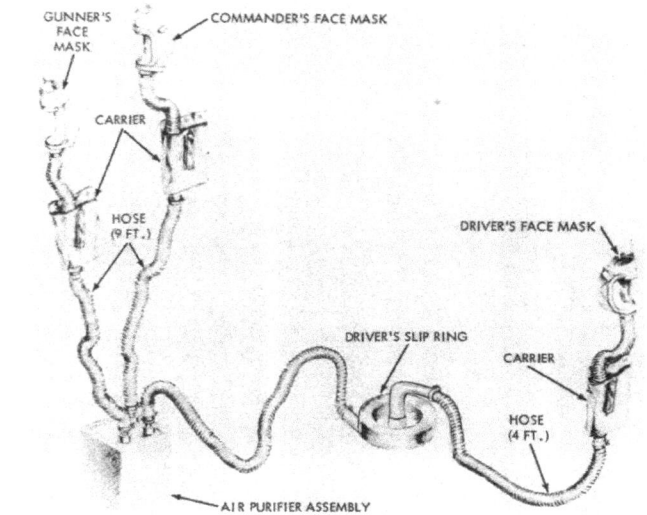

The collective protection system at the right supplied purified air to each of the three crew members. The modular chemical agent detector located in the XM803 turret is shown below. The various systems provided for the protection of the crew and the vehicle are sketched in the diagram at the lower right.

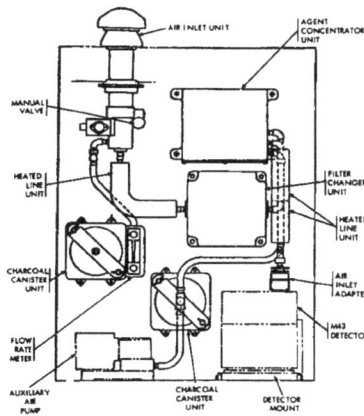

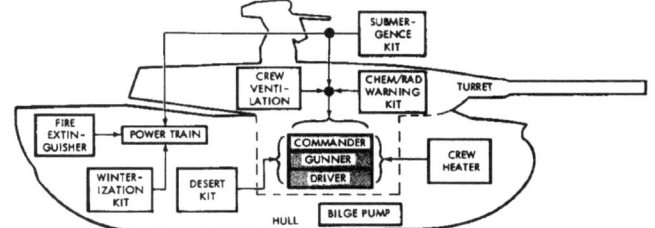

The firepower block diagram at the right outlines the functions of the gunner and the tank commander when operating the weapons and the fire control equipment in the XM803.

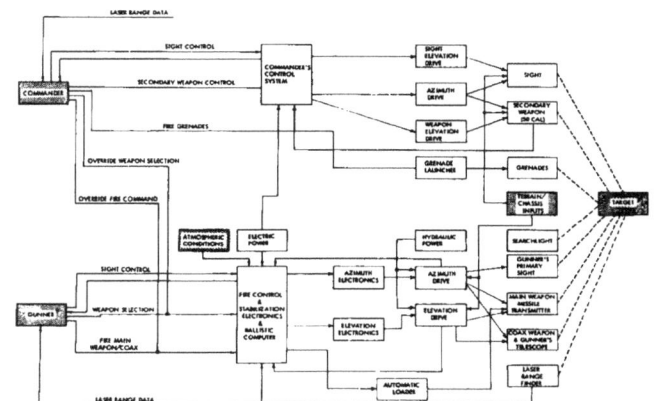

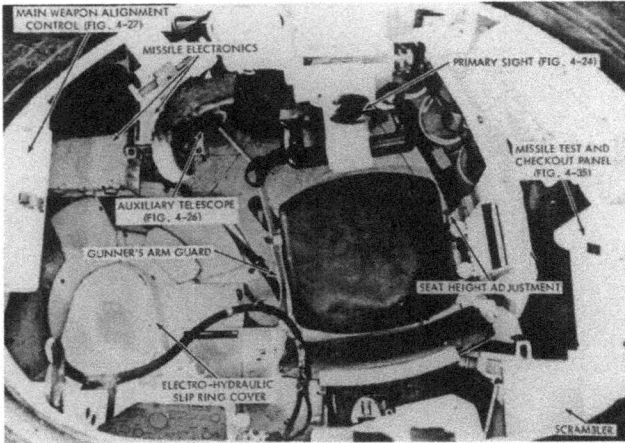

MAIN WEAPON ALIGNMENT CONTROL (FIG. 4-27)
MISSILE ELECTRONICS
PRIMARY SIGHT (FIG. 4-24)
MISSILE TEST AND CHECKOUT PANEL (FIG. 4-35)
AUXILIARY TELESCOPE (FIG. 4-26)
GUNNER'S ARM GUARD
SEAT HEIGHT ADJUSTMENT
ELECTRO-HYDRAULIC SLIP RING COVER
SCRAMBLER

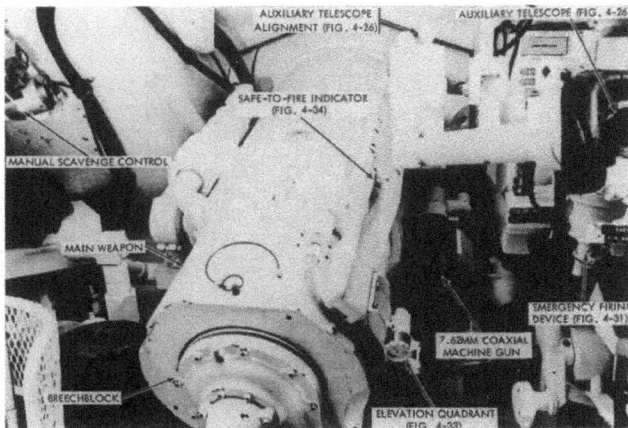

AUXILIARY TELESCOPE ALIGNMENT (FIG. 4-26)
AUXILIARY TELESCOPE (FIG. 4-26)
SAFE-TO-FIRE INDICATOR (FIG. 4-34)
MANUAL SCAVENGE CONTROL
MAIN WEAPON
EMERGENCY FIRING DEVICE (FIG. 4-31)
7.62MM COAXIAL MACHINE GUN
BREECHBLOCK
ELEVATION QUADRANT (FIG. 4-33)

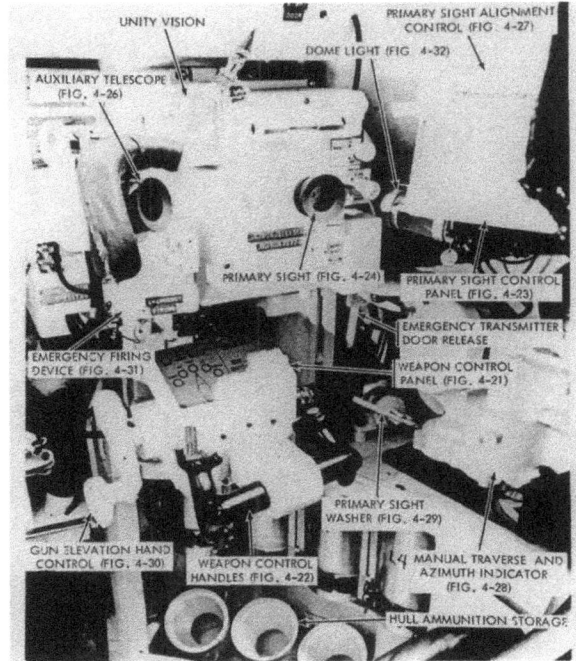

UNITY VISION
PRIMARY SIGHT ALIGNMENT CONTROL (FIG. 4-27)
AUXILIARY TELESCOPE (FIG. 4-26)
DOME LIGHT (FIG. 4-32)
PRIMARY SIGHT (FIG. 4-24)
PRIMARY SIGHT CONTROL PANEL (FIG. 4-23)
EMERGENCY TRANSMITTER DOOR RELEASE
WEAPON CONTROL PANEL (FIG. 4-21)
EMERGENCY FIRING DEVICE (FIG. 4-31)
GUN ELEVATION HAND CONTROL (FIG. 4-30)
WEAPON CONTROL HANDLES (FIG. 4-22)
PRIMARY SIGHT WASHER (FIG. 4-29)
MANUAL TRAVERSE AND AZIMUTH INDICATOR (FIG. 4-28)
HULL AMMUNITION STORAGE

The views above and at the left show the gunner's station in the XM803. Below is a drawing of the gunner's seat.

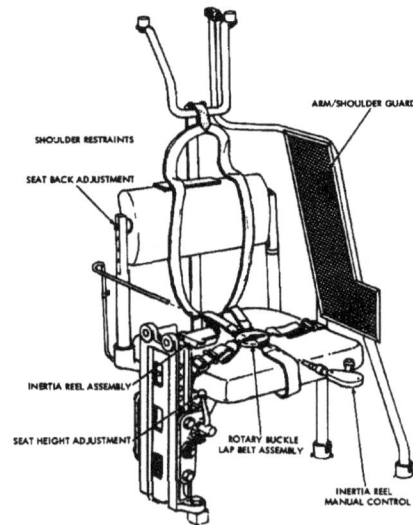

SHOULDER RESTRAINTS
SEAT BACK ADJUSTMENT
INERTIA REEL ASSEMBLY
SEAT HEIGHT ADJUSTMENT
ARM/SHOULDER GUARD
ROTARY BUCKLE LAP BELT ASSEMBLY
INERTIA REEL MANUAL CONTROL

The gunner's stabilized primary sight also incorporated both day and night capability as well as the laser range finder and missile tracker. The missile transmitter was relocated from the left side of the turret on the MBT70 prototypes to the top center of the main weapon mount. This eliminated the need for a separate drive and stabilizer as it followed the main weapon and was stabilized by the gun mount and turret stabilization system. An articulated telescope for emergency use by the gunner was located just above the 7.62mm coaxial machine gun on the right side of the gun mount. Unlike the auxiliary sighting device on the MBT70, it did not include night vision capability. A kit was provided to mount an AN/VSS-3 searchlight on the gun shield to the left of the main weapon. This unit was capable of producing 50 million candlepower during white light operation and a "pink" filter provided near infrared illumination. In the infrared mode, no visible white light was detectable at a distance greater than 25 feet. This searchlight was not waterproof and it had to be removed or encased during fording or deep submergence operations.

At the right is a sketch of the gunner's control handles.

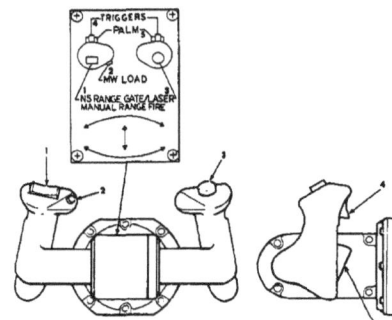

TRIGGERS
PALM
HW LOAD
NS RANGE GATE/LASER MANUAL RANGE FIRE

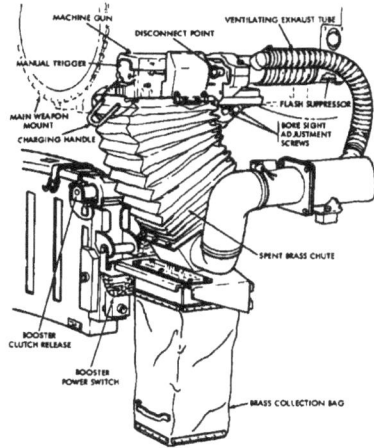

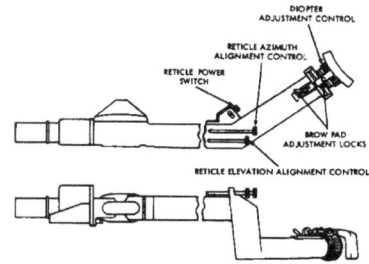

The 7.62mm coaxial machine gun installation is at the left. Above is the gunner's auxiliary telescope. The closed breech scavenging system for the 152mm gun-launcher is shown below.

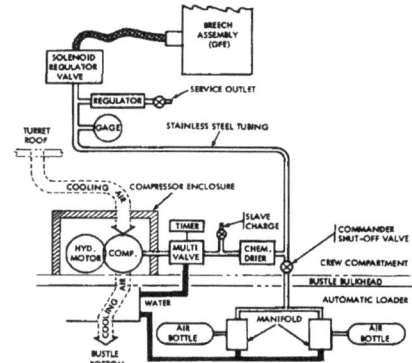

The main weapon in the XM803 was the 152mm gun-launcher XM150E6. It differed in several minor respects from the XM150E5 installed in the MBT70. As a result of the experience with the short barreled gun-launcher in the Sheridan and the M60A2, the new gun-launcher was fitted with a closed breech scavenging system (CBSS). This device used compressed air to blow any smoldering remnants of the combustible cartridge case out of the chamber before the breech was opened. With the CBSS installed, the bore evacuator was superfluous and it was deleted from the later model gun-launcher. An insulated jacket also was fitted around the gun tube to minimize thermal distortion and the gun barrel bend sensor was dropped.

Below are two views showing the installation of the 152mm gun-launcher and its breech mechanism.

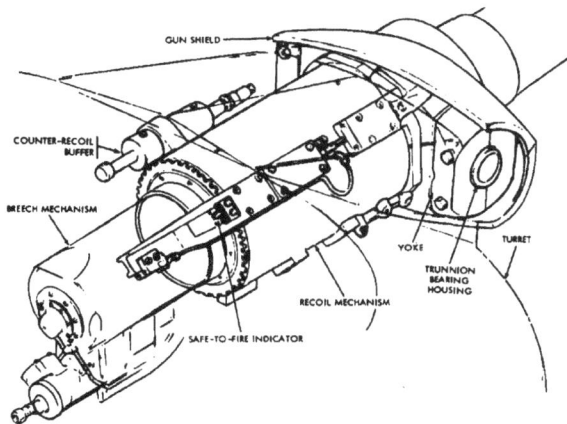

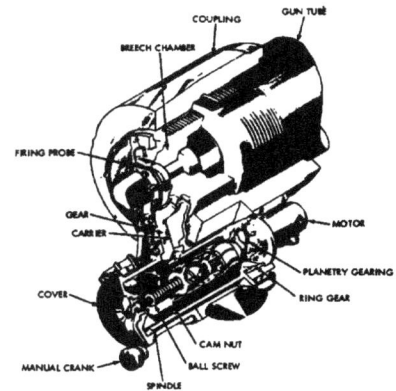

The dimensions of the 152mm gun-launcher XM150E6 can be seen in the sketch below.

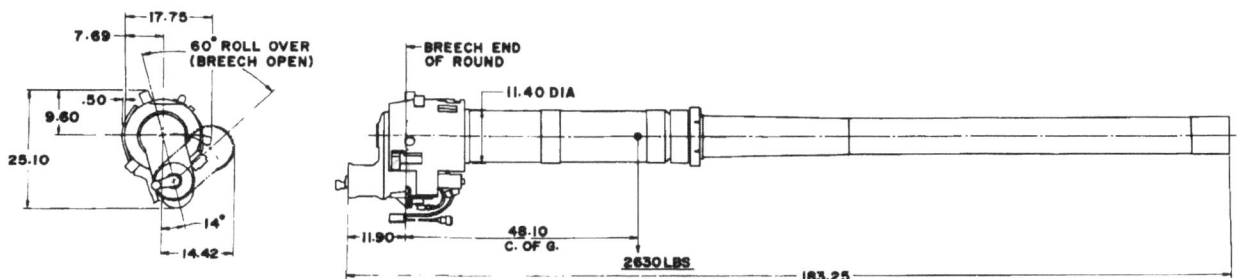

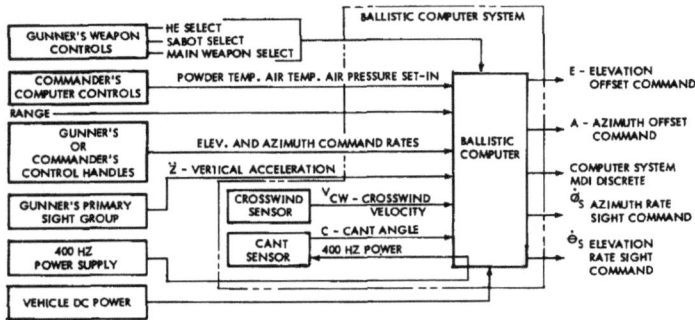

The block diagram at the left illustrates the function of the ballistic computer in the operation of the fire control system.

The ballistic computer in the fire control system determined the main weapon offsets required for firing conventional ammunition from a stationary or moving tank. These were based on the following inputs to the computer: Mode of operation, gunner's primary sight tracking rates, range from the laser range finder, cant angle from the cant sensor, cross wind velocity from the turret mounted cross wind sensor, ammunition type, air pressure, air temperature, and powder temperature. The last three were set into the computer by the tank commander.

The General Motors automatic loader was installed in the turret bustle carrying 24 rounds of 152mm ammunition. An additional seven rounds were stowed in the turret with six in a vertical position and one in a horizontal rack. Nineteen rounds were located in the hull with twelve vertical racks between the rear bulkhead and the turret basket. The remaining seven rounds were in vertical racks attached to the fuel cell access cover on the front bulkhead between the two sets of batteries. All of these racks were provided with spall protection by steel covers except

for the front seven hull mounted rounds. The latter utilized aluminum-nylon-asbestos protective covers. Thus the normal 152mm ammunition stowage totaled 50 rounds.

The driver's counterrotating capsule in the XM803 was similar to that on the prototype MBT70. It could be locked in four positions relative to the chassis. These were facing forward (0 degrees), facing to the rear (180 degrees), and at 56 degrees to the left or right. As on the prototype tanks, these latter positions allowed the use of the side vision blocks for driving in the event that the center block was damaged. A night vision device was interchangeable with the driver's center periscope type vision block. Both types of center vision blocks were fitted with a washer/wiper. The driver's hatch cover was identical to that of the tank commander. In addition to the driver's electrical controls, emergency mechanical controls were provided in the hull which could be connected to the driver's control handles for steering. Like on the MBT70, use of the emergency controls required that both the driver's capsule and the turret be locked in the forward position.

Components and stowage on the XM803 are located on the sketch at the lower right. Below is a drawing of the General Motors automatic loader.

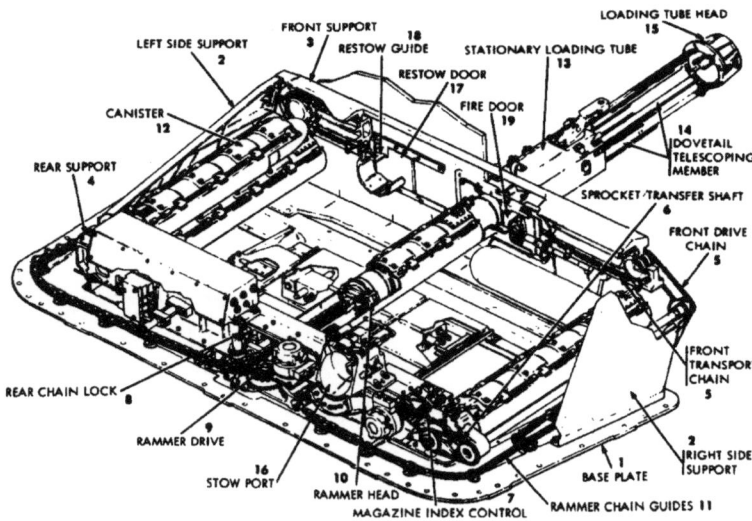

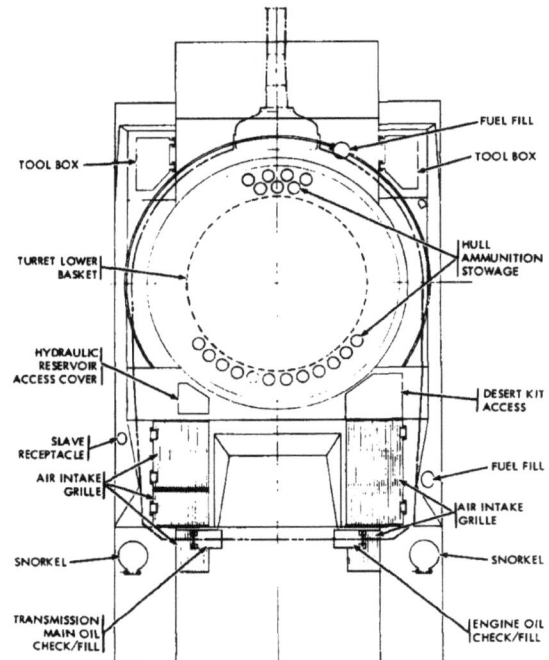

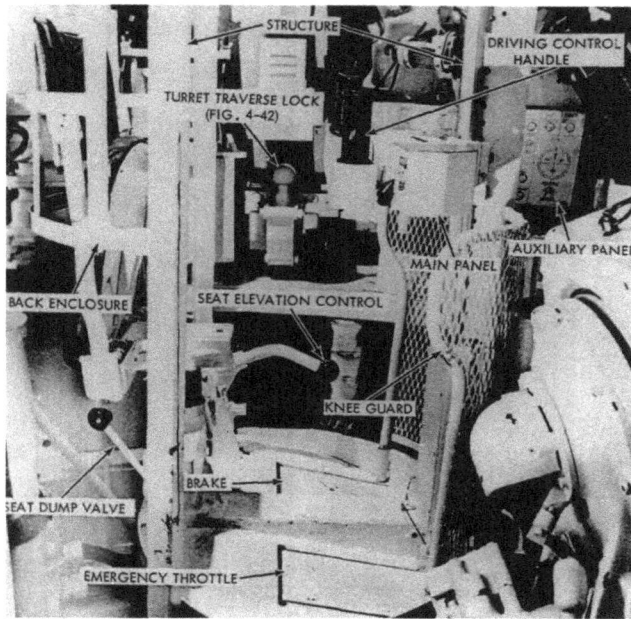

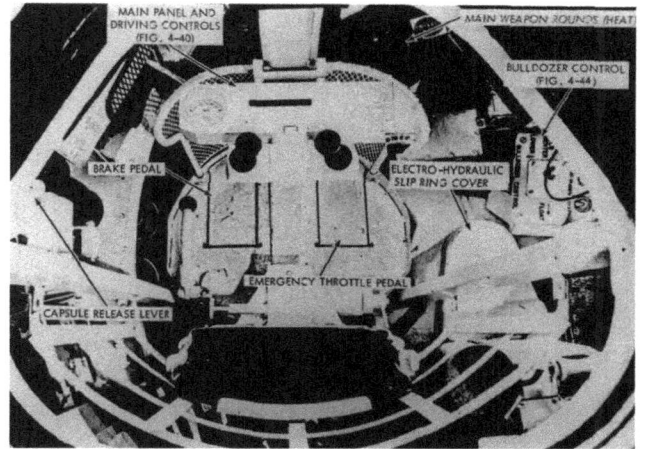

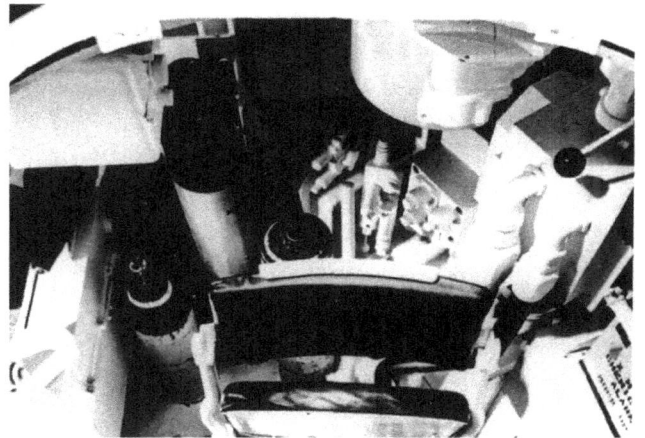

The driver's station in the XM803 can be seen above and at the right. Note the ammunition stowed behind the driver's seat in the latter view.

Below, the driver's controls appear at the left and the drawing at the right shows the method of engaging the emergency controls for the throttle, transmission shift, steering, and brakes.

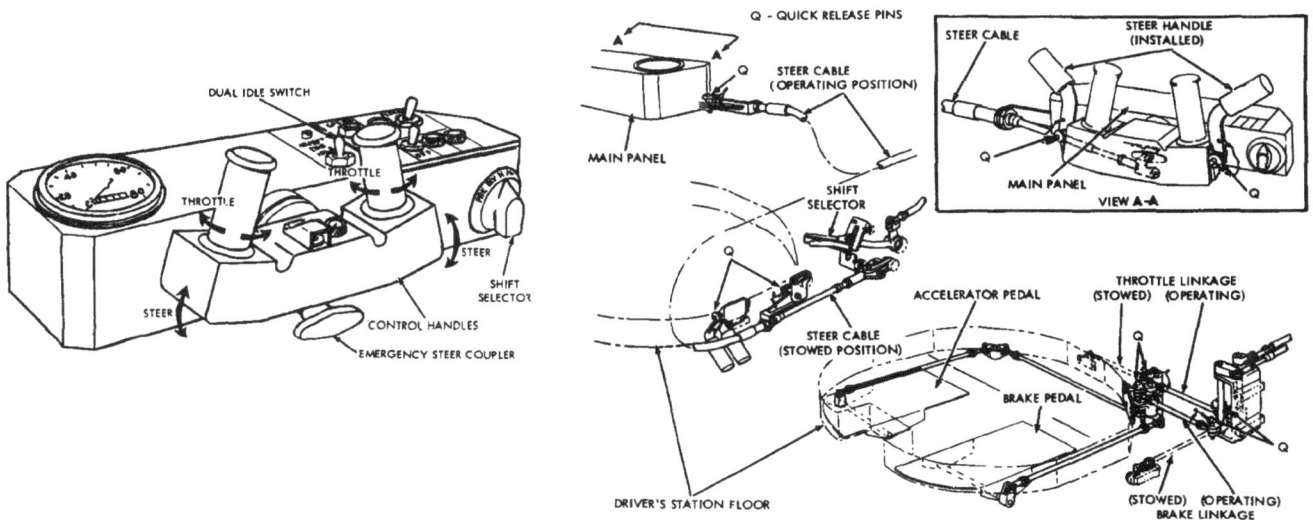

Labels: LIFTING EYE, FINAL DRIVE MOUNTING, ENGINE MOUNT RAIL, SUSPENSION MOUNTING, FENDER STOWAGE BOX, POWER PLANT GUIDE RAIL, CREW COMPARTMENT (ARMOR STEEL FLOOR), SLIP RING MOUNTING, IDLER ARM MOUNTING (CAST ARMOR STEEL), LIFTING EYE, FUEL COMPARTMENT (FRONT TANK)

Labels: POWER PLANT COMPARTMENT, TURRET MOUNTING, FUEL COMPARTMENT (FRONT TANK), TRACK AND SUSPENSION

The general arrangement of the XM803 hull and the location of various components can be seen in these drawings.

Labels: ENGINE EXHAUST GRILLE, GUN TRAVEL LOCK, TRANSMISSION COOLING GRILLE, TRANSMISSION COOLING GRILLE, AIR CLEANER, AIR CLEANER, FINAL DRIVE FILL, TRANSMISSION OIL FILTER, BRAKE ADJUST, TOWING PINTLE, FINAL DRIVE/TRANSMISSION DISCONNECT AND FORDING/SUBMERGENCE DRAIN, FINAL DRIVE/TRANSMISSION DISCONNECT

The XM803 was powered by the air-cooled, variable compression ratio, Continental AVCR-1100-3B engine. This was essentially the same power plant used in the American prototype MBT70. However, it was derated to provide 1250 horsepower at 2600 revolutions per minute to improve the reliability. At this same time, consideration was given to the future installation of a 1500 horsepower gas turbine then under development. This was the AGT-1500 which eventually appeared as the power plant for the M1 tank.

At the right is the mock-up of the Continental AVCR-1100-3B engine. The sketches below show the power pack installation and the engine air filter system in the XM803.

Labels: ENGINE FRONT COOLING FAN, ENGINE REAR COOLING FAN, TURBOCHARGER, EXHAUST, ALTERNATOR, AIR AFTER-COOLER, OIL COOLER, ACCESSORY (FRONT) GEAR TRAIN, LUBRICATING OIL FILTER

Labels: COOLING AIR SHROUDS, ENGINE COOLING FANS, INLET GRILLE, INLET COOLING AIRFLOW, TRANSMISSION, ENGINE COOLING AIR OUTLET GRILLE, TRANSMISSION COOLING AIR OUTLET GRILLE, ENGINE OIL COOLER, TRANSMISSION OIL COOLER, FAN DRIVE, TRANSMISSION/ANCILLARY COOLING FAN, HYDRAULIC OIL COOLER (FUEL COOLER OPPOSITE SIDE), P.T.O., TRANSMISSION/ANCILLARY COOLING AIRFLOW, ENGINE (OIL FILTER AND GENERATOR NOT SHOWN FOR CLARIFY), ENGINE CYLINDERS, ENGINE OIL COOLER, INDUCTION AIR COOLING AIR FLOW, CYLINDER COOLING AIRFLOW, ENGINE OIL COOLING AIRFLOW, INDUCTION AIR COOLER (AFTER COOLER)

Labels: FINE DUST FILTERS, SNORKEL INLET, TO OUTSIDE OF VEHICLE, COARSE DUST FILTERS (PRECLEANER), SCAVENGER BLOWER MOTOR, AIR INLET DOOR, TURBOCHARGER TRANSITION DUCT

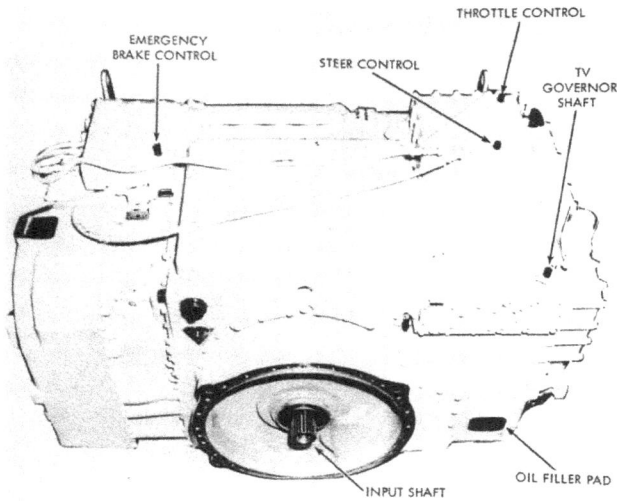

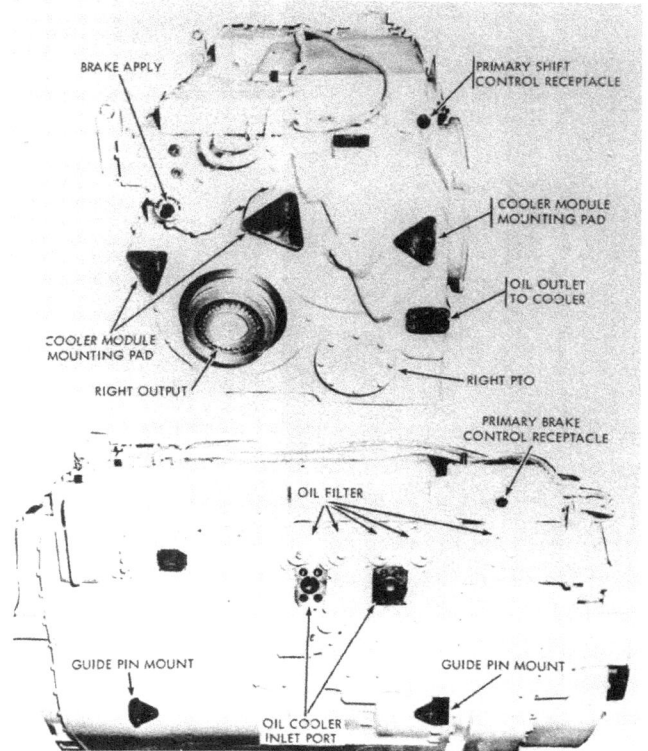

The XHM-1500-2B transmission is shown in the photographs above and at the right. Below at the right is a sketch of the XM803 final drive.

The AVCR-1100-3B engine was coupled to the General Motors XHM-1500-2B transmission. This was a hydromechanical transmission featuring an infinitely variable drive ratio with four forward and four reverse hydrostatic/mechanical phases having equal speeds in both directions. In the first phase, full hydrostatic power was provided. In the second, third, and fourth phases, the power was divided into two paths, one through the propulsion pump and motor unit (hydrostatic) and one through the friction clutches and planetary gearing (mechanical). Combining this power at the outputs provided the hydromechanical type drive. Phase changes were automatic with the driver selecting only forward, neutral, or reverse. Pivot steering was available in neutral. The power was transmitted through the planetary gear final drives to the track sprockets at the rear of the tank.

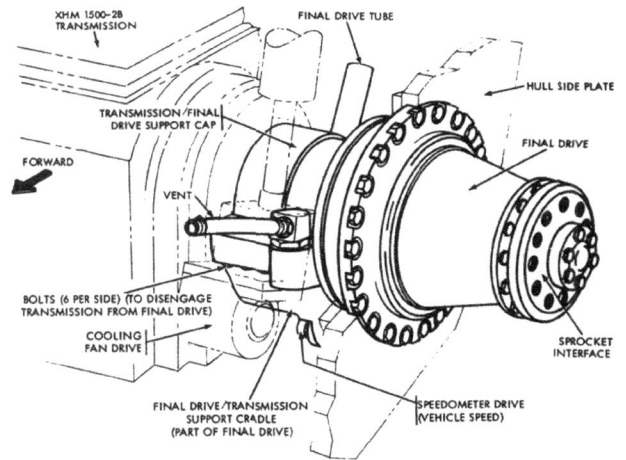

The National Water Lift hydropneumatic suspension was a simplified version of that installed on the MBT70 prototypes. Each of the twelve road wheel stations was fitted with a hydropneumatic suspension unit. These utilized single piston type actuators compared to the double piston actuators on the MBT70 suspension. Also, the four controls on the MBT70 were replaced by three on the XM803.

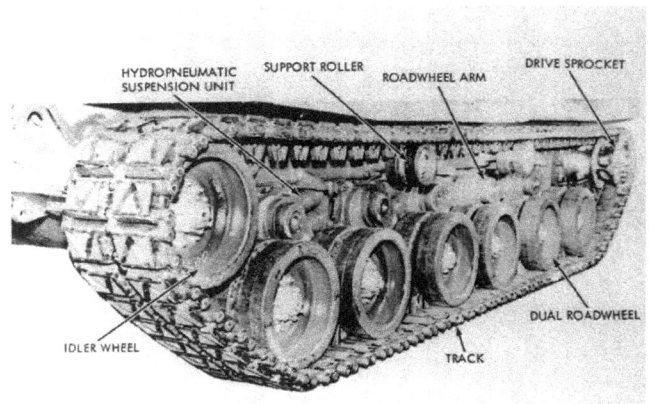

At the right is the XM803 suspension and track. Note that there are only two track support rollers.

155

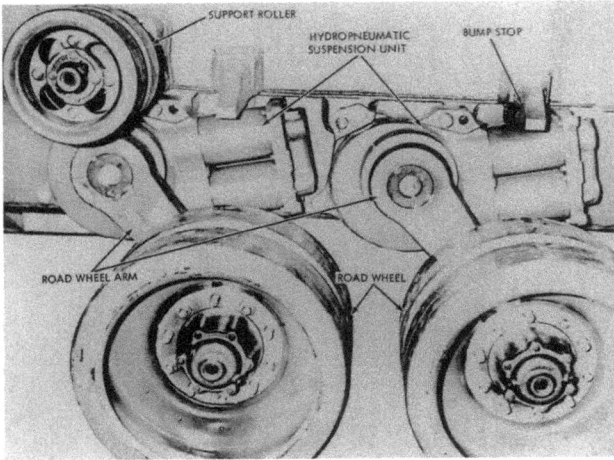

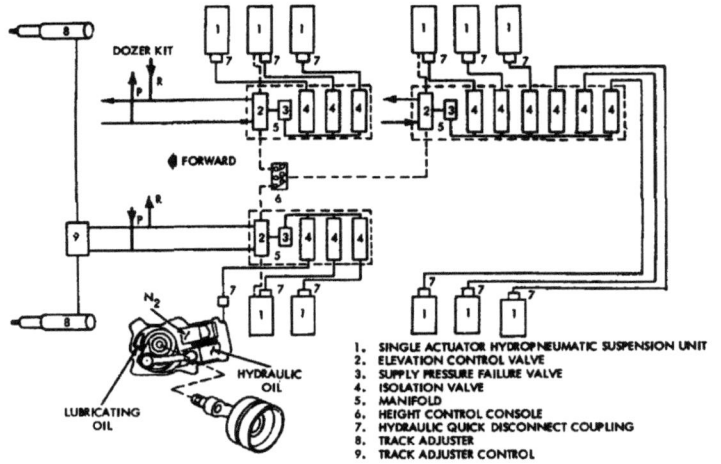

The National Water Lift hydropneumatic suspension on the XM803 is illustrated in the photograph and the schematic diagram above. The suspension control is sketched below. The components are as follows: 1. lock, 2, 3, & 4. elevation controls, 5. track adjuster, 6. light.

1. SINGLE ACTUATOR HYDROPNEUMATIC SUSPENSION UNIT
2. ELEVATION CONTROL VALVE
3. SUPPLY PRESSURE FAILURE VALVE
4. ISOLATION VALVE
5. MANIFOLD
6. HEIGHT CONTROL CONSOLE
7. HYDRAULIC QUICK DISCONNECT COUPLING
8. TRACK ADJUSTER
9. TRACK ADJUSTER CONTROL

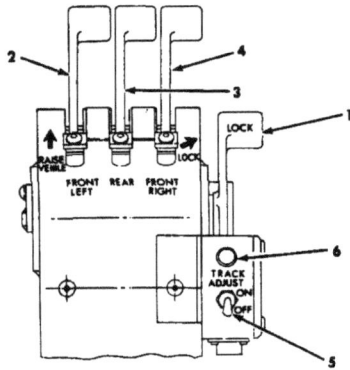

The hydraulic track adjuster is shown below. At the bottom right is a drawing of the road wheel/idler assembly.

Instead of a separate control for the three suspension units in each quadrant of the tank, the XM803 used separate controls for each of the two front quadrants and a single control for the six rear suspension units (three on each side). A single 3000 psi hydraulic system supplied power for the hydropneumatic suspension as well as the drives for the turret and gun mount. The three manually operated suspension control handles were located in the hull at the gunner's station. They were actuated by the gunner on instructions from the tank commander. The ground clearance was adjustable from six to 25 inches and the four corner units incorporated damping valves for greater pitch control. The normal ground clearance was considered to be 21 inches. The hydraulic system also powered the adjustable idler assembly to maintain the proper track tension. The Diehl tracks on the MBT70 were replaced by 24.5 inch wide double pin tracks and the number of track support rollers was reduced from three to two on each side. Armor side skirts were installed to protect the upper suspension components and to permit a reduction in the thickness of the lower hull.

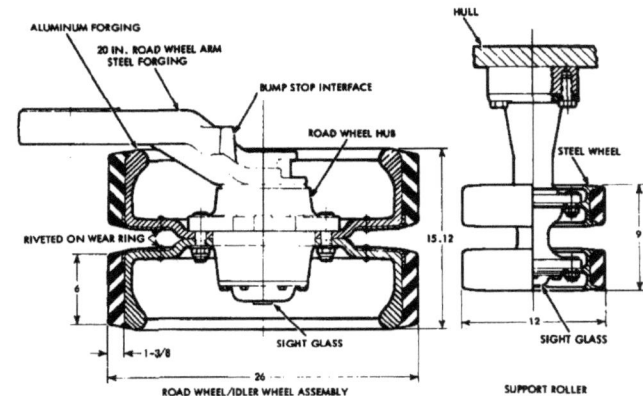

156

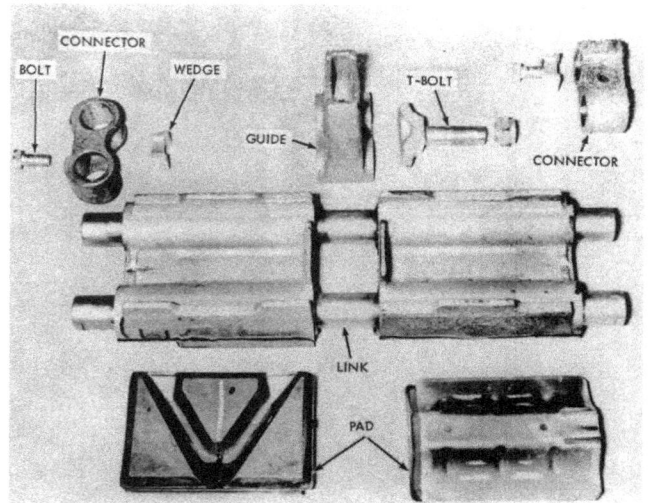

Above are views of the drive sprocket (left) and a disassembled track link (right). Below, the XM803 is sketched in the fording configuration (left) and with the deep water fording kit installed (right).

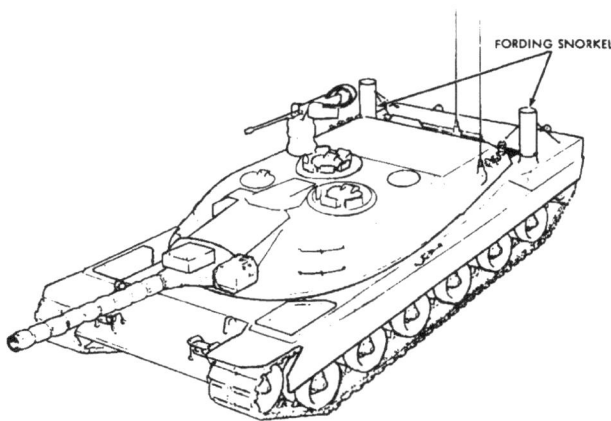

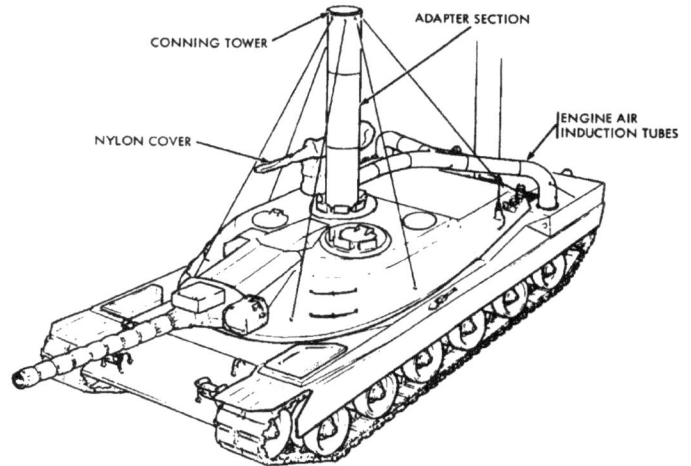

The 57 ton XM803 was capable of fording water depths up to the turret top, if the various seals were in place and the snorkel tubes were erected. For submerged operation, a deep water kit was available. This consisted of a 30 inch diameter sectional conning tower fitted over the commander's hatch and engine air intake tubes installed between the conning tower and the fording snorkel attachment points. Scuba equipment was provided for emergency use by the crew. The engine compartment was flooded during fording operations.

At the right is a sketch of the winterization kit installation for cold weather operation.

Other kits intended for use with the XM803 included a bulldozer, a winterization kit, and a kit for desert operations. The latter required the installation of an air conditioner replacing the small right fuel tank in the engine compartment and two rounds of 152mm ammunition from the crew compartment rear bulkhead.

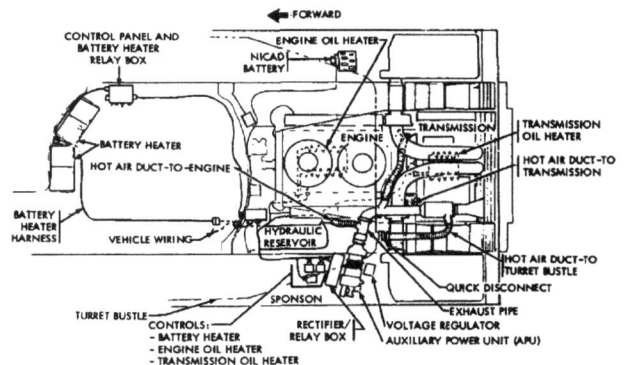

Above is the pilot 152mm gun-launcher tank XM803. Instrumentation has been installed on the tank for the test program.

Initially, two XM803s were authorized for construction, but only one, serial number seven, was finally completed. Despite the simplification, the cost of the tank remained a major problem. Also, opinions were changing among the future users of the vehicle. Problems with the gun-launcher system during operations of the M551 Sheridan and the M60A2 tank resulted in a loss of favor for the combination gun-missile system. Many armor officers preferred a high velocity gun to the more complex armament of the XM803. As a result, Congress, in December 1971, directed that the program be cancelled and that a new project be initiated to develop a less expensive tank. The program was officially deactivated on 30 June 1972, but the single XM803 was utilized, along with several of the MBT70 prototypes, to obtain data for the new tank development program.

Below, another photograph of the XM803 pilot appears at the left. At the right is a view of the XM803 during the development of the fire control equipment. The engineer in the turret hatch is Michael Leu.

PART III

AN AFFORDABLE MAIN BATTLE TANK

Although Congress cancelled the XM803 program in late 1971, the Army's need for a new tank was recognized. To meet this requirement, 20 million dollars were authorized for the procurement of two prototypes of a new main battle tank. This was in addition to the 20 million dollars allocated to phase out the XM803. Although the funds were provided specifically for two industrial prototype tanks, the Army initially preferred to concentrate on component development until a specification could be prepared for a new main battle tank. At that time, there was considerable controversy within the Army itself as to what form the tank should take. In February 1972, the Main Battle Tank Task Force (MBTTF) was established at Fort Knox, Kentucky under the leadership of Major General William R. Desobry. Its objective was to define the characteristics of the new MBT and to prepare the Material Need (MN) document. Technical support for the Task Force was provided by the Advanced Concepts Branch of the Tank Automotive Command (TACOM). Eventually, this work resulted in a series of eight main concept studies prepared by J. B. Gilvydis. The first three of these studies covered the work performed for the Task Force during its period of operation from February to August 1972. The later five were conducted for the Main Battle Tank Project Manager's

Office, after it was established in September, and for the Army Materiel Command (AMC). This work continued until March 1973.

On 5 February 1972, General Henry A. Miley, Jr., the Commanding General of AMC, concluded in a letter to the Department of the Army that tank system concept studies by industry would be of value to the Army and would help meet the directions of Congress. Following this, contracts were awarded to Chrysler Corporation and General Motors Corporation for such design studies to be completed by September 1972. The Chrysler program was to cover new evolutionary design concepts based on the M60A1 tank. The General Motors contract was to study the design of a new MBT using advanced design components based on the experience of the XM803 program. Both projects provided concept studies to support the work of the MBT Task Force. Initially, the new main battle tank was designated as the XM815, but with the revision of the nomenclature system, this was soon changed to the XM1.

TACOM presented the results of their first study to the MBTTF at Fort Knox on 25 February 1972. The initial concept, designated as LK 10322, was basically a conventional arrangement with the engine in the rear, the commander, gunner, and loader centrally located in the turret,

Below is the LK 10322 design concept with heavy protection, rear engine, and a conventional turret.

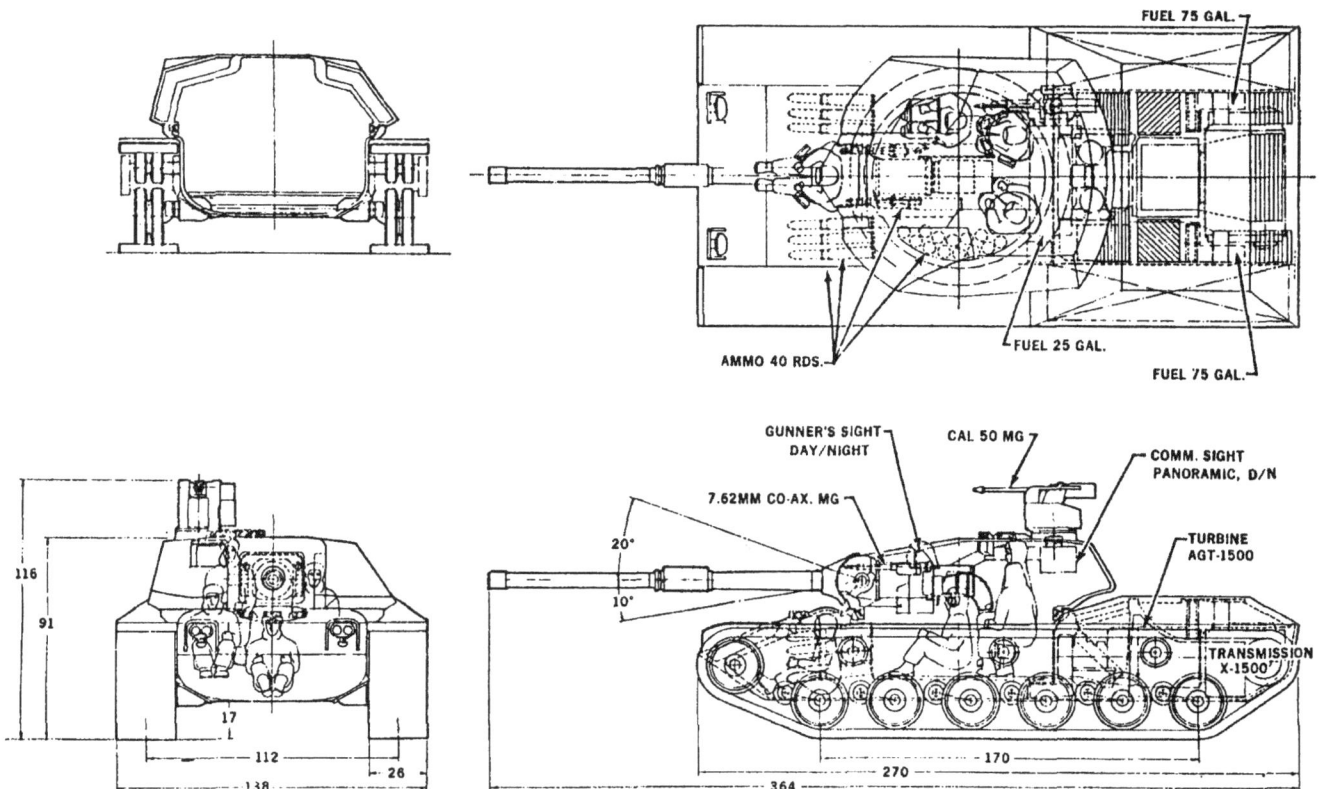

and the driver in the front center of the hull. The total armored volume of this concept was about 540 cubic feet. This compared to an armored volume of approximately 425 cubic feet for the Soviet T55 and 650 cubic feet for the M60A1. Armed with the 105mm gun M68, the LK 10322 concept was used in this study to analyse the effect on the combat weight of the different thicknesses of armor required to defeat various enemy threats. Fourteen different levels of protection were considered in this first study with the estimated vehicle weights ranging from about 34 tons to approximately 57 tons. The armor ranged from all round protection against only 23mm armor piercing projectiles to frontal protection against 120mm shaped charge rounds. Various composite armor arrays developed by the Ballistic Research Laboratory were utilized in different versions of the concept. Since the design of the armor was rapidly developing, frequent changes were required to adapt the constantly changing arrays to the tank concept studies. It is interesting to note that the initial drawing for the LK 10322 concept featured the AGT-1500 gas turbine, the power plant eventually selected for the M1 production tank. With six road wheels per track, a tube-over-bar suspension was specified for the LK 10322 to improve the cross-country performance.

During April and May, a second TACOM study investigated modified versions of the LK 10322 concept in an effort to reduce the weight. Initially, the hull design remained the same, but the turret was slightly modified. The side walls were tapered five degrees to the rear and the slope of the right side was increased from 14 to 20 degrees from the vertical. A new low profile commander's station was introduced with an externally mounted 7.62mm machine gun and the 7.62mm coaxial machine gun was replaced by a .50 caliber weapon. These modifications resulted in estimated weight reductions ranging from a little less than three tons to over three and one half tons depending upon the protection level under consideration. Later, the hull of the modified LK 10322 was lengthened from

270 inches to 275 inches which allowed an increase in fuel capacity from 300 to 350 gallons. This hull also could accept either the AGT-1500 turbine with the XHM-1500 transmission or the AVCR-1100 diesel engine with the X-1100 transmission. The average weight increase with these hull changes was about 600 pounds.

On 20 June 1972, the MBT Task Force requested TACOM to perform a third weight and protection analysis. This study was to evaluate the effect of 72 different component combinations applied to the modified LK 10322 concept. The following component list was supplied by the Task Force for the analysis.

1. Primary Armament
   a. 105mm gun M68
   b. 110mm gun, United Kingdom
   c. 120mm gun, Federal Republic of Germany

2. Fire Control
   a. XM803 system (without commander's day/night sight)
   b. M60A1 (with turret integrated night thermal sight)

3. Power Plants
   a. AVCR-1100 diesel engine with the X-1100 transmission
   b. DB1500 diesel engine with the Renk HSWL 354 transmission
   c. AGT-1500 gas turbine with the XHM-1500-2 transmission

4. Suspension
   a. Tube-over-bar
   b. Piston hydropneumatic

5. Tracks
   a. T142 or Diehl
   b. Lightweight, 28 inch width

This study used two different approaches. The first calculated the armor weight for the hull and turret and determined the protection level obtained with total vehicle weights of 43, 45, 47, and 49 tons. The second approach analysed the effect of 72 combinations of components on the total vehicle weight at a single level of protection. The protection level selected was that required to defeat the combined threat of the Soviet 115mm APFSDS projectile at 800 meters range and the 3.2 inch diameter HEAT round on the front and up to 30 degrees to each flank.

In addition to the 72 variations of the modified LK 10322 concept from the third TACOM study, the MBT Task Force reviewed fifteen additional designs. Numbers 73 through 80 were submitted by General Motors Corporation and 81 through 83 came from Chrysler. Candidates 84 through 87 were product improved versions of the M60A1. In reviewing the 72 candidates from the third

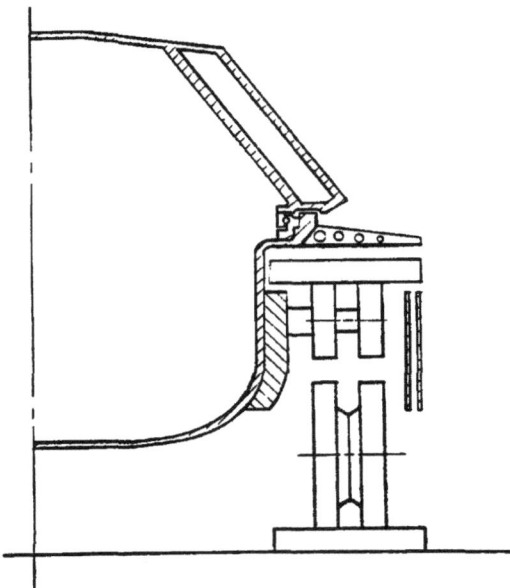

The cross section at the left shows the spaced armor arrangement incorporated in some of the new designs.

The sketch at the right and the drawing at the bottom of the page depict a General Motors tank concept powered by the AGT-1500 gas turbine. The estimated weight of this vehicle was about 47 tons.

TACOM study, the Task Force eliminated those armed with the United Kingdom 110mm rifled gun and the 120mm smooth bore weapon from the Federal Republic of Germany. The 110mm gun was not considered sufficiently superior to the 105mm gun M68 and the 120mm smooth bore was not expected to be available in time to meet the proposed production schedule. At that time, the Task Force also considered the AGT-1500 gas turbine to be a high risk power plant and it was dropped from the concepts under review. This reduced the number of proposals in the TACOM study from 72 to 16. Those remaining differed primarily in the engine and transmission with minor variations in the fire control systems as well as the suspensions and tracks. As a result, the 16 candidates were reduced to two featuring the two power packages considered acceptable. Thus one was powered by the air-cooled Continental AVCR-1100 engine with the X-1100 transmission. The other was fitted with the Daimler Benz (later MTU) DB1500 liquid-cooled diesel with the Renk transmission. These two concepts were included by the Task Force among the five candidates for the final trade off analysis.

Of the eight General Motors proposals, two were powered by the AGT-1500 gas turbine using the X-1100 transmission, one utilized the DB1500 diesel with the Renk transmission, and two were fitted with the AVCR-1100-3B with the X-1100 transmission. The remaining three had a twin engine arrangement consisting of two General Motors 8V71T liquid-cooled diesels, again with the X-1100 transmission. All eight concepts used a high strength torsion bar suspension and were armed with the M68 105mm gun. Estimated weights of these vehicles ranged from 47 to 67 tons depending upon their level of protection. The Task Force rejected the three proposals powered by the twin 8V71T diesels because of anticipated increased maintenance time and reduced reliability. With the elimination of the AGT-1500 in two of the concepts, the number of General Motors candidates was reduced to three. One of these also was rejected because of its excessive weight of 67 tons. One of the remaining two was selected as the General Motors candidate even though it weighed 56 tons because it came closest to meeting the MN requirements for ballistic protection.

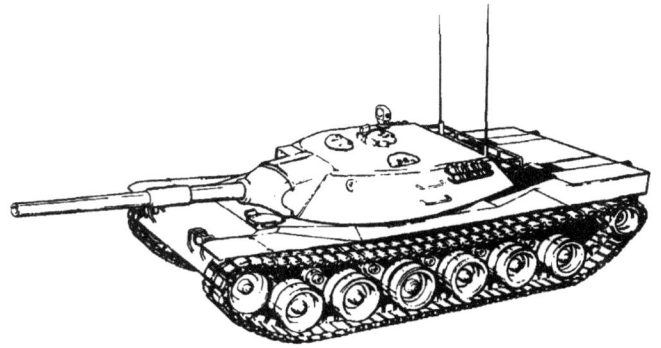

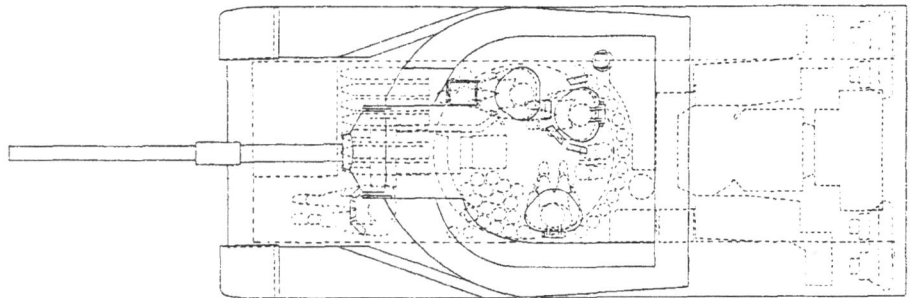

Note: All Dimensions are in Inches.

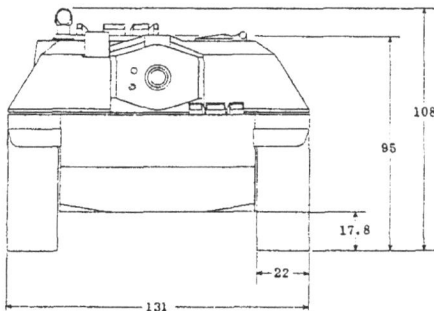

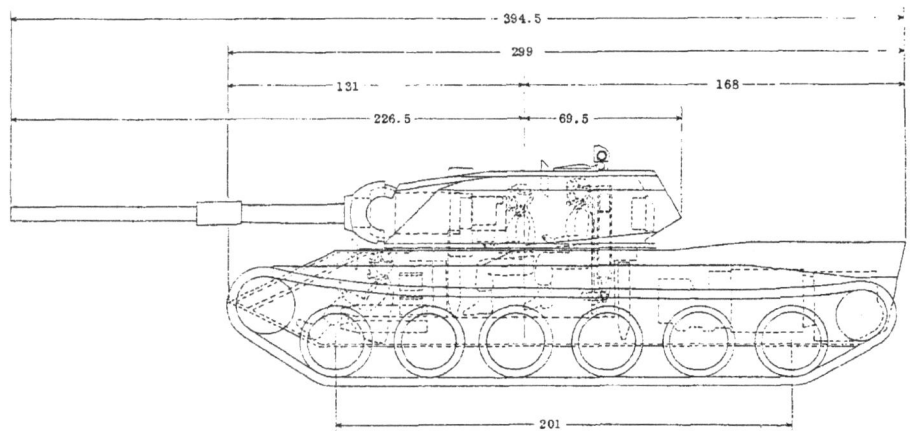

163

One of the three Chrysler concepts was selected as superior to the other two in several important areas and essentially equal in other respects. It was powered by the AVCR-1100 engine with the X-1100 transmission and had an estimated combat weight of 48.5 tons. However, like the other candidates that met the MN weight requirements, it did not provide the desired protection level.

None of the four product improved M60A1s met the MN requirements for ballistic protection or sustained speeds on 10 per cent and 30 per cent grades. Of the four, numbers 84 and 85 provided the highest level of protection. Since number 85 featured the more advanced fire control system, it was selected as one of the five candidates for the final trade off analysis.

The most obvious conclusion from the review of these concept studies was that the level of ballistic protection required by the proposed MN document was incompatible with the weight limitation specified. This problem was outlined in a fact sheet forwarded from the Task Force to the Commanding General of AMC, dated 3 August 1972. This fact sheet concluded that the only practical solution was to either reduce the level of ballistic protection or increase the 49 ton weight limitation in the proposed MN document. The latter alternative was subsequently adopted.

Although General Motors and Chrysler provided support to the MBT Task Force under their study contracts for new tank prototypes, the final reports on both of these contracts were not submitted until October 1972. In their report, General Motors proposed two concepts. They were based on estimated average production unit costs of $400,000 and $500,000, both in fiscal year 1972 dollars. Referred to as the 400K and 500K tanks, the latter was a growth version of the former with improvements in ballistic protection, firepower, and night vision. Both were manned by a crew of four with the driver in the left front hull. The usual arrangement of the three man turret crew was reversed placing the gunner and tank commander on the left side of the cannon with the loader at the right rear.

The turret on both vehicles utilized the large bearing ring from the XM803. Both tanks were armed with a stabilized 105mm gun M68. Forty rounds of 105mm ammunition were stowed in the hull below the level of the turret ring without separate compartmentation.

The 400K tank was protected by welded plate armor arrays on the turret front. The sides and rear of the turret consisted of single plate rolled homogeneous armor. The hull front was a single plate of high obliquity armor. Spaced armor skirts were provided on the forward sides of the hull. The tank was powered by the previously proposed twin General Motors 8V71T diesel engines connected to a modified X-1100 transmission through a transfer case. With this arrangement, the tank could be driven on one engine in an emergency. The twin engine installation was rated at 1200 gross horsepower. The estimated combat weight of the 400K concept was 52 tons. Although the armor protection was superior to the M60A3, it did not meet the requirements for the new MBT. The running gear consisted of an advanced high strength torsion bar suspension with six 30 inch diameter road wheels per side running on a 22 inch wide, double pin, flat track. A 7.62mm coaxial machine gun was located on the right side of the cannon and a 40mm high velocity grenade launcher was installed on the commander's hatch. The independently stabilized gunner's day/night sight included a laser range finder.

The combat weight of the 500K tank was increased to 55 tons providing improved ballistic protection. The front hull now consisted of spaced rolled homogeneous armor and steel/aluminum array skirts were installed on the sides. On the turret, a cast armor trunnion support was joined to frontal spaced armor arrays. The sides and rear also consisted of spaced armor. Although the M68 105mm gun was retained in the 500K tank, the mount was strengthened to permit future adaptation of a 120mm gun firing kinetic energy ammunition. The 7.62mm coaxial machine gun in the 400K tank was replaced by the Bushmaster (20mm-30mm) weapon system then under development.

The General Motors 400K tank proposal is sketched below.

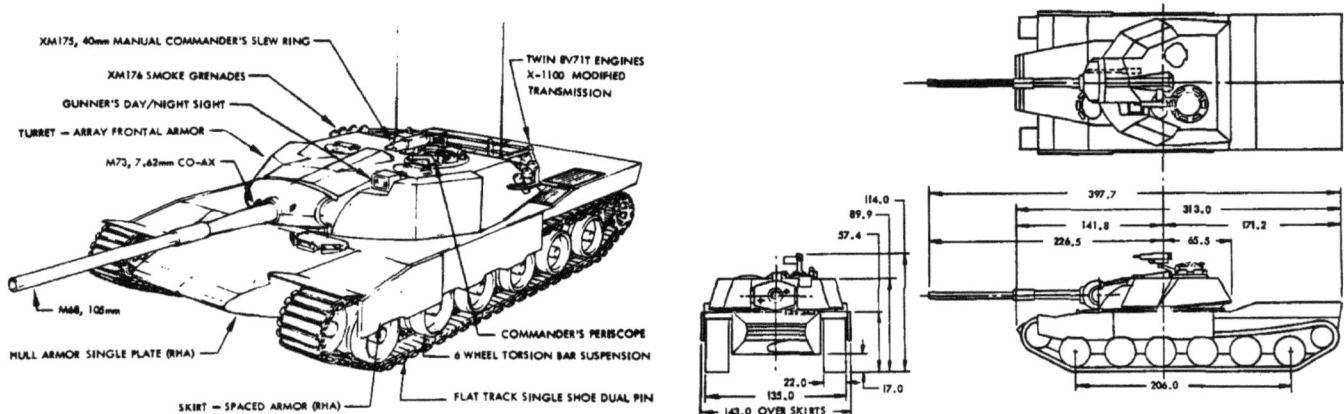

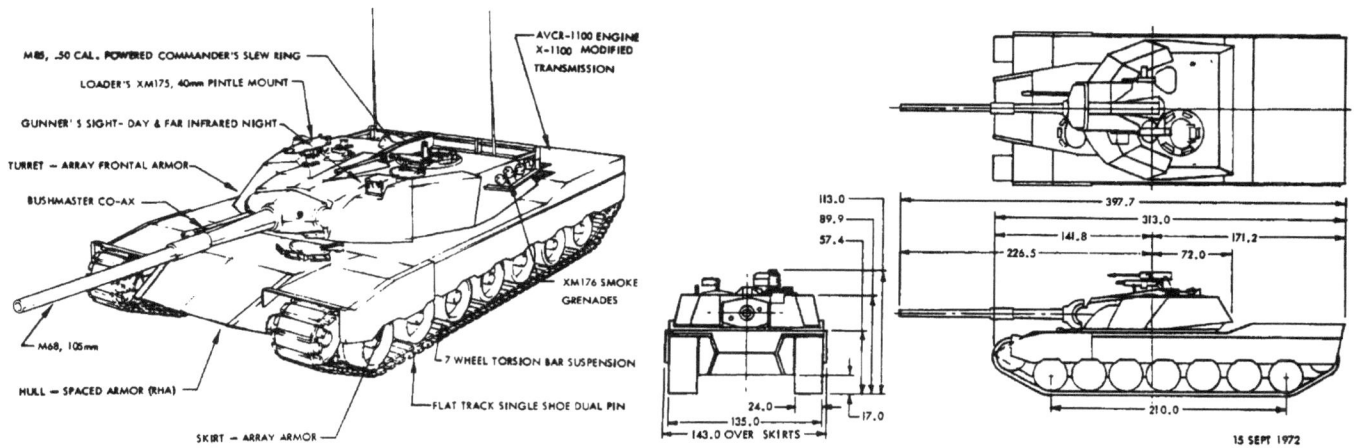

Above is the General Motors 500K tank concept.

The 40mm high velocity grenade launcher was moved to the loader's station and replaced on the commander's hatch by a .50 caliber M85 machine gun. The commander's weapon was now mounted on a powered slew ring. The gunner's stabilized sight with its laser range finder also now included a far infrared imager for night operations. Individual displays were provided for both the gunner and the tank commander. With the heavier weight, the 500K tank was powered by the AVCR-1100-3B engine rated at 1450 gross horsepower using the modified X-1100 transmission. The number of road wheels was raised from six to seven per side and the width of the double pin flat track was increased to 24 inches.

Chrysler's report on the New Evolutionary Tank reviewed eight different vehicle concepts as well as the development status of various design features and components. Main armament on all eight concept studies was the

M68 105mm gun. In one case, this was supplemented by a TOW missile installation. On two other concepts, the Shillelagh II or Swifty gun launched missile was specified. This was a proposed development of the Shillelagh reduced in diameter to permit launching from the 105mm gun. The missile velocity also was increased to reduce the flight time by about 50 per cent and to extend the range by approximately 25 per cent. However, the reduced diameter would have decreased the effectiveness of its shaped charge warhead. The estimated combat weight of the eight Chrysler proposals ranged from about 49 tons to 57 tons. After considering the eight concept studies, the Chrysler report proposed a final configuration combining the best components from two of them. With an estimated combat weight of 57.2 tons, this tank had a conventional layout. The driver was in the center front hull and the three man turret located the gunner and the tank commander on the right side of

Two features incorporated in the Chrysler tank proposals are shown here. The TOW missile installation on the tank turret is at the right. Below, the fuel cells can be seen located on each side of the driver. This arrangement increased the protection, particularly against shaped charge rounds.

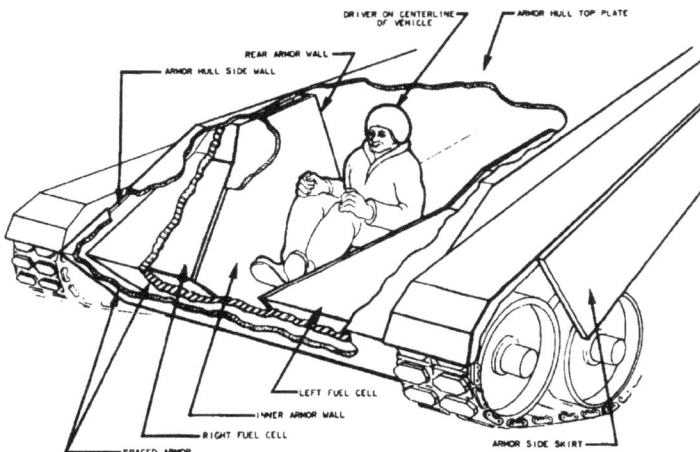

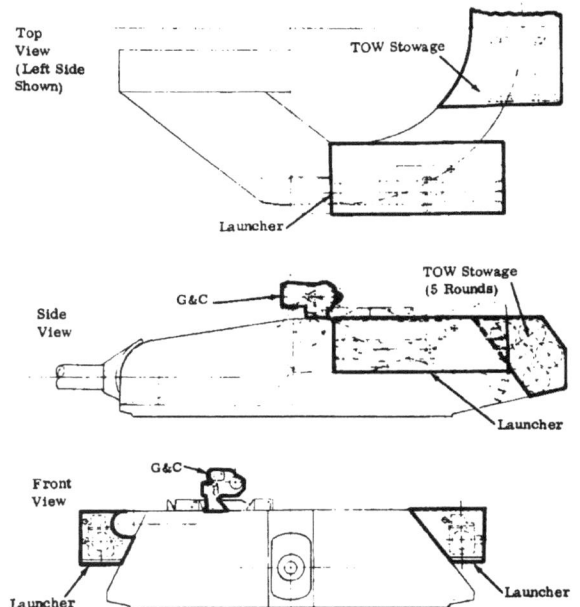

165

Pintle Mount 7.62mm M.G. (M60)

140.0

Spaced Armor

99.5   94.5   75.0

Flat Side Hull   18.0

25.0

112.0

137.0

387.5

211.5   Turret Centerline

43.5   Bustle Ammo Compartment
30 Main Rounds (18 Ready)

Gun Trunnion Center   Hull Ammo
Compartment
20 Main Rounds

Spaced Armor

42.0

34.0   Tube-Over-Bar
Suspension

184.3

305.0

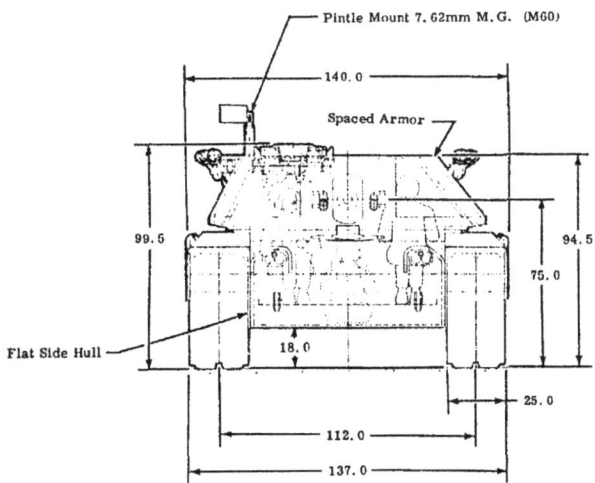

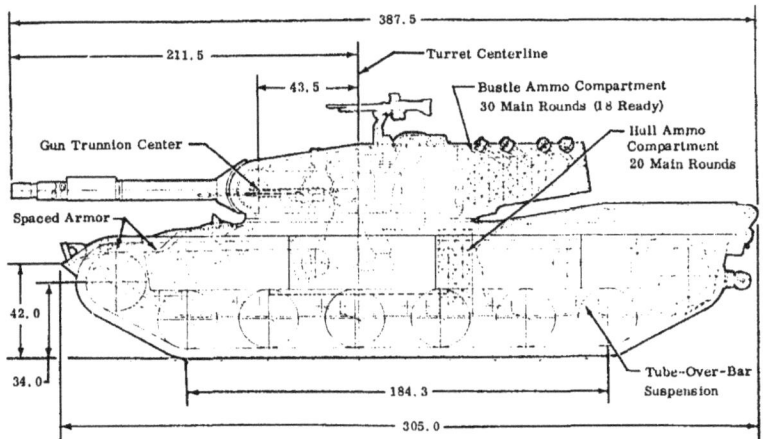

A Chrysler proposal is illustrated in this three view drawing. This tank is powered by the AVCR-1100 diesel engine. It does not include any missile launching capability.

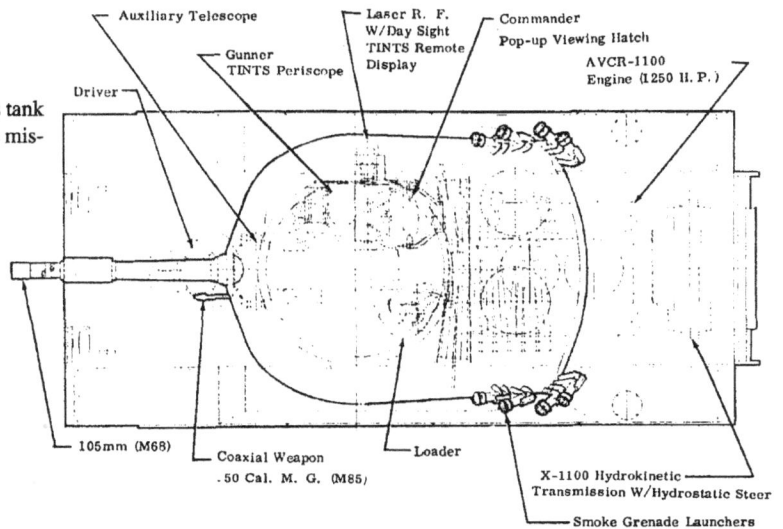

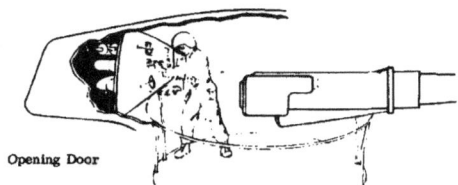

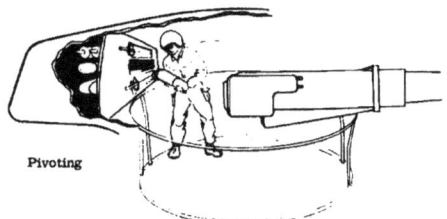

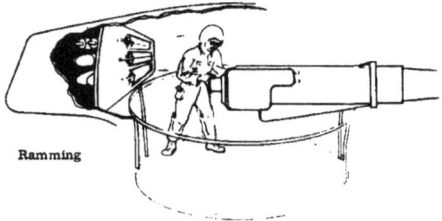

Auxiliary Telescope   Laser R. F.
W/Day Sight
TINTS Remote
Display   Commander
Pop-up Viewing Hatch

Gunner
TINTS Periscope   AVCR-1100
Engine (1250 H.P.)

Driver

105mm (M68)   Coaxial Weapon
.50 Cal. M.G. (M85)   Loader   X-1100 Hydrokinetic
Transmission W/Hydrostatic Steer

Smoke Grenade Launchers

Opening Door

Pivoting

Ramming

the 105mm gun with the loader on the left. The 105mm ammunition was stowed in compartments separated from the crew space. These compartments in the turret bustle and the hull were designed to fail away from the crew space. Thus any ammunition explosion resulting from a penetration of the armor would be directed either to the outside or into the engine compartment away from the crew.

The loading sequence for the rounds stowed behind the armored bulkhead in the turret bustle is shown at the left. Below, this bulkhead protects the crew if the ammunition is detonated.

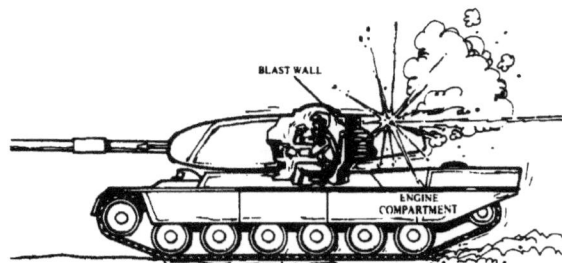

BLAST WALL

ENGINE
COMPARTMENT

166

At the right is the pop-up hatch proposed by Chrysler for the tank commander's station.

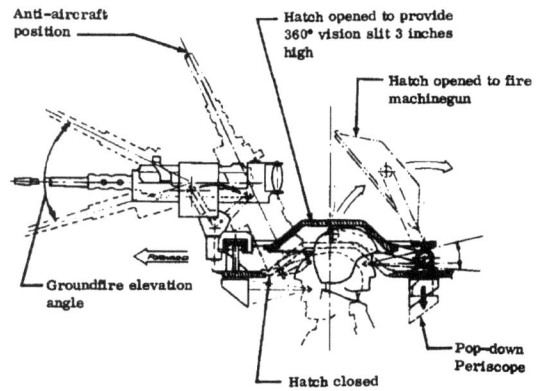

The 105mm M68 gun was selected as the main armament with possible future upgrading through improved ammunition or the introduction of a gun launched missile such as the Swifty. A .50 caliber M85 coaxial machine gun was specified, but the design permitted its replacement by the Bushmaster weapon when it became available. The fire control equipment was a modified M60A3 system with the turret integrated night thermal sight (TINTS). Final selection of the power plant was not made at that time with the German MTU MB 873 Ka 500 (formerly the Daimler-Benz DB1500) 1500 horsepower liquid-cooled diesel, the AVCR-1360-1 air-cooled diesel, and the AGT-1500 gas turbine all being considered. It was proposed to select two of these power plants for test rig evaluation. The X-1100 transmission was to be used in all three cases.

The advanced torsion bar suspension was selected. As mentioned earlier, this system utilized high strength torsion bars permitting a much greater twist before exceeding the elastic limit of the material. Thus a much greater wheel travel could be obtained. As with other proposed tank concepts, it was obvious that the combat weight would greatly exceed the original target if satisfactory protection levels were to be achieved.

Another Chrysler proposal configuration appears in this drawing. This concept was powered by the AGT-1500 gas turbine and featured a hydropneumatic suspension. The tank also was to be armed with the Swifty missile launched from the M68 105mm gun.

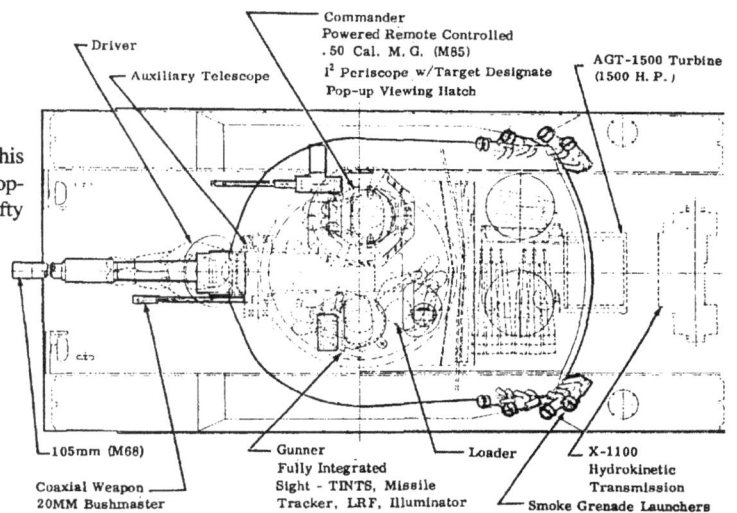

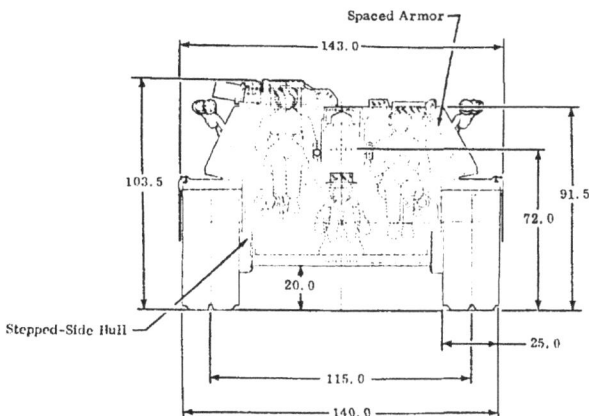

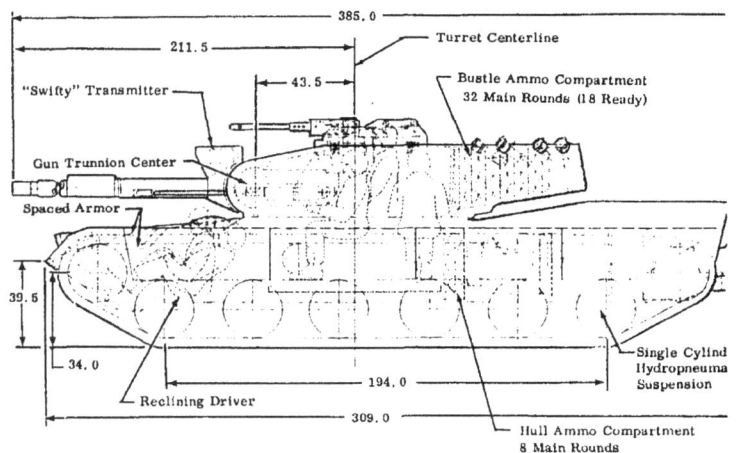

167

A new concept, designated LK 10352, was proposed by TACOM in a fourth study dated July 1972. The main objective of this study was to determine the effect on the gross vehicle weight if the fuel capacity was increased from 300 to 350 gallons. This was similar to the earlier examination of the same increase on the modified LK 10322. The hull was lengthened to 276 inches with the extra six inches between the crew compartment bulkhead and the engine being used to accommodate a tank for the additional 50 gallons of fuel. The lower glacis was redesigned on this new concept and the angle with the vertical increased from 55 degrees to 60 degrees. The upper glacis remained at 65 degrees from the vertical. No changes were made to the turret. The effect of these modifications on the gross vehicle weight was calculated based on the protection required to defeat two threat levels. The greatest of these was the combined threat of the Soviet 115mm APFSDS kinetic energy projectile at a range of 800 meters and a 3.2 inch diameter precision HEAT shaped charge round. The calculation indicated that a vehicle weight of over 52 tons would be required for protection directly from the front out to an angle of 30 degrees to each flank.

In August and September, the LK 10352 concept was modified in a fifth study to determine the effect of second generation armor arrays on the gross vehicle weight. The early steel-glass-steel composite armor was replaced by multiple steel and aluminum plate combinations. For this analysis, three threat levels and the armor arrays required to defeat them were supplied by the Ballistic Research Laboratory. The light threat consisted only of the kinetic energy projectile represented by the Soviet 115mm APFSDS at a range of 800 meters. The medium level of protection was required to defeat both the kinetic energy threat and the 3.2 inch diameter HEAT shaped charge round. A new heavy threat was introduced at this time. It consisted of both the 115mm kinetic energy projectile and a 5.0 inch diameter HEAT round. This was the threat against which the new MBT was to be protected in future TACOM studies. For the front hull armor array, the upper glacis was angled at 35 degrees and the lower glacis at 70 degrees, both from the vertical. The nose also was extended 2.5 inches to give more room to the driver. The turret was modified from the earlier concept by changing the left front from a curved to a flat surface angled back at 30 degrees to the rear and sloped at 60 degrees from the vertical. The slope of the right front was changed from 60 to 55 degrees from the vertical to provide more room for the gunner. This side already had a flat surface angled to the rear at 30 degrees. The turret sides were unchanged, but the bustle was extended by 12 inches to obtain more stowage space. The .50 caliber coaxial machine gun was replaced by the Bushmaster and the .50 caliber weapon was relocated to the commander's hatch. The study indicated that a combat weight of over 59 tons would be required to provide protection against the heavy threat from the front and at 30 degrees to each flank.

A sixth study by TACOM from October 1972 to January 1973 covered the preparation and weight analyses of new design concepts and the building of a full size wooden mock-up. The rapid development of composite armor technology continued to cause repeated changes in the tank design concepts. In fact, according to TACOM personnel at that time, the Ballistic Research Laboratory was cranking out new armor arrays faster than new concept designs based on them could be drawn and analyzed. After a preliminary study, it was decided to draw two different versions of the new tank. One version was to have the main gun ammunition stowed in the crew space as in earlier tanks. In the other, it was to be separated from the crew in ammunition compartments. Two different turret designs were prepared. One turret stowed the ammunition separate from the crew in a bustle compartment. The second turret had a short bustle with no ammunition stowage.

Two designs were considered for the tank with compartmented ammunition. In the first type, 18 out of a total of 40 105mm rounds were stowed in automated racks in the turret bustle behind a one inch thick aluminum alloy bulkhead. These racks were similar to those in the XM803 and the rounds were retrieved through a single round door in the bulkhead. Sixteen rounds were in two revolving drum type racks in the front hull, one on each side of the driver. These rounds were retrieved through small round doors in the one inch thick bulkhead facing the turret basket. The remaining six rounds were stowed in a non-automated rack in a two inch thick box underneath the cannon in the turret basket.

The second design of the tank with the compartmented ammunition eliminated the automated racks. They were replaced by simple lightweight racks and some rounds were relocated. The eighteen rounds remained in the turret bustle, but they were now retrieved through sliding aluminum doors. The six rounds previously stowed in the turret basket were moved to the front hull. The compartments on each side of the driver now contained 11 rounds each and they could be reached through sliding doors.

In the non-compartmented version of the new tank, all 40 rounds of main gun ammunition were stowed in the crew compartment below the turret ring. Twentyeight were in two racks, one on each side of the driver. The remaining 12 were located in the turret basket. These were considered to be ready rounds and were in easy reach of the loader.

The compartmented and non-compartmented concepts were quite similar in external appearance except for slight changes in the turret design. In addition to the shorter bustle, the non-compartmented version retained the five degree taper to the rear on the turret sides. On the compartmented turret, the side walls were parallel to the center line. This provided more stowage space for ammunition in the bustle. Both sides of the turret front were angled back 30 degrees to the rear and sloped at 45 degrees from the vertical. The left and right sides of the turret were sloped at 35 degrees and 20 degrees from the vertical respectively. The roof was flat and the rear of the bustle was angled at 40 degrees from the vertical. As in the previous concepts, the turret ring diameter was 85 inches.

The full size mock-up corresponding to the LK 10372 design can be seen in the photographs on this page. Note the coaxial Bushmaster weapon.

The hull design was identical for both concepts, except for the ammunition racks. The upper and lower front glacises were angled at 35 and 70 degrees from the vertical respectively. Side skirts extended from the front idler to the rear sprocket protecting both the crew and engine compartments. The power plant was the AVCR-1360 engine with the X-1100 transmission. All of the fuel was located in the engine compartment. A tube-over-bar suspension system with six road wheels per side was used with 26 inch wide tracks.

For the commander, a low rotatable station even with the turret roof line was used for both concepts. It was fitted with a .50 caliber M85 machine gun which could be operated from either inside or outside the tank. In addition to the machine gun sight, the commander was provided with six periscopes for all around vision. A Bushmaster weapon system was installed on the left side of the 105mm gun. A day/night sight including a laser range finder was mounted in front of the gunner's station with separate controls for the gunner and the tank commander. Weight analyses of the two concepts armored against the heavy threat indicated that the tank with the compartmented ammunition weighed slightly over 60 tons compared to under 57 tons for the non-compartmented version.

The tank with the compartmented ammunition and armored against the heavy threat was selected for the construction of a full size wooden mock-up. The mock-up was modified in several ways from the original concept drawings. The day/night sight for the gunner and the tank commander was relocated to the right side of the turret. This was done to eliminate a weak ballistic area on the turret caused by the previous mounting of the sight. To provide protection against HEAT rounds, armor arrays extending alongside the entire crew compartment were installed on the hull side areas over the sponsons. The overhanging frontal sections of the turret were lowered by about four inches leaving only a minimum clearance for the driver's head when driving with his head exposed. This was to improve the ballistic protection by reducing the shot trap in this area. The turret roof was slightly crowned to increase its rigidity and to eliminate any possibility of buckling. This was achieved by raising the center line one inch and sloping the roof toward the sides at five degrees from the horizontal. After completion of the mock-up, a new concept drawing, designated LK 10372, was prepared incorporating all of the changes.

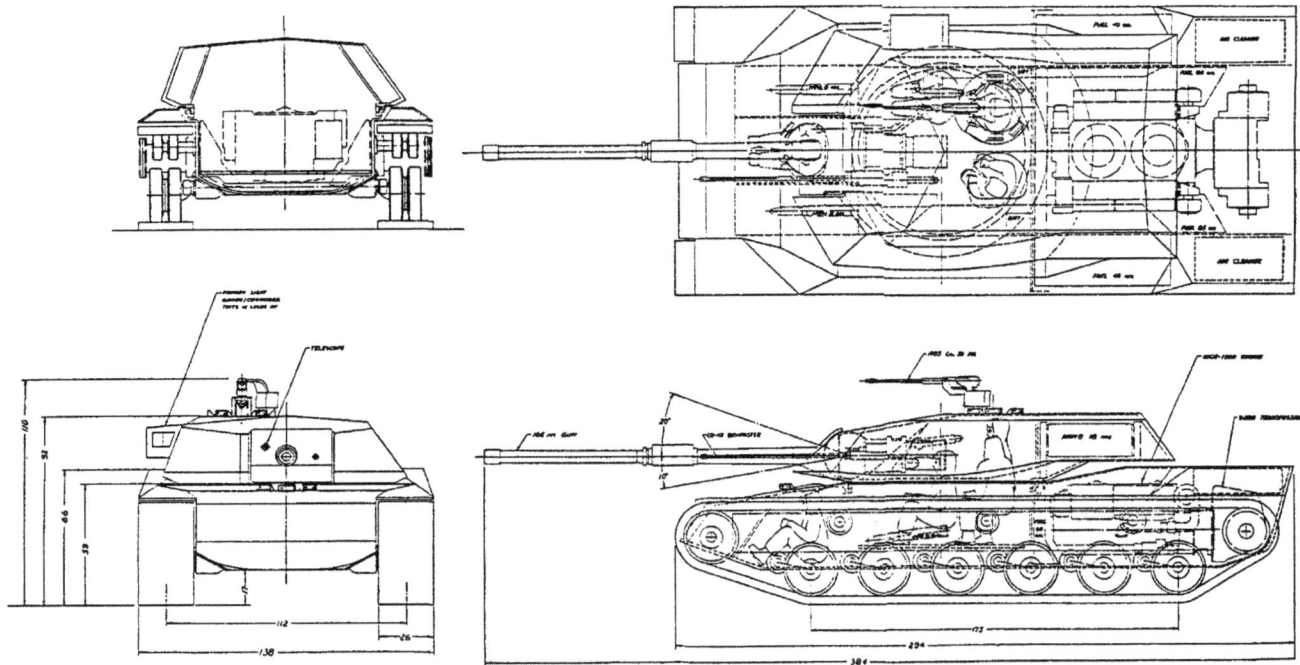

Above is the LK 10372 concept drawing reflecting the modifications incorporated in the full size mock-up.

Data on improved armor arrays were received from the Ballistic Research Laboratory at the conclusion of the previous study. To evaluate the effect of these new arrays on the vehicle weight, a new concept was prepared including the changes applied to the mock-up as well as several new features. This design, designated as LK 10379, was covered in the seventh study from TACOM in February 1973. From this point on, the analyses considered only the protection level required to defeat the heavy threat. This was still the Soviet 115mm APFSDS projectile at 800 meters and the 5.0 inch diameter HEAT round. The 40 rounds of 105mm ammunition in the LK 10379 concept were located in the hull below the turret ring and were not stowed in separate compartments. The Bushmaster weapon system was removed from inside the turret and remounted on the left side exterior near the bustle. The day/night sight for the gunner and the tank commander was installed through the right side of the turret as on the mock-up. The gun shield was narrowed from a width of 38 inches to 12 inches. The five degree taper to the rear of the turret side walls was eliminated making the sides parallel to the turret center line. This provided more space for the tank commander and simplified the design of the Bushmaster mount. The slope of the left and right sides of the turret remained at 35 and 20 degrees from the vertical respectively. The angle of the turret frontal sections toward the rear was increased from 30 to 40 degrees, but they remained sloped at 45 degrees from the vertical. The turret roof was crowned as on the mock-up. The hull front on the LK 10379 was

redesigned with a 55 degree angle from the vertical on both the upper and lower glacises. This increased the hull length making the six road wheel suspension marginal for good mobility. Therefore, a new tube-over-bar suspension was specified with seven road wheels per track. As on the mock-up, armor arrays were installed on the hull sides above the sponsons. The combat weight of the LK 10379 concept was calculated to be a little over 60 tons. The analysis showed that by reducing the protection on certain areas of the hull and lowering the fuel capacity from 350 to 300 gallons, the weight could be reduced to about 58 tons.

The eighth and last in the main series of new XM1 concept studies was performed by TACOM in March 1973. Its objective was to explore ways to reduce the weight of the proposed tank to below 58 tons. To achieve this goal, a new concept drawing, LK 10382, was prepared with several changes from the previous design. However, the same armor arrays were used to permit accurate comparisons. The front hull was changed back to the earlier design with the upper and lower glacis plates sloped at 35 and 70 degrees from the vertical respectively. This reduced the length of the hull by 15 inches and lowered the weight by about 900 pounds. The hull height was reduced from 42 to 40 inches and the sides above the sponsons were sloped at a greater angle from the vertical. The fuel tank behind the crew compartment bulkhead was removed and the engine was moved four inches forward. This dropped the fuel capacity back to 300 gallons. The turret ring diameter was reduced from 85 to 83 inches which also shortened

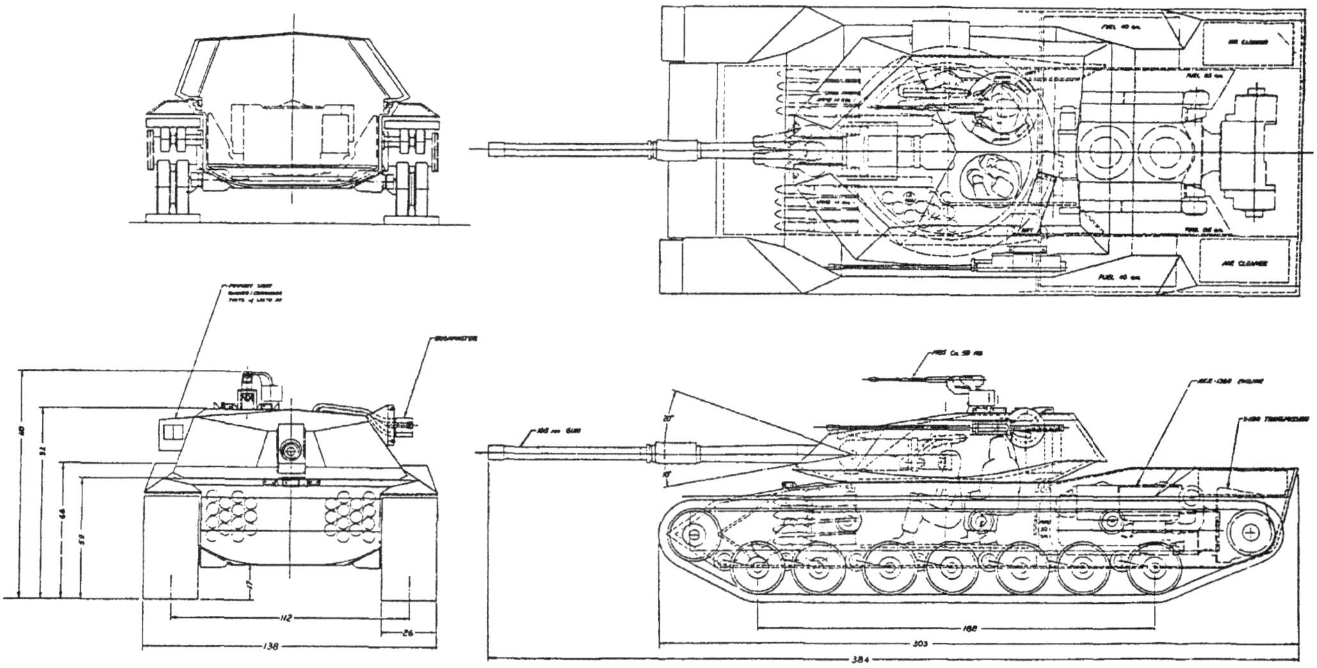

Concept drawing LK 10379 appears above. Note that the 105mm ammunition is not isolated from the crew space and the Bushmaster weapon has been relocated to the left side of the turret. This tank also is fitted with seven road wheels per track.

The suspension in design concept LK 10382 below has returned to the use of six road wheels per track and the ammunition feeding system for the Bushmaster has been simplified.

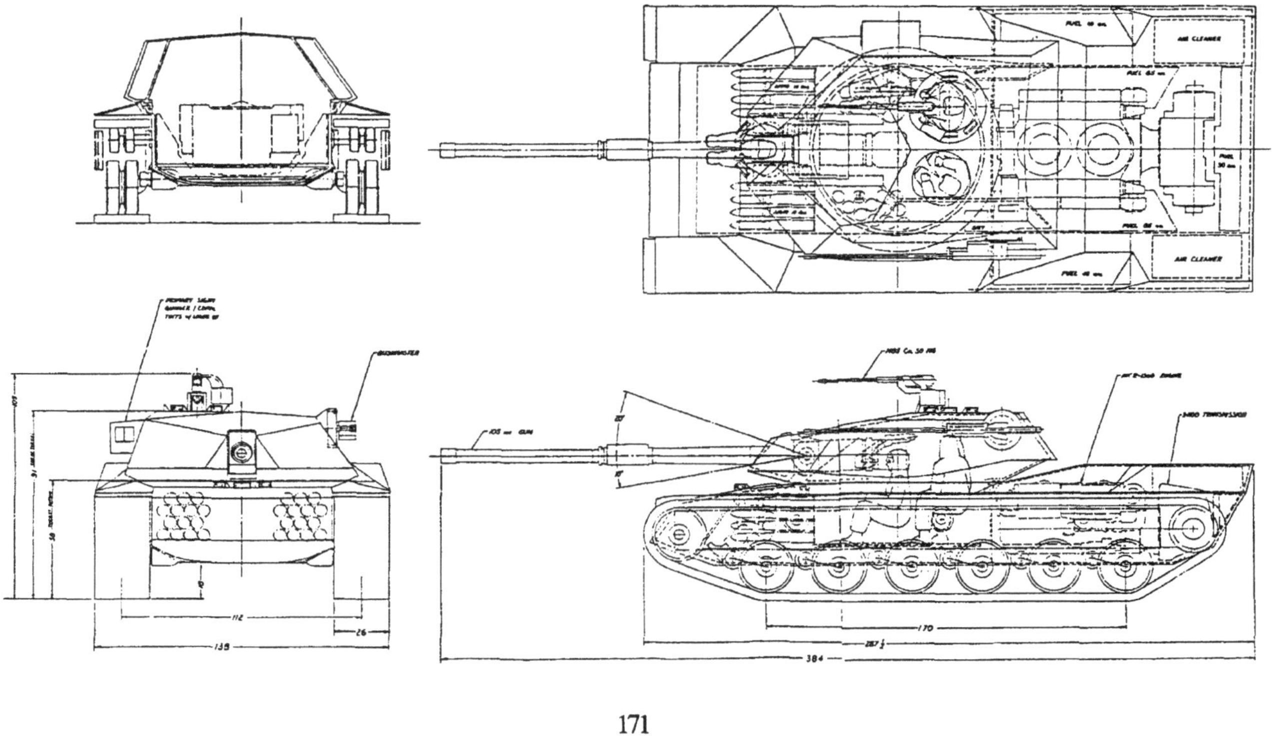

the hull by an additional two inches. With the shorter hull, the suspension was changed back to six road wheels per side. The ground clearance was increased from 17 to 18 inches for better mobility and mine protection. Except for the smaller ring, the turret design remained the same, but the Bushmaster was moved farther to the rear, lowered, and mounted closer to the turret wall. The ammunition feeding system for the Bushmaster was changed so that it fed directly through the turret side armor. This eliminated the partially exposed ammunition chutes which formerly were routed through the turret roof. The estimated combat weight of the LK 10382 design was between 57 and 58 tons depending upon weights assumed for the suspension and the possible reduction in thickness of the hull bottom plate under the engine compartment.

After the complexity and high cost that resulted in the cancellation of the XM803, Congress required that these factors be tightly controlled in the new tank program. Thus a design to unit cost ceiling was established at the beginning of the project. The specification requirements in the Materiel Need (MN) document were presented as ranges permitting the contractors some leeway in meeting the design objectives. As the Army later observed during a Congressional hearing in April 1974, "We would like to have the maximum performance, but we are sticking to a design to unit cost and we therefore have to give the contractor the ability to make trade offs to meet the design to cost goal".

As mentioned earlier, the weight limitation in the original MN document was incompatible with the protection level required. Since crew survivability had the highest priority on the list of characteristics for the new tank, the maximum weight was raised to 58 tons to permit adequate armor. Later, this would be increased to 60 tons. During the initial studies in 1972, the design to unit cost goal was set at $400,000. However, by the time that the request for proposal was issued, this had increased to $507,790, also in 1972 dollars. This compared to estimated unit costs of $339,000 for the M60A1, $432,000 for the M60A3, and $611,000 for the cancelled XM803. The new program was divided into three phases. The first of these was a competitive advanced development validation phase during which contracts would be awarded to two sources. Each of these contractors would design, build, and test one complete prototype tank and one automotive test rig. They also were to furnish an additional hull and turret for ballistic evaluation. This 34 month advanced development phase would end with a combined developmental/operational test (DT/OT-I) by the Army. These tests would be used to validate the contractors' proposals for phase 2, which would be a sole source full scale engineering development (FSED) program. Phase 3 would put the new tank into production.

The artist's concept at the right shows the Chrysler proposal for the XM1 validation phase tank.

The various tank concepts being considered by the Army at this time reflected a change in thinking since the period of the MBT70 and the XM803. Although some sources regarded the appearance of the long range anti-tank missile as sounding the death knell of the tank, this was not the case. The main effect of the new missiles was to release the tank from the defensive role since other units were now capable of defending themselves against armored attack. The tank could now resume its place as the key offensive weapon in the combined arms team. In that role, the missile firing gun-launcher system was less attractive as the main tank weapon than the fast shooting 105mm gun, particularly with improved ammunition. Thus when proposals were requested for the development of the new tank, the 105mm gun with its quick reaction, multitarget performance, and immediate availability was the obvious candidate despite its shorter range.

Proposals for the advanced development validation phase were received from Chrysler and General Motors in May 1973. Originally, Ford Motor Company had been expected to participate, but they declined to enter the competition. As a result, after review, contracts were awarded to Chrysler and to the Detroit Diesel Allison Division of General Motors on 28 June 1973. As mentioned earlier, these contracts required the design, construction, and testing of a completely integrated prototype tank. It was to include all subsystems except for the night vision equipment. An automotive test rig also was to be constructed for use in the test program and an additional ballistic hull and turret were to be provided for ballistic tests at Aberdeen Proving Ground. These cost plus incentive fee (CPIF) contracts eventually amounted to $68,999,000 for Chrysler and $87,969,000 for General Motors.

Chrysler proposed a conventional arrangement for its prototype tank with a three man turret crew and the driver in the center front hull. The gunner and the tank commander were in the usual location on the right side of the M68 105mm gun with the loader on the left. A coaxial Bushmaster weapon was fitted on the left side of the cannon. An M85 .50 caliber machine gun was mounted on the tank commander's station and either a 7.62mm M60D machine gun or a 40mm high velocity grenade launcher was installed on the turret roof in front of the loader's hatch.

The model in these photographs represents the proposed Chrysler XM1 validation phase tank. Note the coaxial Bushmaster weapon. The blow off panels for ammunition stowed in the turret bustle can be seen in the views from the top front.

The side skirt sections on the original proposal were designed to swing either up or out as indicated on the model above and the sketches below.

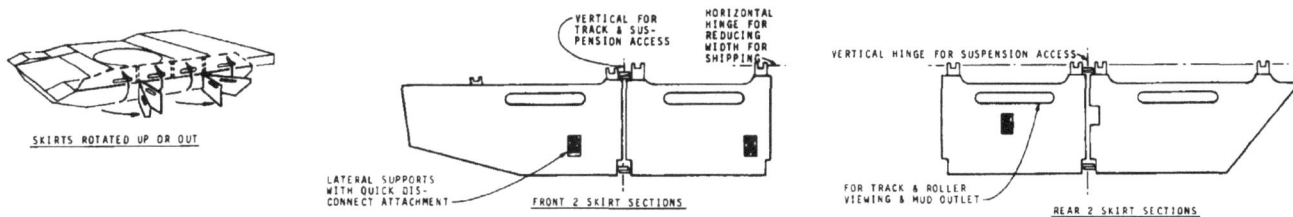

SKIRTS ROTATED UP OR OUT

VERTICAL FOR TRACK & SUS- PENSION ACCESS

HORIZONTAL HINGE FOR REDUCING WIDTH FOR SHIPPING

LATERAL SUPPORTS WITH QUICK DIS- CONNECT ATTACHMENT

FRONT 2 SKIRT SECTIONS

VERTICAL HINGE FOR SUSPENSION ACCESS

FOR TRACK & ROLLER VIEWING & MUD OUTLET

REAR 2 SKIRT SECTIONS

173

Another view of the Chrysler XM1 validation phase model appears above and the evolution of the turret design is illustrated at the right. Various components of the turret and hull are shown in the sketches below.

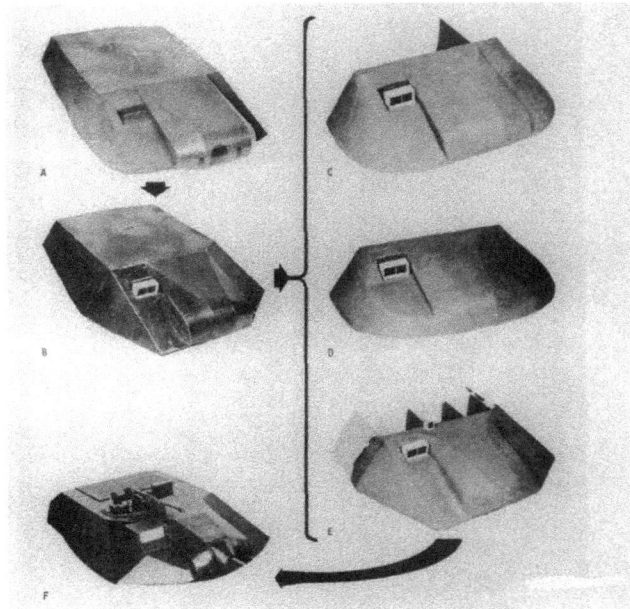

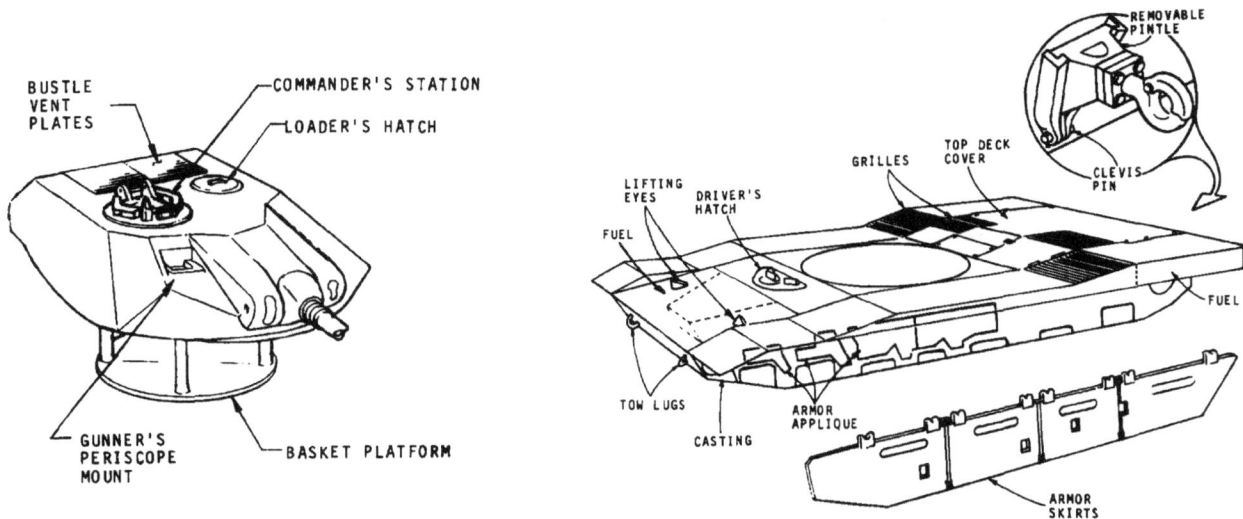

A simplified line of sight stabilization system was expected to provide performance equal to at least 90 per cent of a dual axis system with a cost saving of about $3000. The 105mm ammunition was stowed in separate compartments vented away from the crew space. New composite special armor was incorporated into the frontal areas of the tank and side skirts extended from the front idler to the rear sprocket partially shielding the hydromechanical suspension. This was actually a tube-over-bar suspension with seven road wheels per track using rotary hydraulic shock absorbers. The term hydromechanical suspension was selected to avoid confusion with the earlier tube-over-bar suspension installed experimentally on the M60A1 which lacked the performance of the new design. It also was noted that the same name could be applied to the advanced high

strength torsion bar suspension if it replaced the tube-over-bar design in the future. At the time of the proposal, the advanced torsion bar was considered a high risk item, but the option of using high strength steel or titanium torsion bars was kept open for the future.

In regard to the power plant, Dr. Philip W. Lett Jr., the leader of the Chrysler team, made a bold decision. Despite the fact that the Army MBT Task Force had considered the gas turbine to be a high risk item, he selected the AGT-1500 with the X-1100 transmission to power their prototype. In retrospect, this appears to have been a crucial decision in the design of the new tank. Subsequent events were to prove that the AGT-1500 would develop into a highly reliable power plant and it had several advantages over its diesel competitors. Although both the AGT-1500

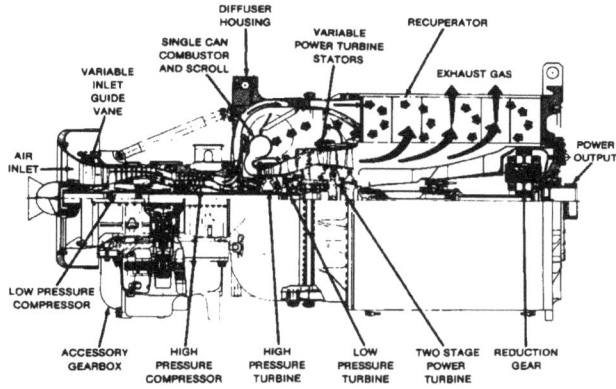

The Lycoming AGT-1500 gas turbine can be seen in the drawing and photograph above and the engine and its various accessories are sketched at the bottom of the page. At the right, Major General Baer (far left) and Dr. P. W. Lett (at right) are with a group observing the AGT-1500.

and the AVCR-1360 diesel were rated at 1500 gross horsepower, the former utilized only 30 horsepower for cooling compared to 160 horsepower for the latter. Also, a growth potential to 2000 gross horsepower was estimated for the turbine compared to 1800 gross horsepower for the variable compression ratio diesel. The less complex design of the turbine engine with its fewer parts was expected to provide higher reliability and reduced maintenance in the future. Its lighter weight also permitted increased armor protection without exceeding the overall weight limit for the tank. The gas turbine did not generate smoke during operation and it was capable of using a wider range of fuels than the diesel. Also, the turbine was easier to start at low temperatures. The greatest disadvantage of the gas turbine was its higher fuel consumption. This was particularly obvious at low vehicle speeds. However, the small size of the power plant allowed space for an increase in fuel capacity to achieve the specified cruising range. Another problem was the high volume of combustion air used by the gas turbine. This required a filter system with approximately three times the capacity of that necessary for the diesel engine. The turbine, of course, did not require the high volume of cooling air used by the air-cooled diesel.

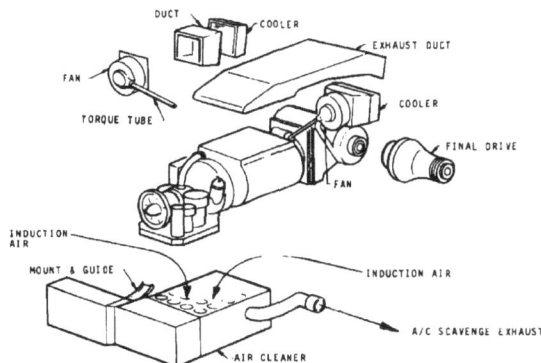

The development of the gas turbine tank engine resulted from recommendations of the Committee on Gas Turbine and Electric Propulsion for Tanks, chaired by Mr. James C. Zeder. When presented in 1965 to the Mobility Advisory Group, the recommendations of the Zeder Committee were initially rejected because of cost. However, the program was subsequently approved. After the evaluation of several proposals from industry, the Lycoming Division of AVCO was selected to develop the new turbine, designated as the AGT-1500. At that time, it was considered as a possible power plant for the MBT70. However, after the selection of the diesel engine for that program, attempts were made to cancel the AGT-1500 and to transfer its funds elsewhere. These attempts were prevented by the efforts of Lieutenant Colonel John W. Wiss, Chief of the Army Tank Automotive Command Laboratories, and Mr. Wayne S. Anderson who directed the Propulsion Systems Laboratory. Thus the AGT-1500 was available when needed for the XM1 program and its state of development was not far from that of the AVCR-1360 variable compression ratio diesel. Although the initial cost of the gas turbine exceeded that of the diesel engine, it was anticipated that the life cycle cost would be competitive because of reduced maintenance requirements.

Above is an early General Motors concept drawing of the XM1 validation phase tank.

The early General Motors concept for the XM1 validation phase tank also was armed with the 105mm gun M68 and a Bushmaster. However, the latter was located in a separate mount on the left side of the turret. The turret crew rode in the usual positions with the gunner and tank commander on the right side of the cannon and the loader on the left. A .50 caliber M85 machine gun was provided for the tank commander and a 7.62mm machine gun was installed at the loader's position. The main gun ammunition was separated from the crew space in compartments. The driver was located in the left front hull and special armor was incorporated in the frontal areas of the tank. The Continental AVCR-1360 diesel engine drove the tank

through the General Motors X-1100 transmission. As mentioned earlier, the AVCR-1360 diesel was a 1500 horsepower development of the AVCR-1100 used in the MBT70 and the XM803. The suspension on the General Motors tank was fitted with six road wheels per track. This was a hybrid suspension with road wheels 1, 2, and 6 fitted with the National Water Lift hydropneumatic suspension units. Road wheels 3 and 5 utilized high strength torsion bars.

In early July 1973, Major General Robert J. Baer, the XM1 Project Manager, and personnel from the Ballistic Research Laboratory escorted representatives from Chrysler and General Motors to the United Kingdom for an update on the British developed Chobham special armor.

The Continental AVCR-1360 diesel engine is shown in the sketch and photograph below. The 120 degree angle between the cylinder banks of this engine gave it a low silhouette.

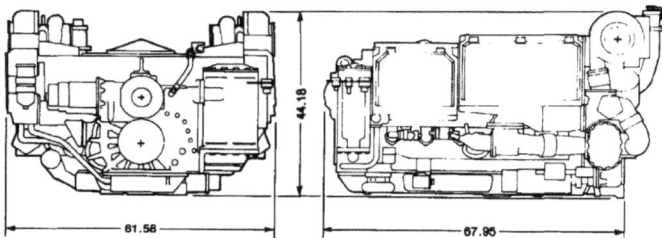

176

The model of the proposed General Motors XM1 validation phase tank can be seen in the photographs on this page. Unlike the Chrysler design, the Bushmaster weapon is mounted externally on the left side of the turret.

The six road wheel suspension and the vertical turret armor of the early General Motors XM1 concept are visible in the views below and at the left.

GUNNER'S SIDE-MOUNTED PRIMARY SIGHT

COMMANDER'S STATION
(M85 .50 CAL MACHINE GUN ON A POWERED STATION)

LOADER'S STATION
(M60D MACHINE GUN PINTLE MOUNTED)

COAXIAL WEAPON

DRIVER'S STATION

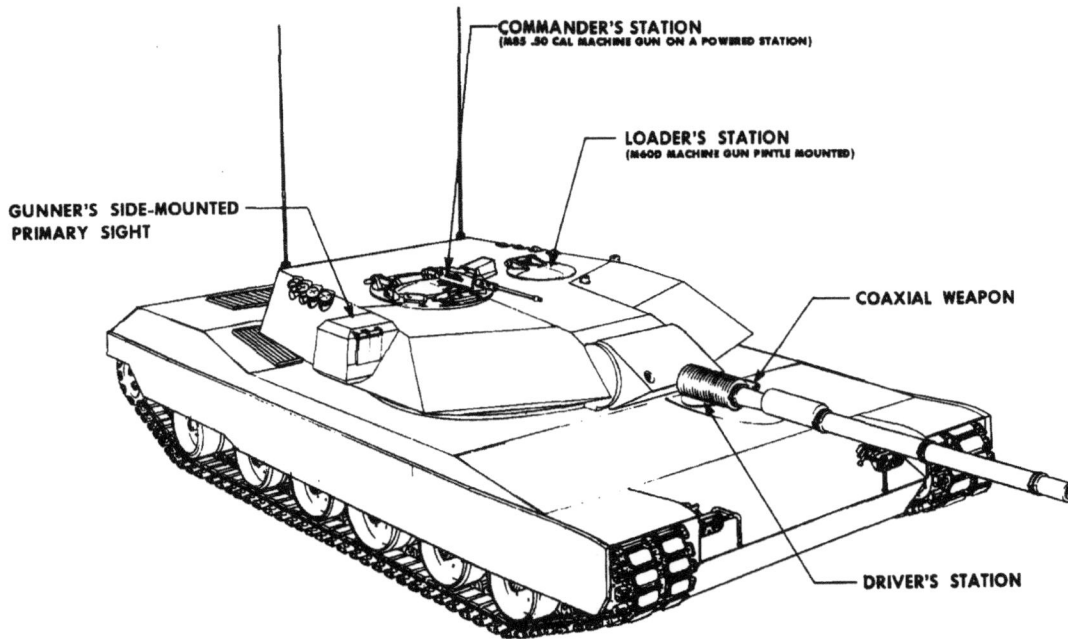

Above is a drawing of the redesigned configuration for the General Motors XM1 validation phase tank. Note that the externally mounted Bushmaster has been eliminated in favor of a coaxial weapon.

During this trip they observed the proposed design of a new United Kingdom vehicle utilizing this special armor as well as the manufacturing processes required for its production. Based upon the newly obtained data, both contractors reevaluated their proposed armor configurations. Chrysler made a number of changes, but they retained the basic design in their original proposal. However, General Motors made major changes in the configuration of their proposed tank. The most obvious change was in the turret where the vertical front and side walls were replaced by sloped armor. The modifications by both contractors required additional work by the Ballistic Research Laboratory to develop new special armor combinations. By Janu-

ary 1974, both Chrysler and General Motors completed the final design of their special armor configurations and some of the hull and turret armor structures had been delivered to Aberdeen Proving Ground for ballistic tests.

Some changes in the specifications for the new tank resulted from a study of the Israeli experience during the October 1973 war in the Middle East. One change which was applied to both of the validation phase prototypes was the replacement of the Bushmaster weapon system with a 7.62mm coaxial machine gun. The original intention had been to utilize the Bushmaster against lightly armored vehicles thus reducing the stowage requirement for the expensive main gun rounds. However, battle experience showed

The Chrysler turret design also underwent modification. Below, the Chrysler validation phase tank appears at the left with its original turret and at the right with the modified design.

Above, the Chrysler and General Motors XM1 validation phase prototypes are compared with the 105mm gun tank M60A1. The low silhouette of the new vehicles is obvious.

that the tank crews would invariably use the main gun against these vehicles. Also, there was a need for a coaxial machine gun for use against infantry and to suppress antitank weapons. Elimination of the bulky Bushmaster and its ammunition provided space to increase the number of 105mm rounds to 55. Some of the modifications resulting from the study of battlefield experience were applied at a later date to the full scale engineering development (FSED) tank proposals. These included changes in the ammunition compartments to accommodate the larger number of 105mm rounds and the use of a less flammable hydraulic fluid.

The General Motors XM1 prototype can be seen above and below. A .50 caliber M85 machine gun is provided for the tank commander and a 7.62mm M60D machine gun is installed at the loader's hatch.

The General Motors XM1 prototype is shown in these photographs during the evaluation program. Test instrumentation has been installed on the turret roof and the gun barrel.

The complete validation phase prototypes were delivered to Aberdeen Proving Ground for the combined developmental/operational test (DT/OT-I) which ran from 31 January to 7 May 1976. During this same period, the competing automotive test rigs were driven more than 3000 miles over a variety of terrain. Both prototypes were displayed to the public for the first time on 3 February 1976. The results of the DT/OT-I indicated that both the Chrysler and the General Motors prototypes met the specified requirements for the new tank. The Chrysler tank with its gas turbine power plant showed slightly greater acceleration, but the performance was satisfactory for both vehicles. The operational portion of the tests included two weeks of simulated combat operations with the tanks manned by crews from Fort Knox and Fort Hood. After the evaluation of the competing prototypes, the selection of the contractor for the full scale engineering development phase was scheduled for July 1976.

Additional views of the General Motors XM1 prototype appear here. The secondary armament has not been installed on the tank during this part of the test program.

Sleeping Bags & Covers (3)
Antenna Bag & Mast Sections
Flag Set

Single Bit Axe &
Double-Faced Hand Hammer

Shovel

Track Connecting Fixtures (2)

Pintle Assembly

Mattock Pick & Handle

Pinch Point Crowbar

Tow Cable Assemblies (2)

Artillery Cleaning Staff
Sections (5) W/Stowage Box

Sleeping Bag & Cover

Bustle Stowage Compartment
W/Provision for Padlocks

Combat Meals (24) W/Stowage Bag

2-3/4-Lb. Portable Fire Extinguisher

.50 Cal. Ammo - 340 Rnds.
(85 Rnds./Can)

Tarpaulin

External stowage on the General Motors XM1 can be seen above and at the left. Below, an exploded view of the main gun mount is at the left and the fuel and ammunition stowage is sketched at the right.

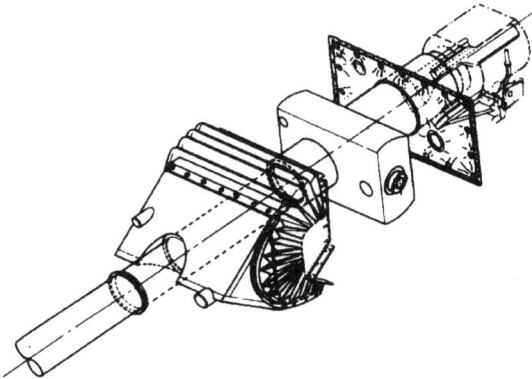

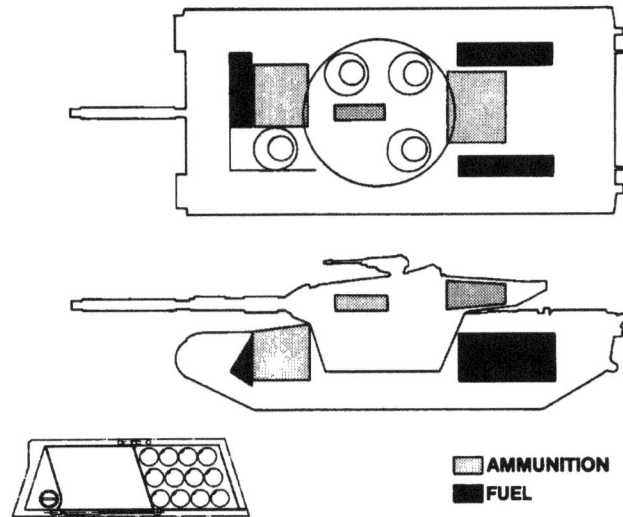

AMMUNITION
FUEL

Below, the stations in the General Motors XM1 for the gunner and the tank commander appear at the left and right respectively. The use by the gunner and the tank commander of the gunner's primary sight is illustrated in the sketch at the right.

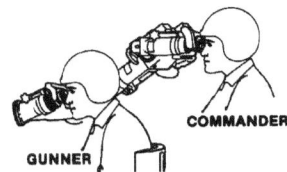

GUNNER

COMMANDER

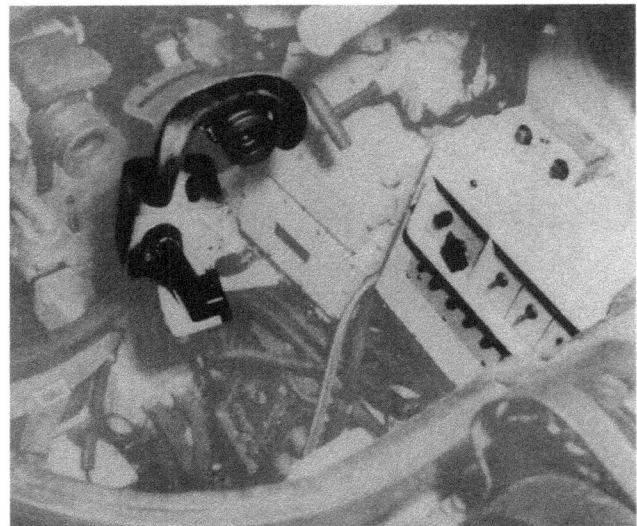

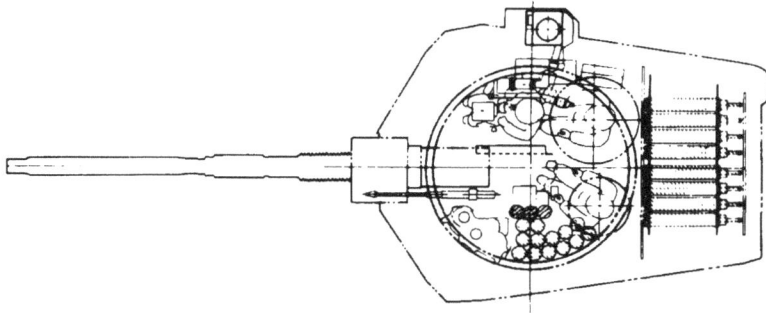

Above, the turret layout in the General Motors XM1 can be seen at the left and at the right is a view of the tank commander's hatch. The sketches below show the gunner's sighting equipment at the left and the operation of the commander's hatch and his relay to the gunner's primary sight at the right.

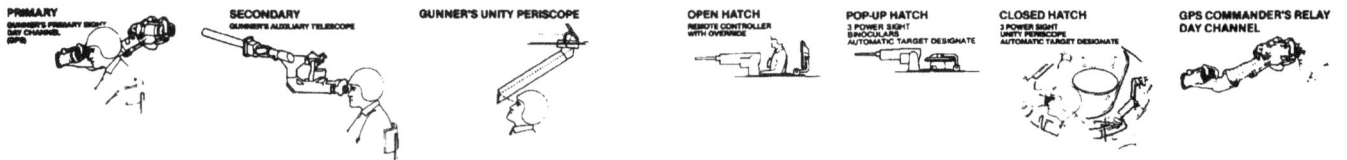

PRIMARY
GUNNER'S PRIMARY SIGHT DAY CHANNEL (GPS)

SECONDARY
GUNNER'S AUXILIARY TELESCOPE

GUNNER'S UNITY PERISCOPE

OPEN HATCH
REMOTE CONTROLLER WITH OVERRIDE

POP-UP HATCH
3 POWER SIGHT BINOCULARS AUTOMATIC TARGET DESIGNATE

CLOSED HATCH
3 POWER SIGHT UNITY PERISCOPE AUTOMATIC TARGET DESIGNATE

GPS COMMANDER'S RELAY DAY CHANNEL

LOADER'S PERISCOPE

At the left is a view of the 7.62mm machine gun installed at the loader's hatch and a sketch of the loader's periscope.

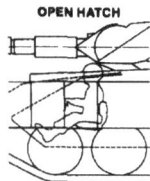

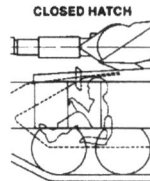

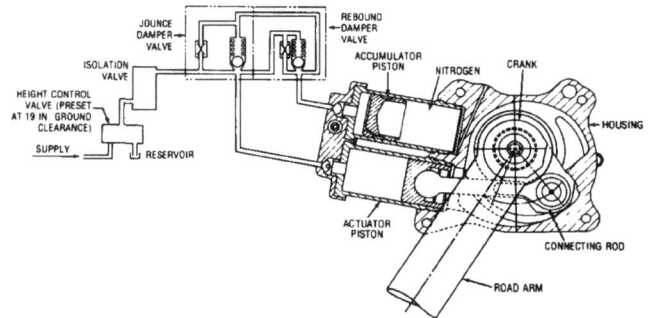

OPEN HATCH

CLOSED HATCH

At the left, the driver's station in the General Motors XM1 is shown below and his position with the hatch open and closed appears above. At the right is a diagram of the National Water Lift suspension unit (above) and an artist's cutaway drawing showing the hydropneumatic and torsion bar components of the hybrid suspension (below).

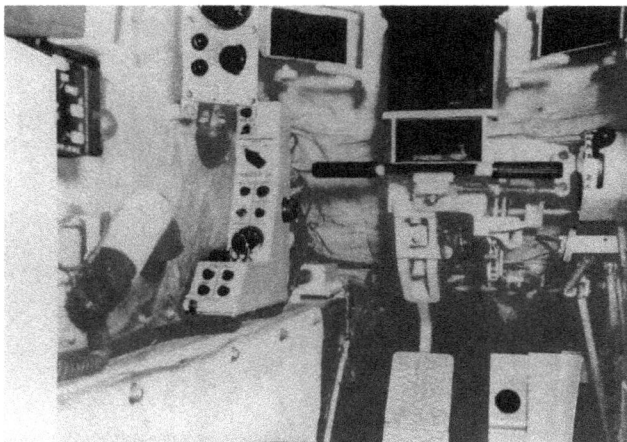

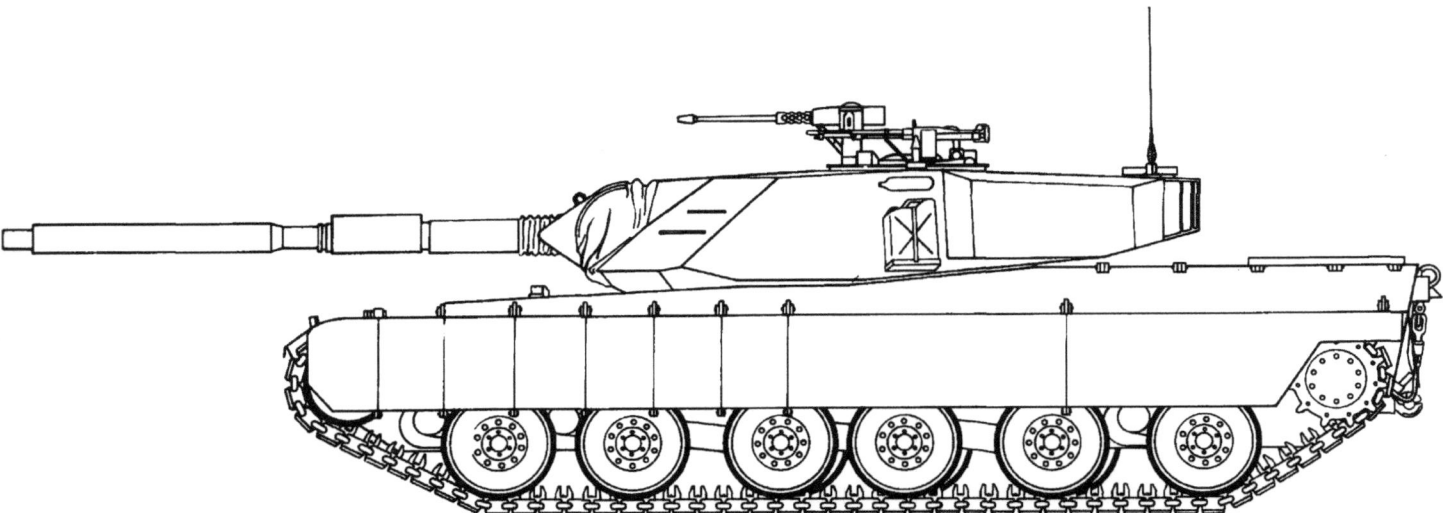

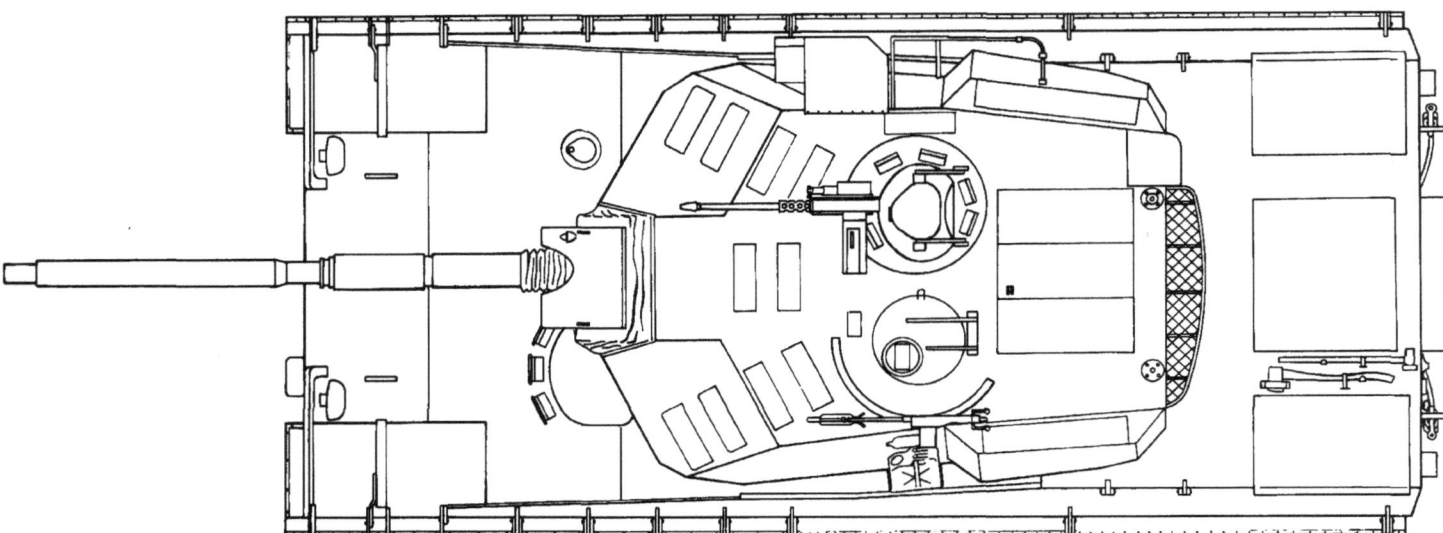

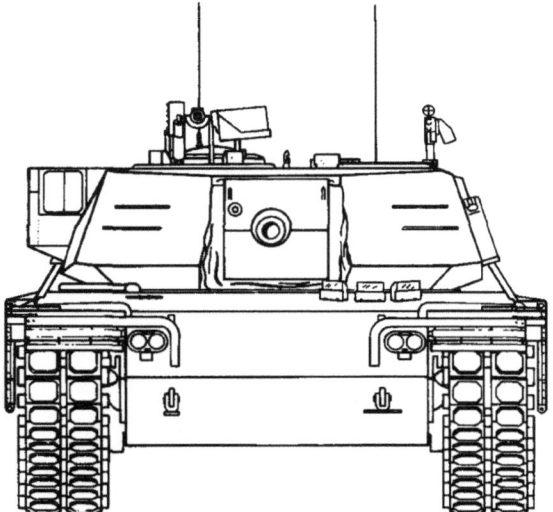

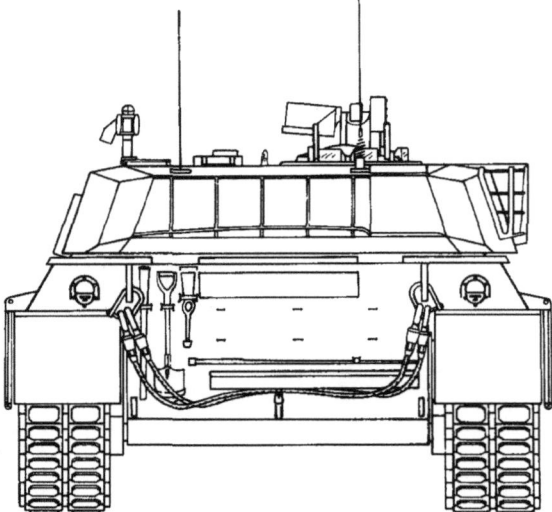

Scale 1:48

© D.P. Dyer

105mm Gun Tank XM1, General Motors Prototype

184

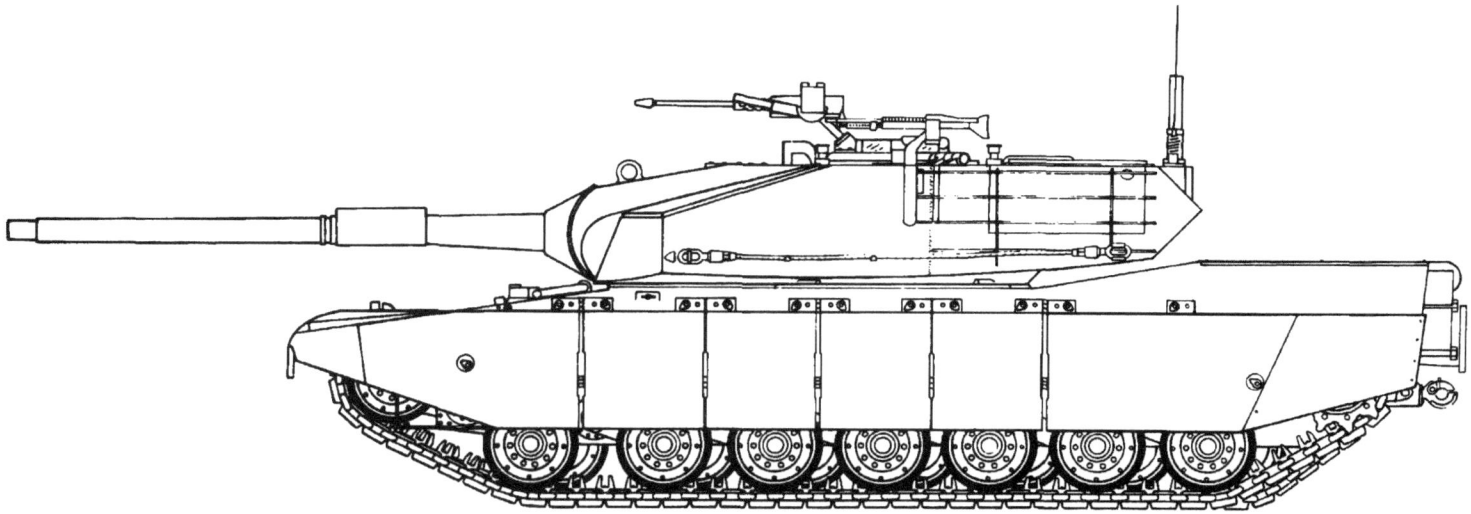

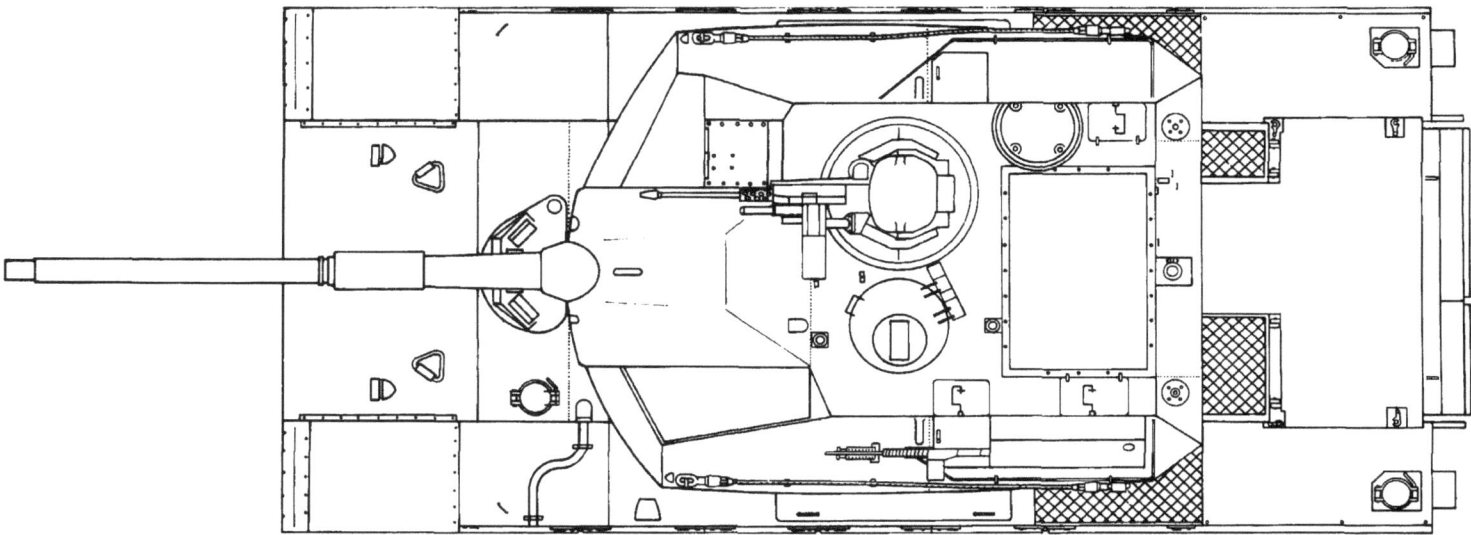

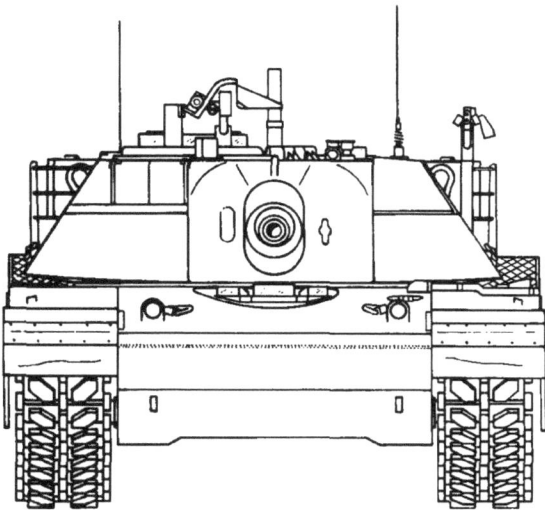

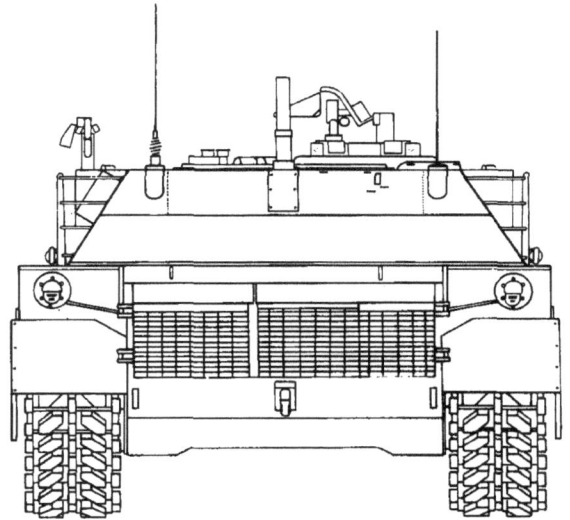

105mm Gun Tank XM1, Chrysler Prototype

185

The Chrysler XM1 validation phase prototype, registration number JE0001, appears in the photographs above and below. Like the General Motors prototype, it features a .50 caliber M85 machine gun for the tank commander and a 7.62mm M60D machine gun for the loader.

Details of the Chrysler XM1 validation phase prototype can be seen in these photographs. Note that there is only a single blow off panel on the turret bustle roof.

The views on this page show the Chrysler XM1 prototype during test operations. Note that the rear sections of the side skirts have been removed in several of the photographs.

Below, the Chrysler XM1 prototype can be seen at the left during the automotive tests and, at the right, the tank is firing its 105mm gun.

Below are two photographs of the Chrysler XM1 automotive test rig, registration number JE0002.

After the cancellation of the MBT70 program, the Federal Republic of Germany continued with the development of their new Leopard II tank. In line with efforts to standardize materiel in NATO, Germany proposed that the Leopard II be modified to meet American requirements. In December 1974, a Memorandum of Understanding between the United States and the Federal Republic of Germany was signed in which the U.S. agreed to evaluate the Leopard II under the same test conditions as the XM1 candidates. At that time, it was estimated that the Leopard II would cost about one million 1975 dollars compared to approximately $750,000 for the XM1. The modified version, designated as the Leopard II AV was armed with the 105mm gun. It had improved armor protection and a simplified fire control system to reduce costs. Tests of the Leopard II AV were carried out at Aberdeen Proving Ground from September to December 1976. Previously, on 22 July 1975, FMC Corporation had been awarded a contract for a study to determine the costs and technical problems involved in manufacturing the Leopard II AV in the United States. The tests at Aberdeen indicated that the Leopard II AV met or exceeded most of the U.S. requirements. The accuracy of its fire control system was slightly higher than that of the XM1, but it was more expensive. The armor protection, ammunition compartmentation, and gun movement were considered somewhat inferior to that on the American tank. Rather than continue the evaluation of the complete tank, it was agreed in January 1977 that only certain major components would be considered for adoption to enhance NATO standardization.

Above, the Chrysler XM1 automotive test rig becomes airborne during a high speed test.

The German Leopard II AV armed with the 105mm gun appears above and below. Note the similarity of the turret configuration to that of the early General Motors proposal.

189

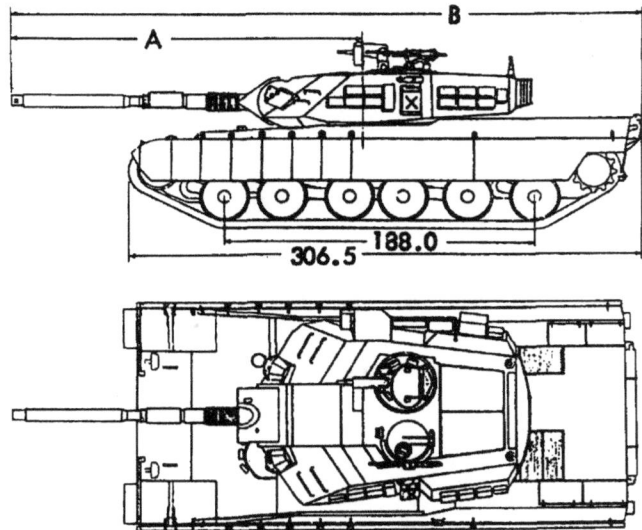

	Dimensions (inches)	
	A	B
Basic GM XM1 105mm (diesel)	213.5	382.0
Standardized GM XM1 105mm (turbine)	213.5	382.5
Standardized GM XM1 FRG 120mm (turbine)	215.0	384.0
Standardized GM XM1 UK 120mm (turbine)	219.13	388.13

Top and side views of the General Motors XM1 are sketched at the right. Above, the dimensions of the diesel tank are compared with those of the vehicle powered by the AGT-1500 gas turbine and armed with the M68 105mm gun, the German 120mm smooth bore gun, and the British 120mm rifled cannon.

Shortly before the end of the XM1 validation test program, a new requirement was introduced for inclusion in the proposals for the full scale engineering development/producibility engineering and planning (FSED/PEP) contract. These proposals were now to include firm price ceilings for the first two production options. These were for 110 low initial rate production (LIRP) tanks in fiscal year 1979 and 352 tanks in fiscal year 1980. The winner of the competition for the XM1 FSED/PEP contract was selected on schedule in July 1976, but the results were not released by the Army. Information published later in the press indicated that General Motors won with a bid of 208 million dollars compared to Chrysler's bid of 221 million dollars. However, these bids were based upon the Chrysler and General Motors tanks being powered by the AGT-1500 turbine and the AVCR-1360 diesel respectively. The Army, by this time, favored the gas turbine and although awarding the contract to General Motors, planned to change the power plant to the AGT-1500. However, this approach was rejected by the Office of the Secretary of Defense. As the Deputy Secretary of Defense later testified before Congress: "I was unalterably opposed to making these changes in a sole source environment after the contractor was selected. I wanted to know what those unit costs were, what the delays in the program might be, if any, and have those costs determined competitively between the two contractors." Also, a decision had been made to arm the new tank with a 120mm gun at some time in the future. Therefore, the turret had to be capable of mounting either the 105mm gun M68 or a 120mm gun without extensive modification. The 120mm guns under consideration were the British rifled weapon and the German Rheinmetall smooth bore gun used in the Leopard II. As a result of the new requirements, a revised request for proposal was issued, with the award of the FSED/PEP contract now rescheduled to November 1976. Both contractors submitted proposals by 28 October for versions of their tanks with a hybrid turret capable of mounting either a 105mm gun or a 120mm gun. In the interest of NATO standardization, provision was made for the use of the gunner's auxiliary sight from the Leopard II as well as metric fasteners at crew serviced interfaces. General Motors presented a version of their tank powered by the AGT-1500 gas turbine and Chrysler proposed an alternate design using the AVCR-1360 diesel engine. However, they recommended their turbine driven tank as the primary choice. All of the proposed tanks were fitted with British developed smoke grenade launchers.

As at the top of the page, the dimensions of the General Motors XM1 diesel tank are compared below and in the sketch at the right with those of the redesigned version powered by the gas turbine and fitted with three types of armament.

	Dimensions (inches)		
	A	B	C
Basic GM XM1 105mm (diesel)	95.0	108.5	113.15
Standardized GM XM1 105mm (turbine)	96.05	109.55	114.20
Standardized GM XM1 FRG 120mm (turbine)	96.05	109.55	114.20
Standardized GM XM1 UK 120mm (turbine)	96.05	109.55	114.20

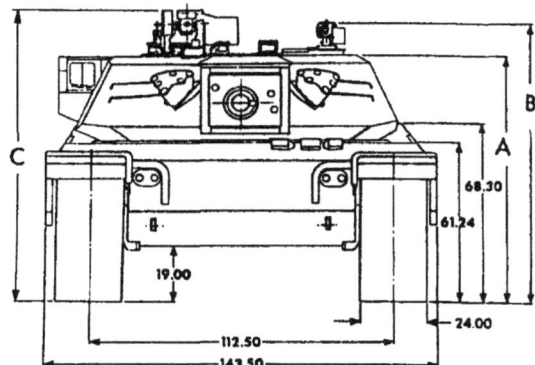

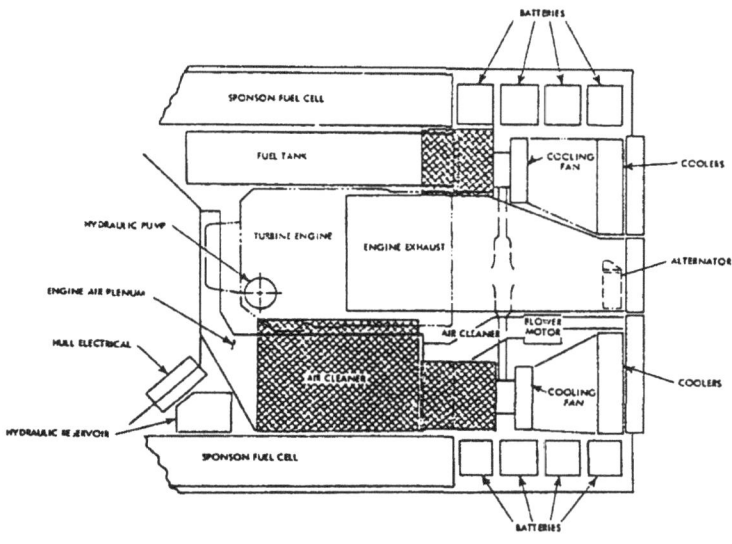

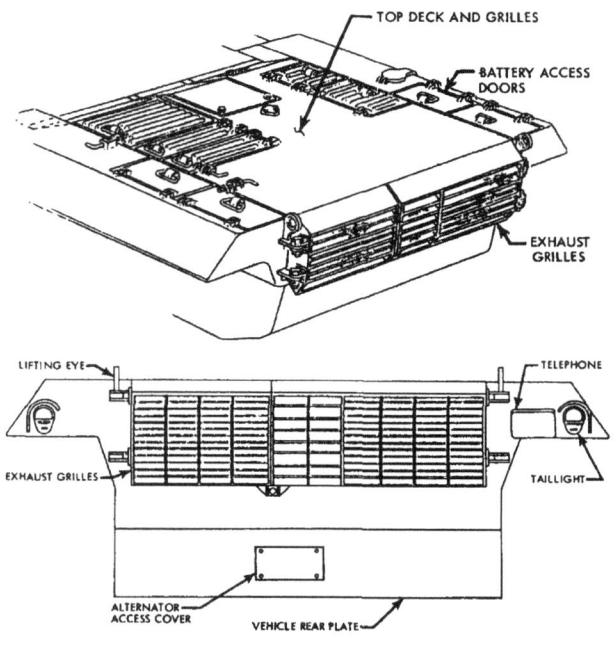

Changes in the General Motors XM1 when powered by the AGT-1500 gas turbine are shown in these drawings. The power pack itself appears at the right. Below, the dimensions and silhouette area are compared between the basic diesel powered General Motors XM1, the turbine powered version with the 105mm gun, and the M60A3 tank.

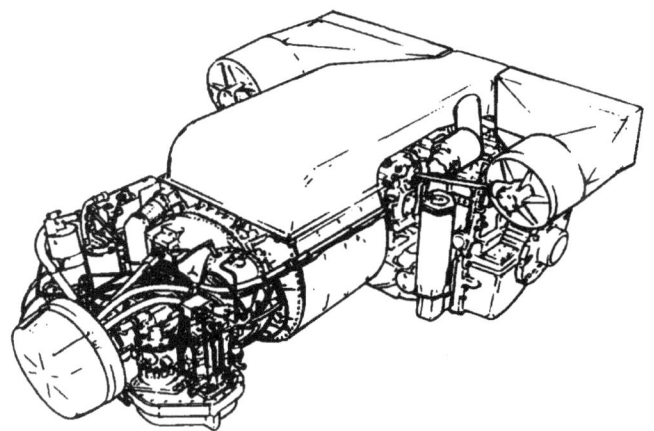

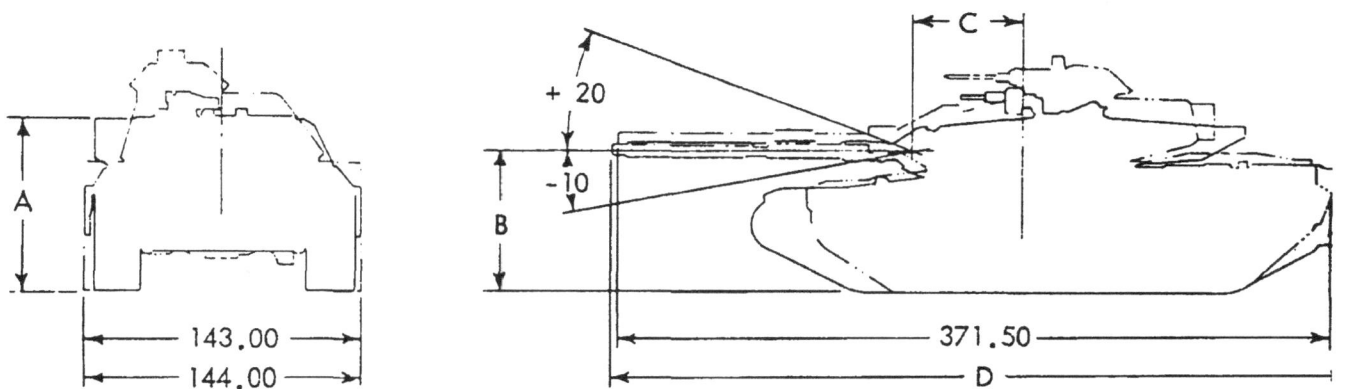

DIMENSIONS

	Basic GM XM1	Standardization GM XM1
A	95.0	96.05
B	77.0	78.50
C	50.0	50.00
D	382.0	382.50

SILHOUETTE AREA (square feet)

	Frontal	Side
Basic GM XM1	77.7	163.8
Standardization GM XM1	78.6	170.1
M60A3	86.1	166.0

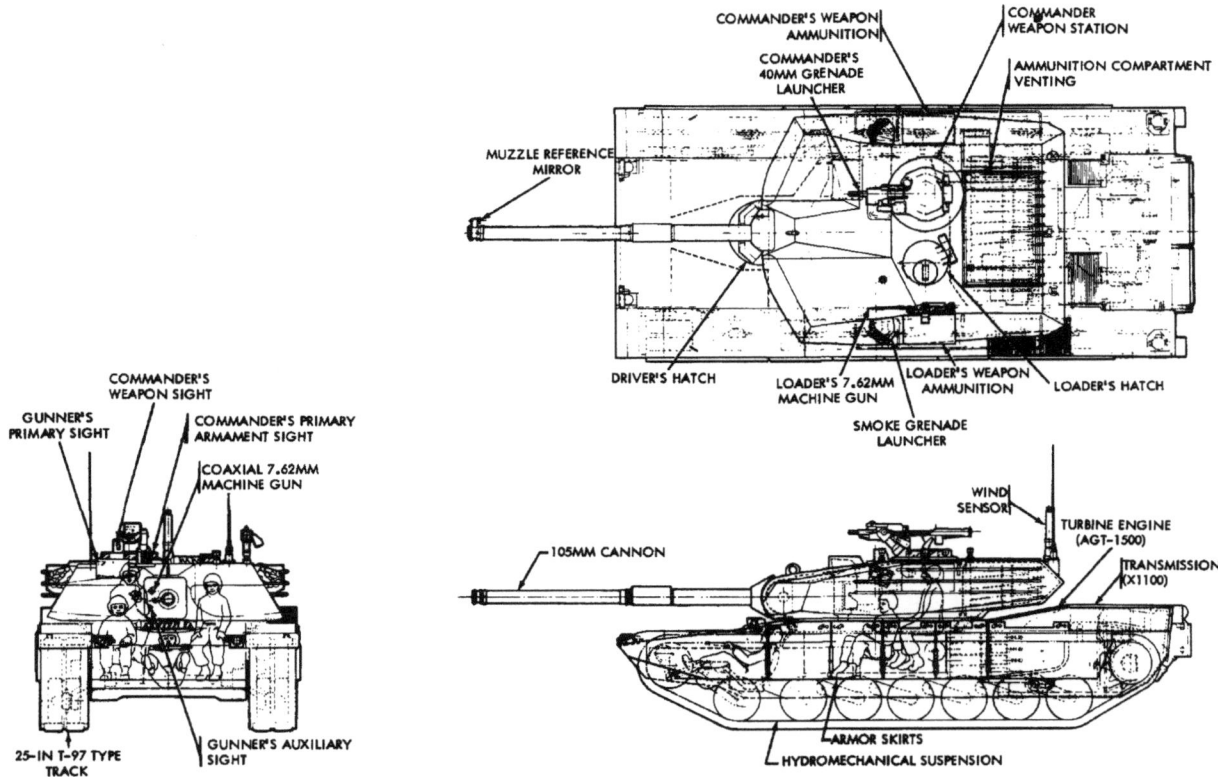

COMMANDER'S WEAPON AMMUNITION
COMMANDER WEAPON STATION
COMMANDER'S 40MM GRENADE LAUNCHER
AMMUNITION COMPARTMENT VENTING
MUZZLE REFERENCE MIRROR
DRIVER'S HATCH
LOADER'S 7.62MM MACHINE GUN
LOADER'S WEAPON AMMUNITION
LOADER'S HATCH
SMOKE GRENADE LAUNCHER

COMMANDER'S WEAPON SIGHT
GUNNER'S PRIMARY SIGHT
COMMANDER'S PRIMARY ARMAMENT SIGHT
COAXIAL 7.62MM MACHINE GUN
25-IN T-97 TYPE TRACK
GUNNER'S AUXILIARY SIGHT

WIND SENSOR
TURBINE ENGINE (AGT-1500)
TRANSMISSION (X1100)
105MM CANNON
ARMOR SKIRTS
HYDROMECHANICAL SUSPENSION

The early (above) and late (below) Chrysler proposal drawings for the full scale engineering development program (FSED) can be seen here. Compare the new special armor gun shield below with the cast version in the top drawing.

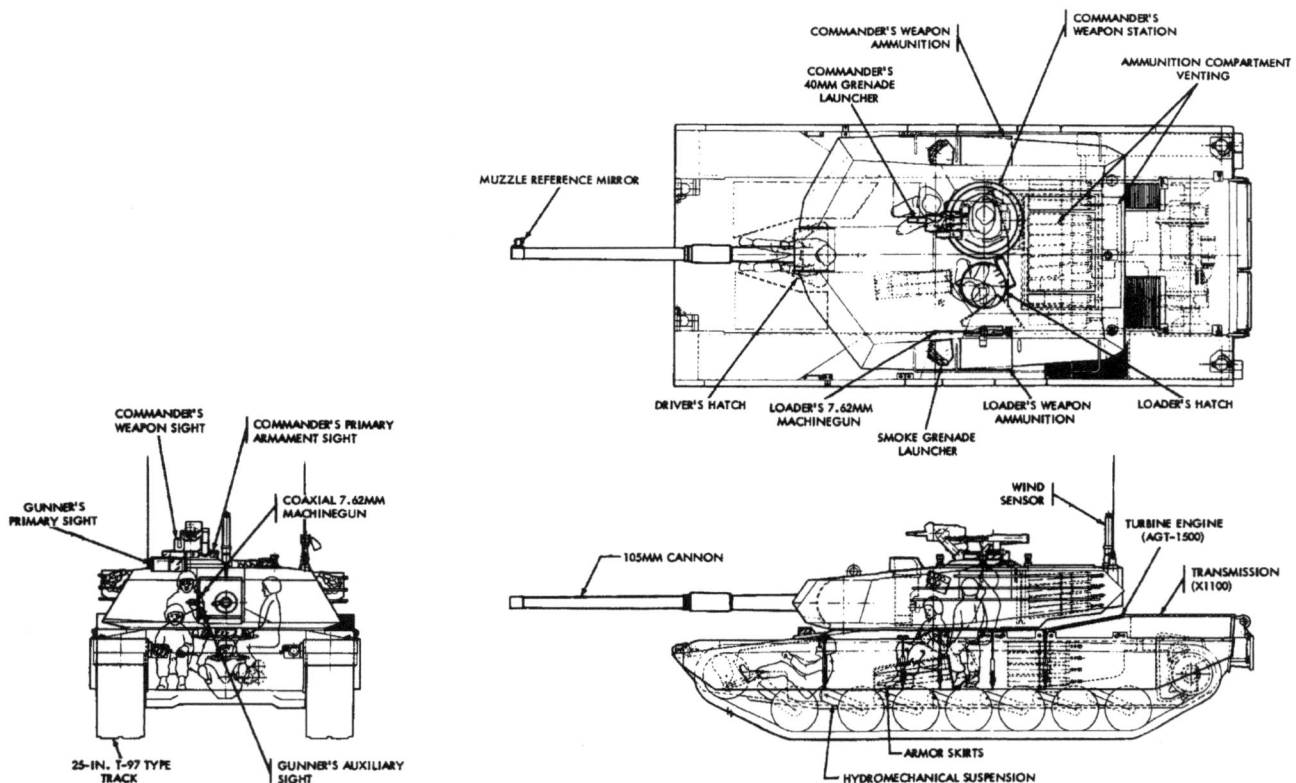

COMMANDER'S WEAPON AMMUNITION
COMMANDER'S WEAPON STATION
COMMANDER'S 40MM GRENADE LAUNCHER
AMMUNITION COMPARTMENT VENTING
MUZZLE REFERENCE MIRROR
DRIVER'S HATCH
LOADER'S 7.62MM MACHINEGUN
LOADER'S WEAPON AMMUNITION
LOADER'S HATCH
SMOKE GRENADE LAUNCHER

COMMANDER'S WEAPON SIGHT
COMMANDER'S PRIMARY ARMAMENT SIGHT
GUNNER'S PRIMARY SIGHT
COAXIAL 7.62MM MACHINEGUN
25-IN. T-97 TYPE TRACK
GUNNER'S AUXILIARY SIGHT

WIND SENSOR
TURBINE ENGINE (AGT-1500)
TRANSMISSION (X1100)
105MM CANNON
ARMOR SKIRTS
HYDROMECHANICAL SUSPENSION

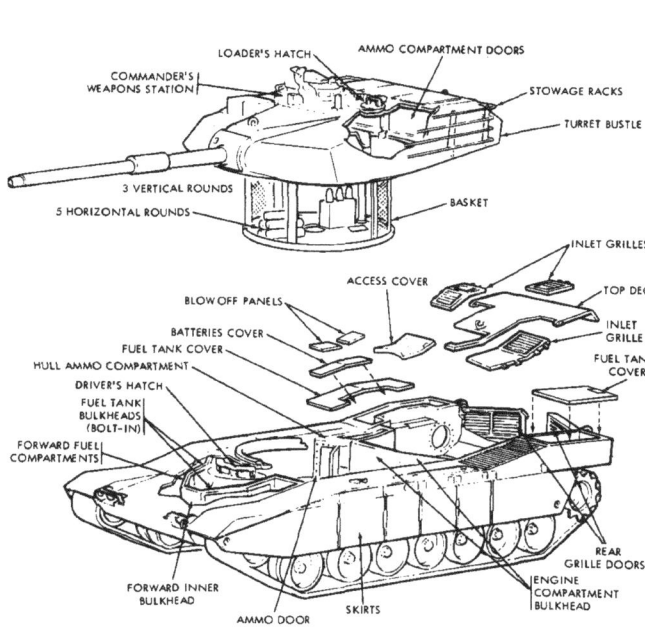

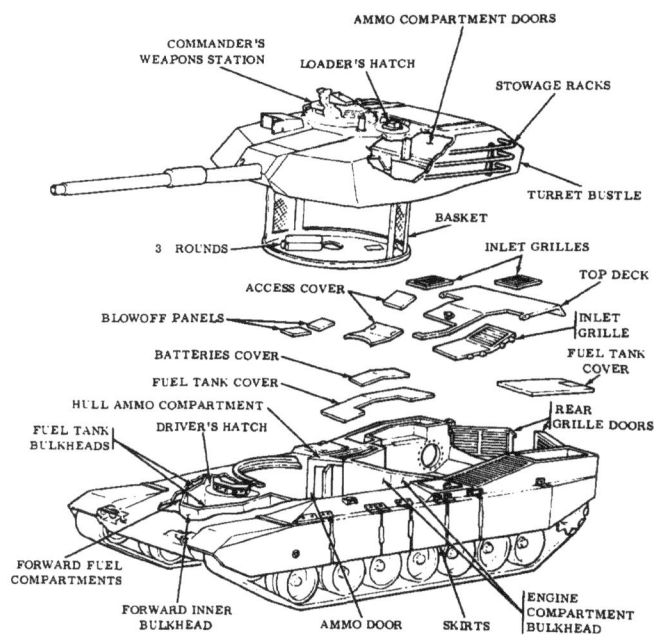

Various components in the early (left) and late (right) Chrysler full scale engineering development proposals can be compared above. The changes are listed in the sketch below at the right.

Prior to the submission of the final proposals on 28 October, Chrysler made a number of additional changes to enhance the survivability of their candidate and to reduce the cost. The arrangement of the special armor was modified and a special armor gun shield replaced the casting on the earlier proposal. A new minilaser range finder was incorporated and an extension from the gunner's primary sight was provided for the tank commander removing the need for a separate sight. Other cost reductions included the elimination of the power elevation for the commander's weapon and the driver's built in test equipment (BITE) panel. The latter was replaced by pop-up indicators and other gages in the engine compartment and at the crew stations.

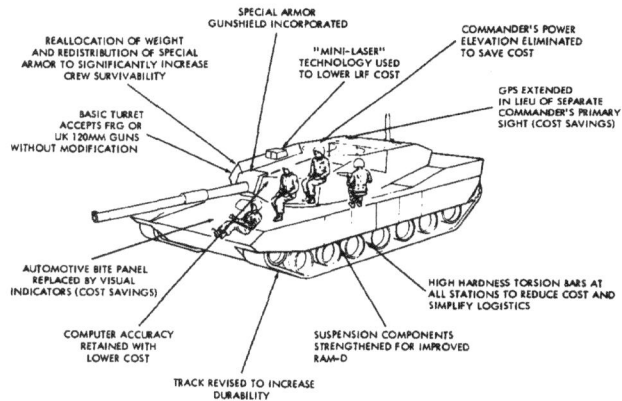

Below, the sketches show the fire control equipment for the early (left) and late (right) Chrysler proposals. Note the change in the commander's primary armament sight.

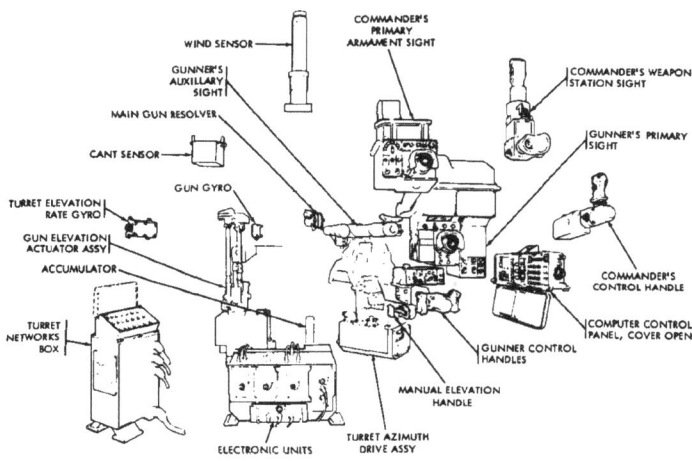

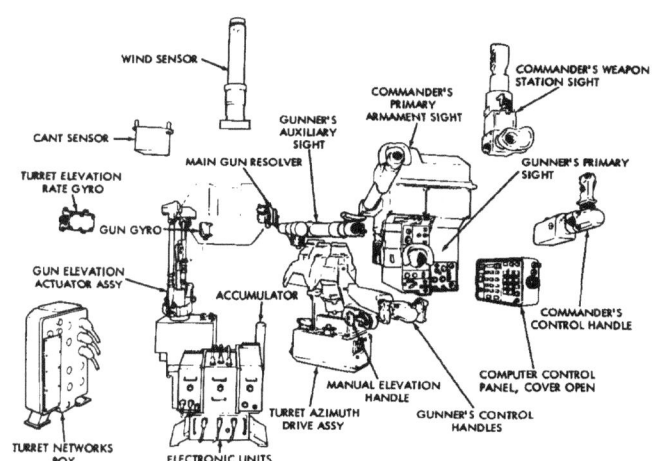

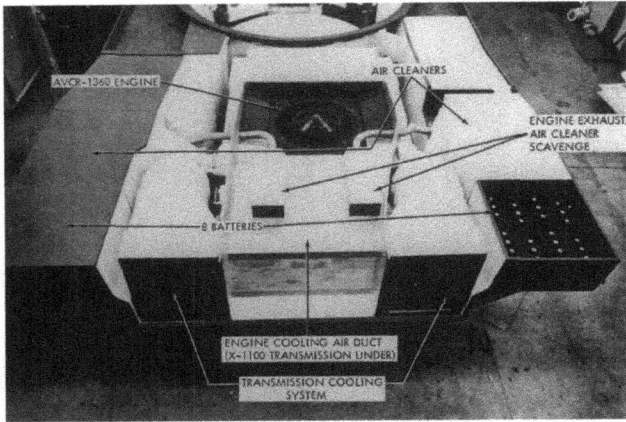

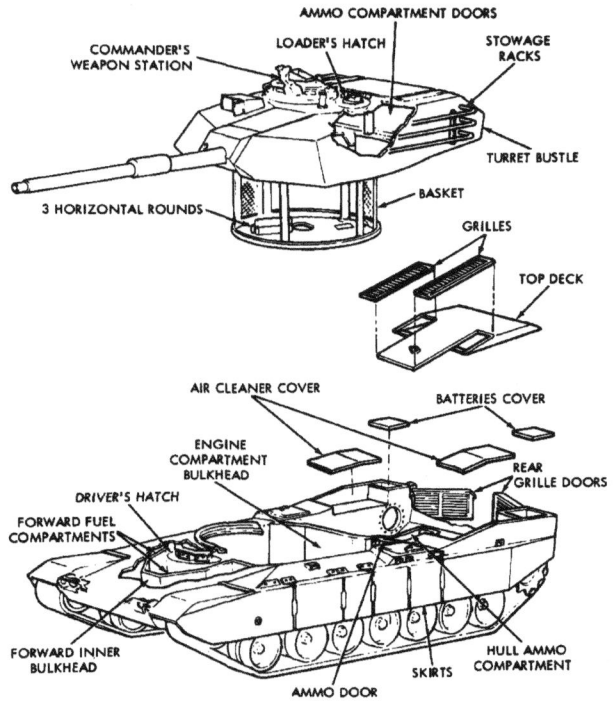

Above, the mock-up for the installation of the AVCR-1360 diesel engine in the Chrysler XM1 can be seen in the photograph. The sketch at the right shows the hull and turret components of the proposed Chrysler diesel powered tank.

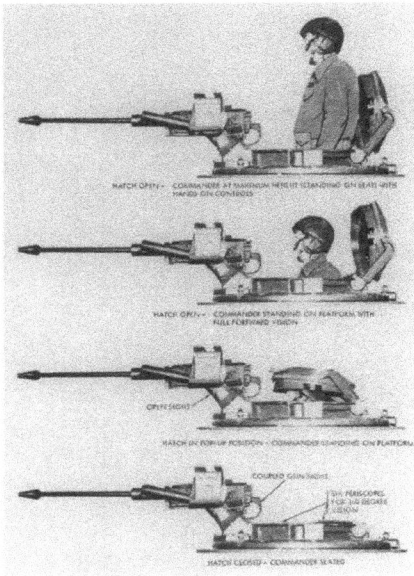

The various positions of the tank commander's hatch cover on the Chrysler XM1 prototype are illustrated at the left. This version is armed with a .50 caliber M85 machine gun. In the sketch of the hatch below, the commander is provided with a 40mm grenade launcher.

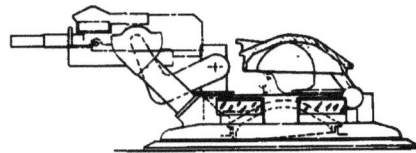

Below, models of the Chrysler full scale engineering development (FSED) tank are armed with the M68 105mm gun at the left and the German 120mm smooth bore cannon at the right.

The full size mock-up of the XM1 FSED tank appears above. A .50 caliber M2 machine gun and a 7.62mm M240 machine gun are now provided for the tank commander and the loader respectively.

FULL SCALE ENGINEERING DEVELOPMENT

At a news conference on 12 November 1976, it was announced that Chrysler had won the FSED/PEP contract. Published reports indicated that the cost reduction efforts at Chrysler had reduced their bid to 196 million dollars compared to a 232 million dollar bid from General Motors. The increase in the latter was, no doubt, partially due to the higher initial cost of the AGT-1500 turbine compared to the AVCR-1360 diesel engine. However, Chrysler's diesel powered proposal for the FSED/PEP program was priced at 186 million dollars. The five month delay in awarding the contract had permitted further refinement of the design

and better determination of the manufacturing costs. For example, the design to unit cost for Chrysler's tank decreased from $574,000 for the original FSED version to $422,000 for the final turbine powered tank proposal. The unit cost for Chrysler's diesel tank design was $376,000. All of these prices were in 1972 dollars for comparison with the earlier estimates. The actual unit price of the winning tank was expected to be about $754,000 in 1976 dollars.

Under the 36 month FSED/PEP contract, Chrysler was to manufacture 11 new pilot tanks incorporating the latest changes resulting from the validation phase tests and

Below at the left, work continues on the full size XM1 FSED mock-up and, at the right, the assembly of the pilot tanks is in progress at the Detroit Army Tank Plant.

These photographs were taken at the rollout of the first XM1 pilot tank at Detroit in February 1978. In the view above, the two center figures are Dr. P. W. Lett (left) and Eugene Trapp (right).

The first XM1 FSED pilot is shown above and below after the rollout in Detroit.

other experience. The validation phase prototype and the automotive test rig also were to be refurbished for use in the test program. Evaluation of these vehicles would determine the details of the final design for low initial rate production (LIRP). All of the 11 new pilot tanks were manufactured at the Detroit Army Tank Plant with the first one being delivered in February 1978. The last of the 11 was completed the following July. Developmental Test II (DT-II) extended from February 1978 to September 1979 at Aberdeen Proving Ground. Operational Test II (OT-II) at Fort Bliss, Texas lasted from April 1978 to February 1979 with the tanks manned by the 2nd Squadron, 3rd Armored Cavalry Regiment. Three of the tanks, after testing at Fort Bliss, were refurbished at Detroit, fitted with the latest configuration hardware, and shipped to Fort Knox in June 1979 for additional durability and reliability testing. This program was conducted by H Company, 2nd Squadron, 6th Cavalry under the direction of the Armor and Engineer Board. During these tests, the reliability, availability, maintainability-durability (RAM-D) goal was 272 mean miles between failures (MMBF). The XM1 reached 326 MMBF. Initially, Chrysler personnel provided the maintenance, but in August the organizational maintenance was taken over by the soldiers of H Company for the remainder

The XM1 FSED pilot appears in these photographs with the cannon aimed forward and to the rear.

These photographs show additional details of the XM1 pilot tank. Note the muzzle reference device installed on the gun tube.

Side views of the XM1 pilot tank appear below. In the right-hand photograph the tank is climbing a vertical obstacle.

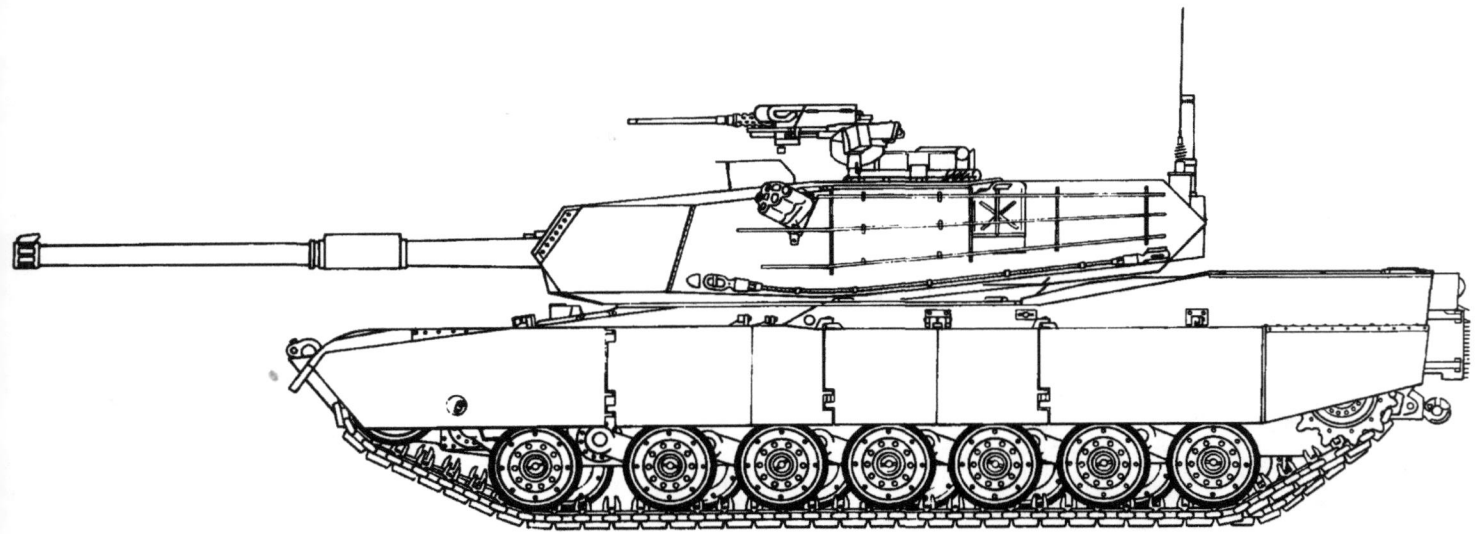

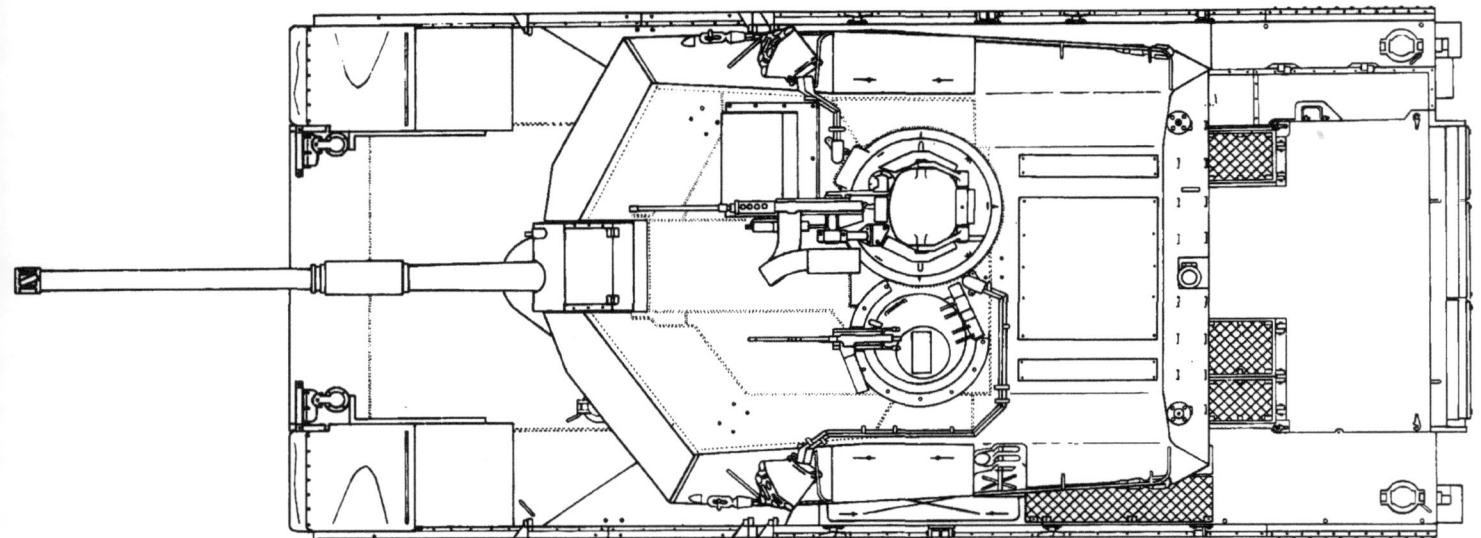

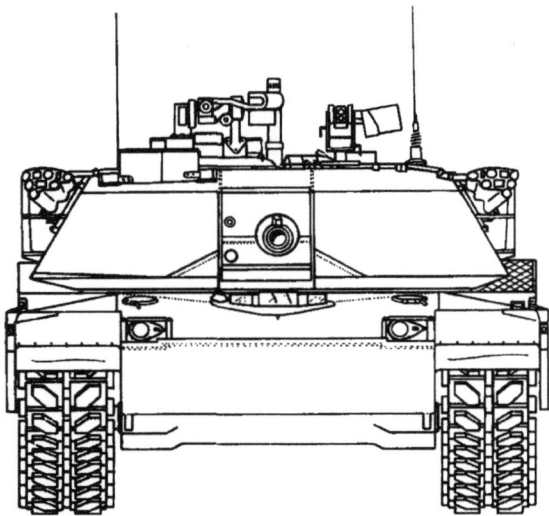

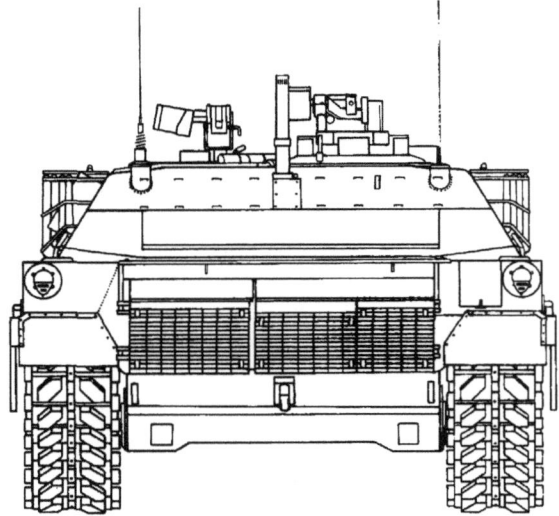

© D.P. Dyer

105mm Gun Tank XM1, FSED Pilot

Above, the XM1 pilot is operating at the Detroit Tank Plant. Below, a closeup of the tank commander's station is at the left and the tank is loaded on a railway flat car at the right.

of the test program. During the OT-II at Fort Bliss, many of the power train failures were caused by dust being ingested into the engine. After modification of the seals in the air filtration system, there were no further failures from this cause during all of the tests at Fort Knox.

The tests at Fort Bliss also revealed a tendency for the tracks to be thrown by the buildup of soil between the sprocket hub and the bottom of the rear sponson. Some tracks also were thrown by rapid backing and turning when changing firing positions. This problem was solved by the combination of some additional suspension hardware and the redistribution of the ground pressure with an increase in track tension. The hardware change consisted of a sprocket hub with discharge ports, a mud scraper, steel blocks welded to the lower rear hull to prevent inside throws, a steel plate mounted above the final drive to maintain track alignment, and a retaining ring on the outside of the sprocket. Satisfactory results were obtained with these modifications during further tests at Fort Bliss in October 1979.

Below, the XM1 is exposed to heavy dust clouds during the tests at Yuma, Arizona. At the right are the hardware modifications intended to prevent the track from being thrown.

Above and below are photographs of the first production XM1 tank delivered at the Lima Army Tank Plant on 28 February 1980. It was named "Thunderbolt", the same as the World War II Sherman tank commanded by General Creighton Abrams.

The survivability of the XM1 was demonstrated during DT-II at Aberdeen using pilot vehicle 11 (PV 11). The tank was fully loaded with fuel and ammunition with dummies placed in the various crew positions. Sensors were installed throughout the tank and, with the engine idling, it was fired upon using various types of ammunition at typical combat ranges. The tank was not destroyed and it was driven away under its own power after the tests. PV 11 also was subjected to a blast from an antitank mine which badly damaged the suspension. However, after short tracking one side of the vehicle, the tank was started and driven away from the test site.

On 7 May 1979, the XM1 was approved by the Secretary of Defense for low initial rate production (LIRP) at the newly equipped Lima Army Tank Plant. As mentioned previously, this was for 110 tanks with the first two being delivered at a special acceptance ceremony on 28 February 1980. At this time, the new tank was named the Abrams in honor of the late General Creighton W. Abrams, an outstanding armor commander and former Army Chief of Staff. The General's wife and three sons attended the ceremony and Mrs. Abrams christened the first production tank.

The LIRP tanks began developmental testing (DT-III) in March 1980 and it continued until September 1981. The major part was conducted at Aberdeen Proving Ground for RAM-D testing. Other work was at Yuma, Arizona for desert tests, Eglin Air Force Base, Florida and the Alaska

Below, an early production Abrams tank is on the test track at the Lima Army Tank Plant. Note the flexible cable step attached to the front side skirt section compared to the rigid metal step on the XM1 pilots.

Above, the 105mm gun in the turret of a new Abrams tank is being aligned at the Lima Army Tank Plant. Below, the XM1 undergoes low temperature testing in the cold room.

Cold Region Test Center for cold room and arctic environment testing, and the White Sands Missile Range, New Mexico for electromagnetic radiation and nuclear vulnerability evaluation. The operational tests (OT-III) started at both Fort Knox and Fort Hood during September 1980 and continued until May 1981. At Fort Hood, the tanks were operated and maintained by the 2/5th Cavalry and the 27th Maintenance Battalion. The new tank was enthusiastically received by the troops who greatly appreciated its speed and agility. The operational tests revealed that the cruising range of the production tank was about 20 to 25 miles less than the objective of 275 miles at 25 miles per hour. This was attributed partially to the suspension modifications made to reduce the track throwing experienced during DT/OT-II. However, it was noted that the turbine operated more efficiently at higher speeds and that an increase in the tank test speed from 25 to only 27 miles per hour produced an improvement in cruising range. Thus by operating at higher speeds, the required cruising range could be obtained.

The photographs below show the Abrams in the hands of the troops during its evaluation program.

Above, the stowage racks are being welded to the turret at the left and at the right is a view of the turret assembly line. Below, the hull is shown at the left before installation of the tracks and the turret is being fitted to the hull at the right.

In February 1981, production was approved for 7058 tanks at a rate of 30 per month. The vehicle was classified as standard on 17 February 1981 as the 105mm gun full tracked combat tank M1.

Production problems at AVCO Lycoming resulted in their failure to meet the delivery schedule and also produced a large number of defects in the engines that were delivered. Poor quality control and shortages of skilled personnel were the primary causes. Eventually these problems were solved and the number of defective engines was minimized as production increased. Early tank production also suffered from delays in obtaining the thermal imaging systems and from some fabrication difficulties with the hull and turret at the Lima plant. These problems also were solved and the production of the new tank soon reached the required level.

Below, the rear of the hull is seen during assembly (left) and the AGT-1500 gas turbine power packs await installation in the new tanks (right).

The welded hull structure can be seen above from the front (left) and the rear (right).

Unlike the M48 and M60 tank series, the M1 did not make use of large armor steel castings. The hull and turret were assembled by welding sections of rolled homogeneous steel armor plate to which the special armor was attached. These steel sections were cut to size using precision flame cutting equipment. The hull was divided into three sections with the crew compartment in the center separated from the fuel cells in the front and the engine compartment in the rear by two bulkheads. In addition to the two fuel cells adjacent to the driver's station, two were located in the engine compartment and two in the rear sponsons. A Halon fire extinguisher system with infrared sensors was installed in both the crew and engine compartments. It could detect a fire in 4-5 milliseconds and

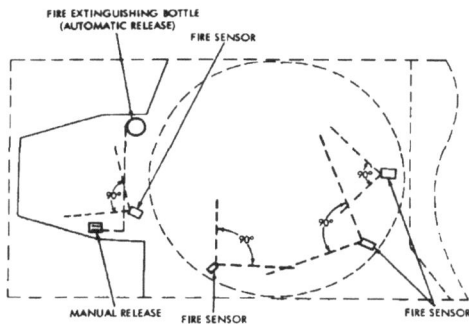

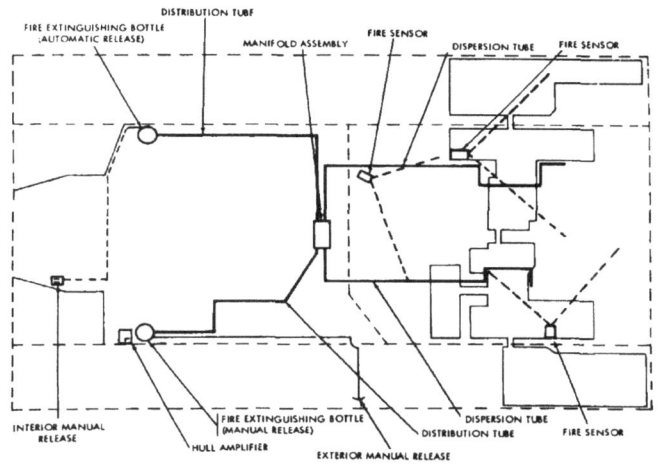

The view angle of the infrared fire sensors in the crew space (left) and the engine compartment (right) is shown in the sketches above. Below, a new production M1 tank is under test.

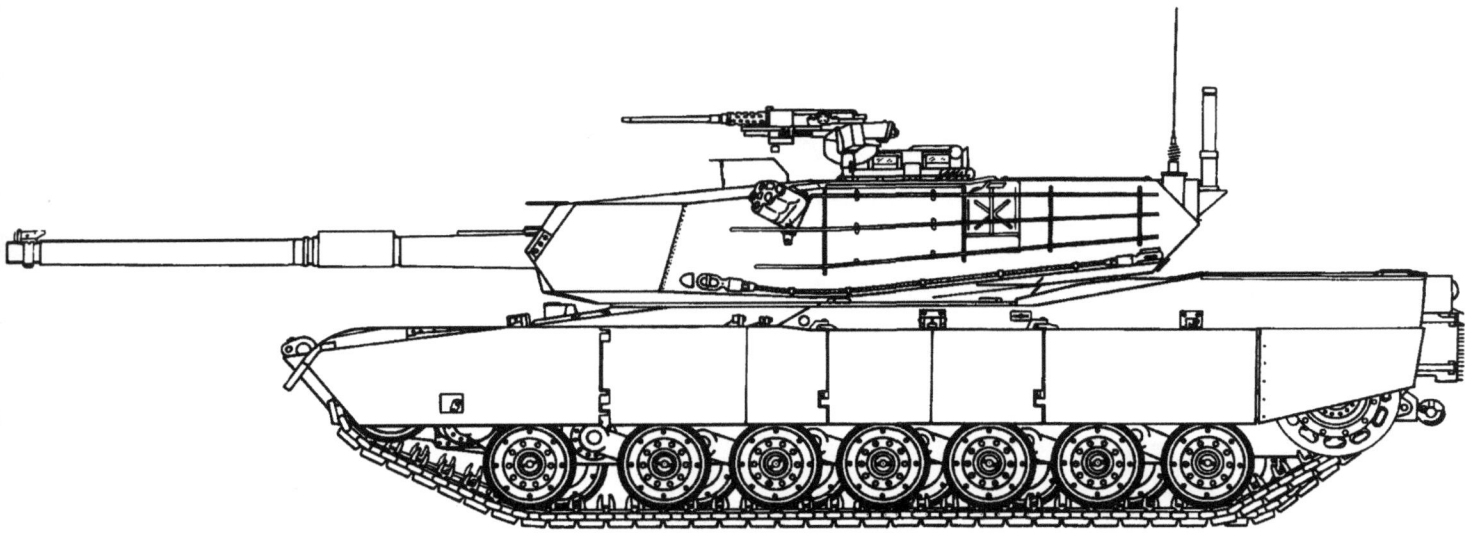

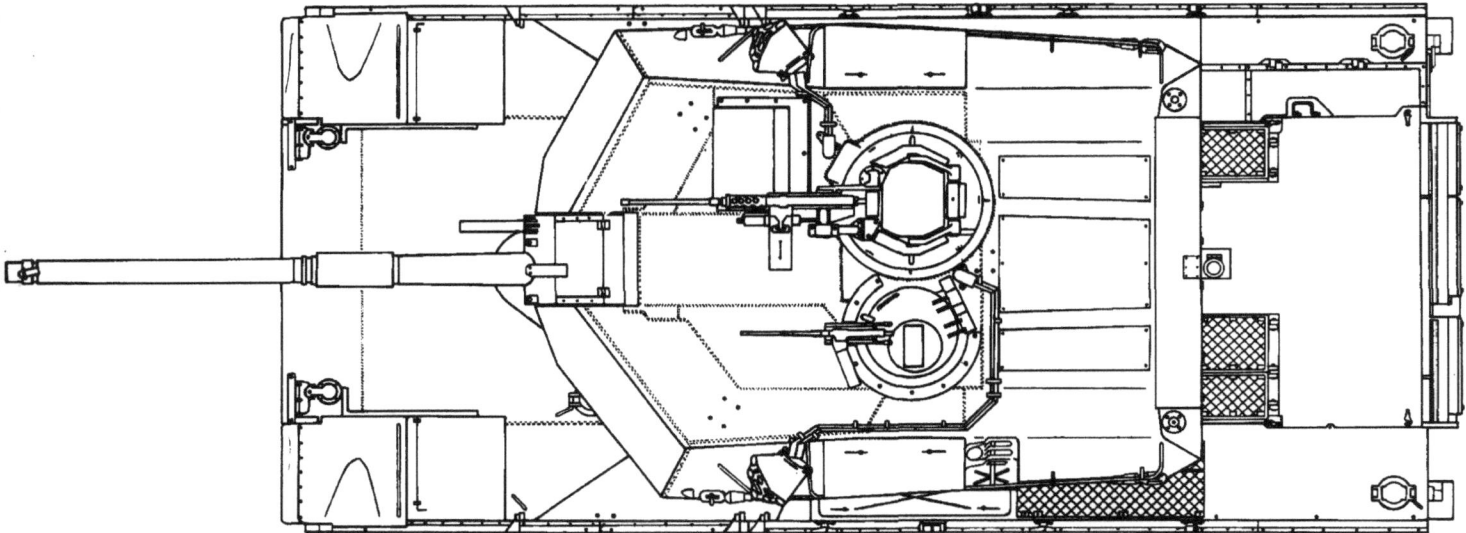

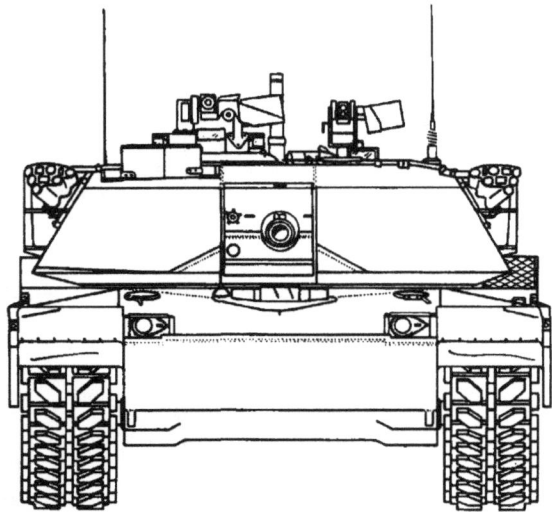

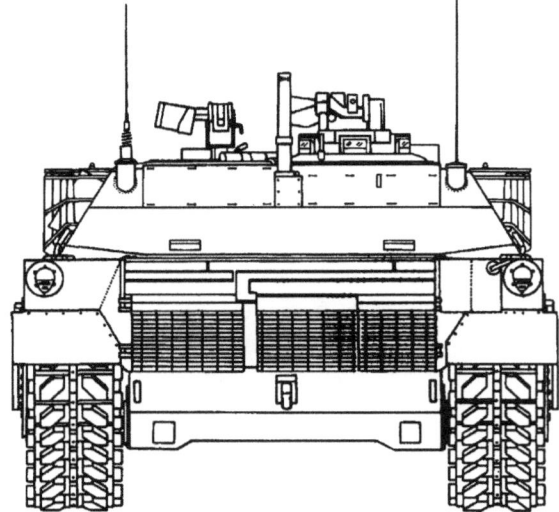

© D.P. Dyer

105mm Gun Tank M1

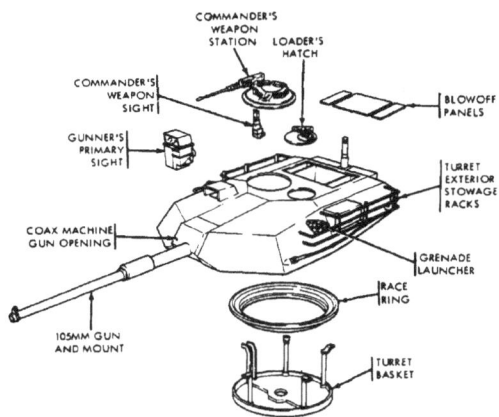

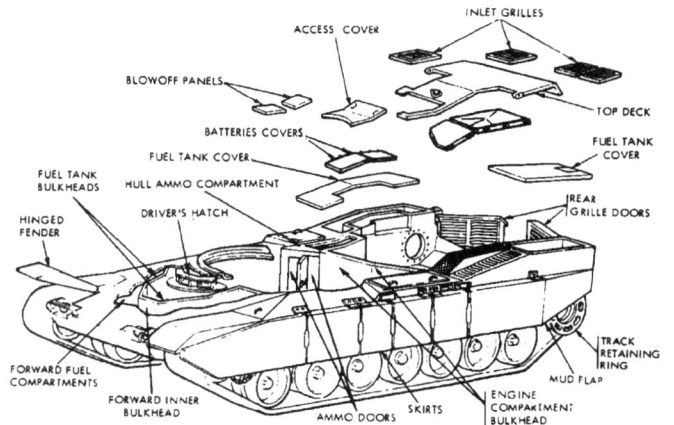

Above, various components of the production M1 tank are identified in the exploded views of the turret and hull. Below are schematic diagrams of the fuel system (left) and the CBR protection system (right).

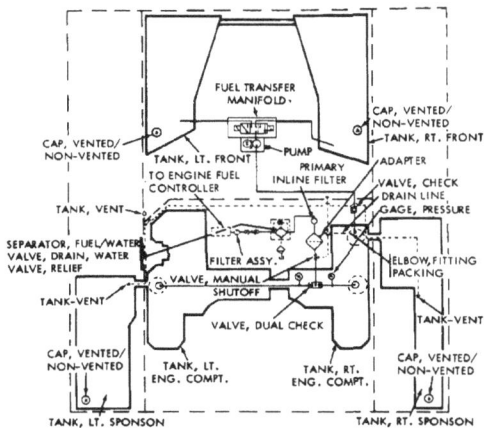

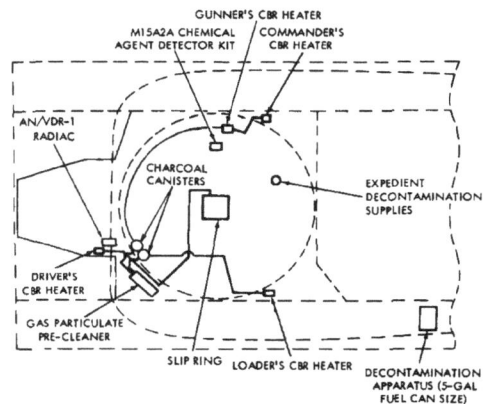

Below, the exterior components of the production M1 tank are identified in this three view drawing.

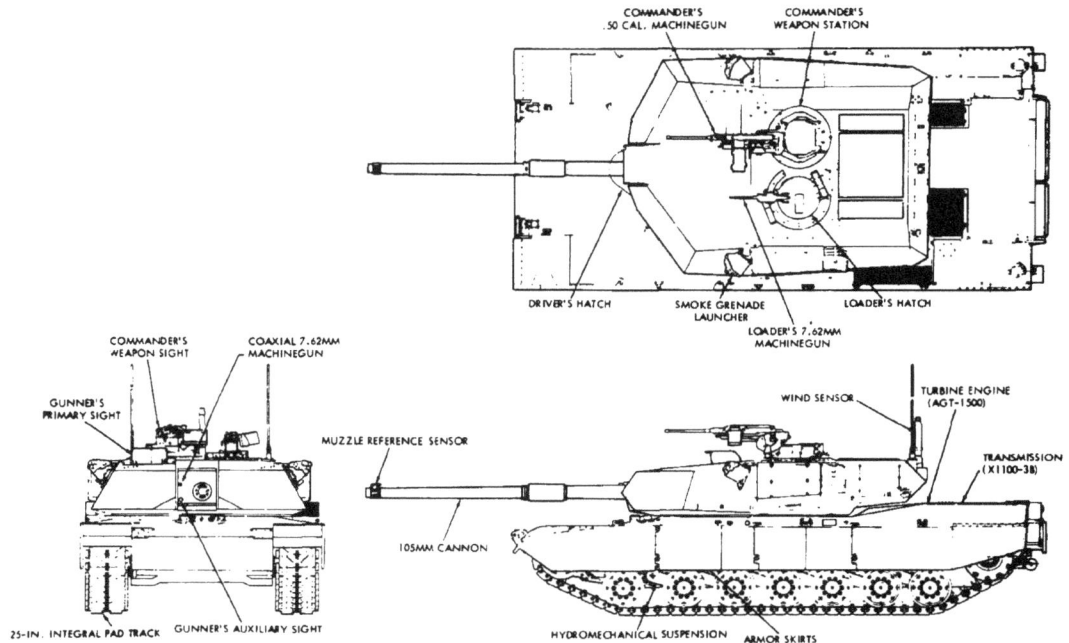

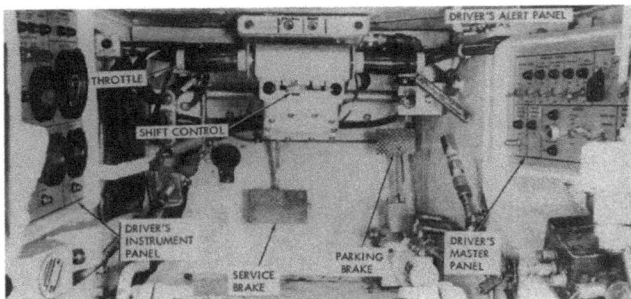

The front and right side of the driver's station can be seen in the two top photographs. The sketch at the lower right shows the driver's position with the hatch open and closed.

extinguish it in about two tenths of a second. The hatch above the driver's seat was fitted with three periscopes providing an overlapping field of vision exceeding 120 degrees. Like on the XM803, there was no floor escape hatch. The driver's T-bar steering control incorporated motorcycle type hand throttles eliminating the need for a floor accelerator pedal thus providing more foot space. The transmission shift selector and push buttons for the intercom system also were located on the T-bar assembly. The service brake pedal was in the center of the floor allowing braking with either foot. The driver's seat was adjustable for driving with the hatch either open or closed and the seat height could be set to four different positions. A large knob on an adjustable lower lumbar support allowed the driver to control the amount of lower back support.

The aluminum turret basket floor was suspended from five posts, two of which mounted the seats for the tank commander and the loader. A door in the basket floor provided access to components in the bottom of the hull. The commander's station permitted observation in the seated position through both the extension from the gunner's primary sight and the sight for the .50 caliber machine gun. Six periscopes permitted 360 degree vision. The three positions of the commander's platform allowed him to observe standing with the hatch in the protected open position or at two heights with the hatch fully open. The commander's .50 caliber machine gun was electrically powered in azimuth, but it was elevated manually.

The loader at the left rear of the cannon had an eight position swiveling seat that was adjustable in the vertical direction. An eight round armored hull ammunition compartment was located on the right side of the tank behind the engine compartment bulkhead. Blowout panels in the top and bottom of the hull vented any ammunition explosion to the outside and away from the crew. Fortyfour 105mm rounds of ammunition were stowed in the turret bustle behind sliding armor plate doors. Twentytwo of these rounds could be reached by the loader without leaving his seat. A knee switch operated the bustle compartment door

At the right is a view of the turret basket before being attached to the turret.

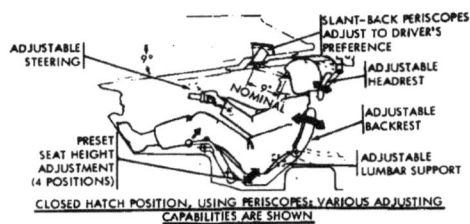

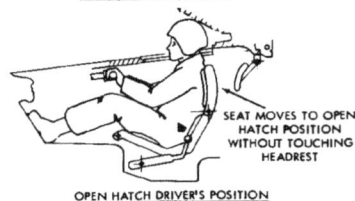

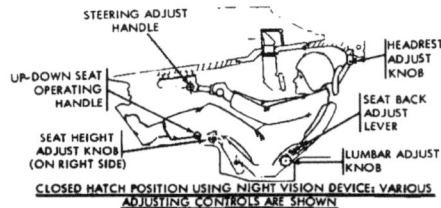

208

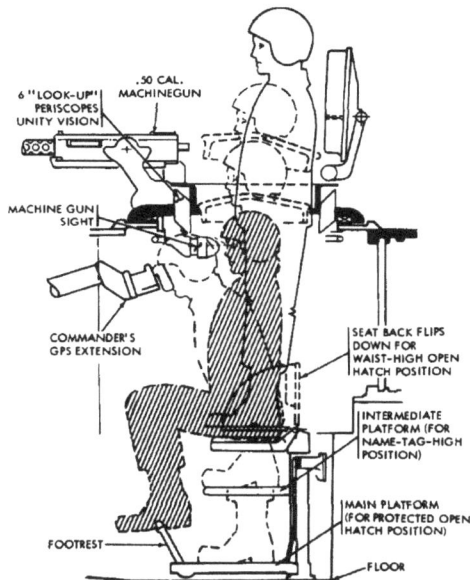

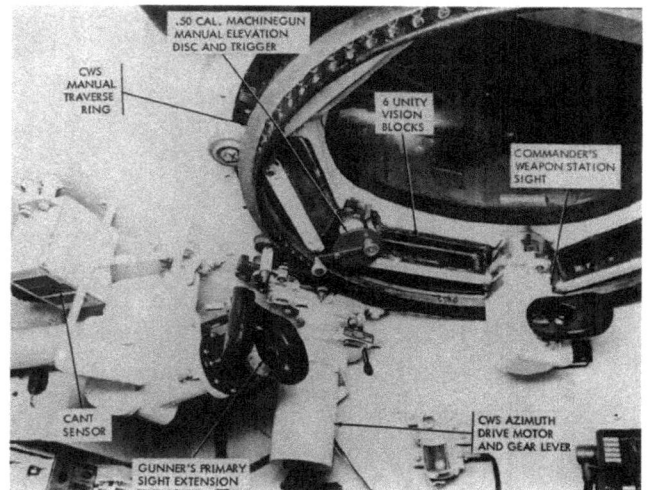

The various positions for the tank commander are sketched at the left and an interior photograph of his station is above.

covering these 22 ready rounds on the left side of the turret. The door also could be operated manually in an emergency. Blowout panels in the bustle roof vented any ammunition explosion to the outside. Three additional 105mm rounds were stowed on the turret basket floor to the left of the cannon in spall protection covers. This brought the total 105mm ammunition stowage to 55 rounds.

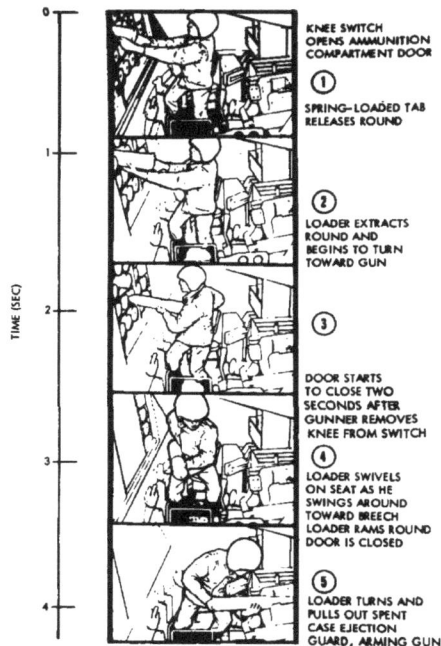

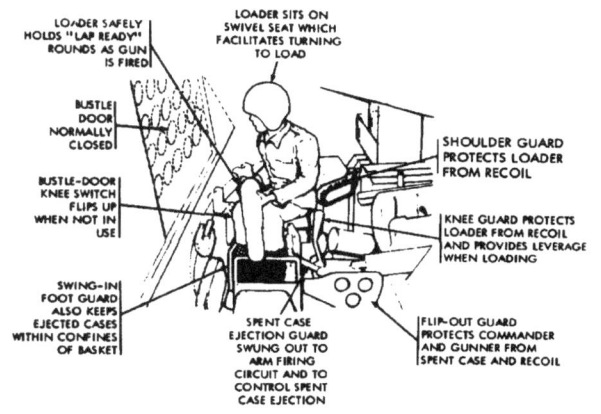

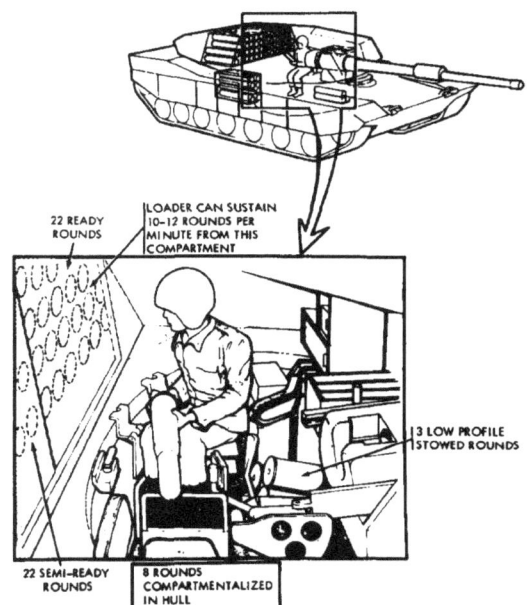

The loading sequence and the ammunition stowage is illustrated in the sketches above and at the right.

The machine gun has not been installed in the view of the loader's hatch above. The 7.62mm M240 machine gun appears at the top right and it is mounted at the loader's hatch in the sketch at the right.

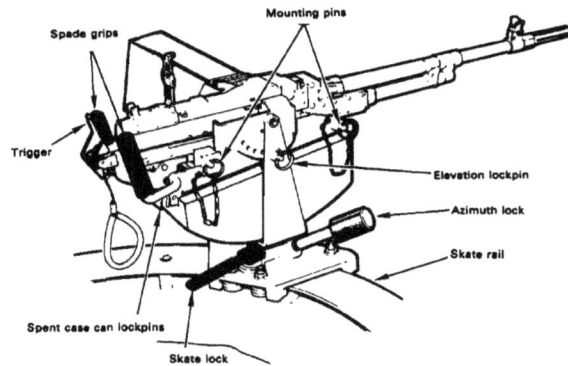

The loader's hatch cover was fitted with a rotatable periscope mount. A 7.62mm M240 machine gun was skate mounted around the outer edge of the loader's hatch. Another 7.62mm M240 machine gun was installed coaxially on the right side of the 105mm gun M68E1 above the gunner's auxiliary sight.

The gunner's seat was pedestal mounted on the turret basket floor and it was adjustable in both the vertical and horizontal directions. No adjustments were required when the gunner shifted from the primary sight to the auxiliary sight. The gunner's primary sight with its extension for the tank commander was the main optical sighting instrument. It also included the laser range finder, the thermal vision system, and the gyro stabilized line of sight platform. The sight was protected by an armor steel cover with doors operable from within the turret. The gunner's auxiliary sight was an eight power articulated telescope installed in the right side of the gun mount below the coaxial machine gun. The heart of the fire control system was the digital computer which received signals from the sensors as well as inputs from various controls. Data from the laser range finder, the cant sensor, and the wind sensor were received automatically, but other inputs were entered into the computer manually. The gunner used the muzzle reference system to determine tube bend corrections which were then entered into the computer. Atmospheric temperature and pressure as well as ammunition temperature and tube wear also were manual inputs to the computer.

The left front of the turret interior appears at the right showing the 105mm gun and the loader's station.

The 105mm gun M68E1 can be seen in its mount below.

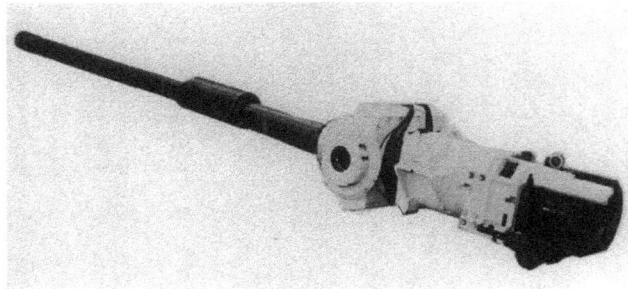

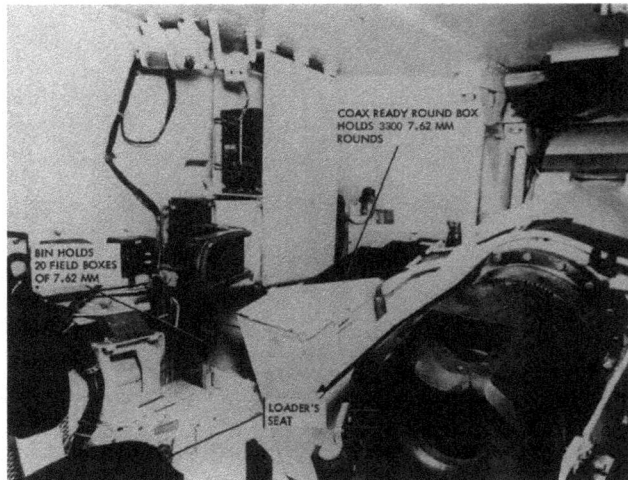

210

Top-left sketch labels:

GUNNER BENDS WRAP AROUND BROW PAD TO FIT FACE

COMPUTER PANEL AT GUNNER'S IMMEDIATE RIGHT

GUNNER ADJUSTS CHEST REST AS FAR REARWARD AS IS COMFORTABLE

GUNNER LEANS INTO AND PULLS AGAINST CHEST REST TO HOLD TORSO SOLIDLY IN POSITION. THIS WILL MINIMIZE SENSITIVITY TO VEHICLE VIBRATIONS AND SEVERE JOLTS.

GUNNER'S CONTROLS INTEGRAL WITH SIGHT BODY

GUNNER PULLS ON HANDLES

GUNNER WEDGES HIMSELF IN

RIGHT KNEE GUARD

LEFT KNEE GUARD

GUNNER ADJUSTS SEAT VERTICALLY AND HORIZONTALLY IN CONJUNCTION WITH BEST CHEST REST POSITION

FOOT REST

REMOTE INTERCOM SWITCH

NOTE: TILT FOOT ONTO FOOT REST BUT NOT ON REMOTE INTERCOMM SWITCH

Top-right photograph labels:

HYDRAULIC SYSTEM PRESSURE GAGE

GUNNER'S PRIMARY SIGHT

COMPUTER CONTROL PANEL

GUNNER'S AUXILIARY SIGHT

GUNNER'S CONTROL HANDLES

7.62 MM COAX MACHINEGUN

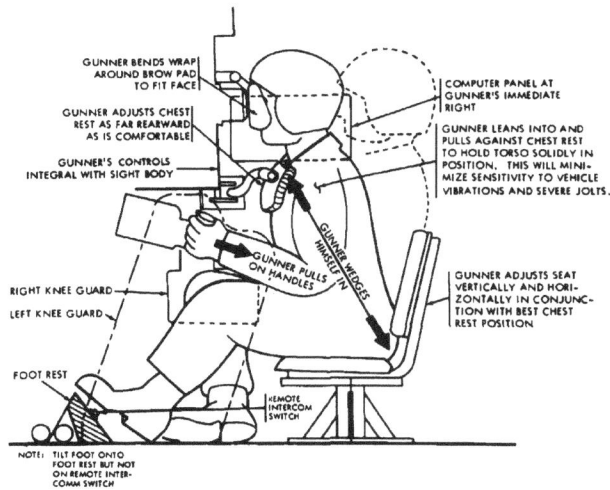

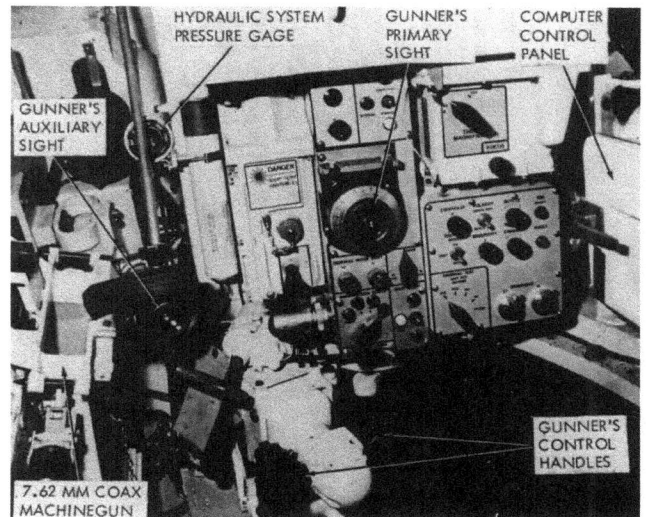

Above is a sketch and a photograph of the gunner's station. Below, the fire control system is diagramed at the left and its various components are illustrated at the right.

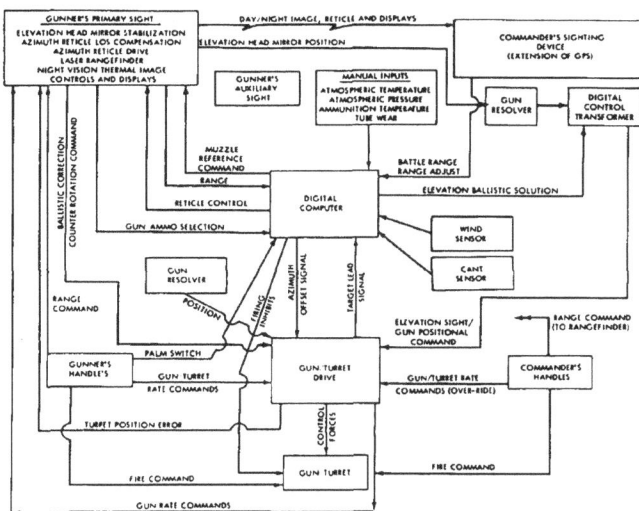

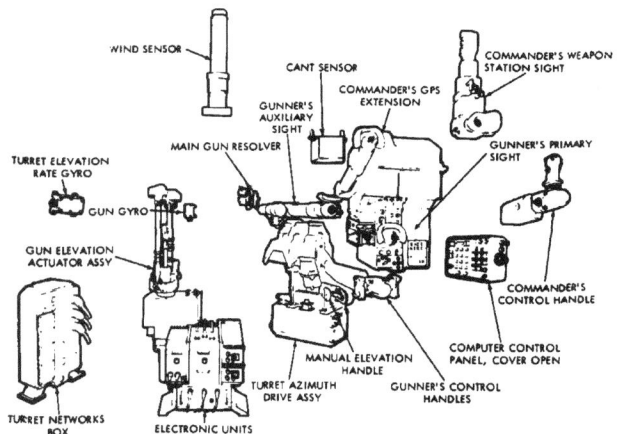

Fire control system diagram (left):

GUNNER'S PRIMARY SIGHT — ELEVATION HEAD MIRROR STABILIZATION, AZIMUTH RETICLE LOS COMPENSATION, AZIMUTH RETICLE DRIVE, LASER RANGEFINDER, NIGHT VISION THERMAL IMAGE, CONTROLS AND DISPLAYS

DAY/NIGHT IMAGE, RETICLE AND DISPLAYS

ELEVATION HEAD MIRROR POSITION

COMMANDER'S SIGHTING DEVICE (EXTENSION OF GPS)

GUNNER'S AUXILIARY SIGHT

MANUAL INPUTS — ATMOSPHERIC TEMPERATURE, ATMOSPHERIC PRESSURE, AMMUNITION TEMPERATURE, TUBE WEAR

GUN RESOLVER

DIGITAL CONTROL TRANSFORMER

BALLISTIC CORRECTION

COUNTER ROTATION COMMAND

MUZZLE REFERENCE COMMAND

BATTLE RANGE RANGE ADJUST

RANGE

ELEVATION BALLISTIC SOLUTION

RETICLE CONTROL

DIGITAL COMPUTER

GUN AMMO SELECTION

WIND SENSOR

GUN RESOLVER

CANT SENSOR

POSITION

FIRING COMMAND

AZIMUTH OFFSET SIGNAL

TARGET LEAD SIGNAL

RANGE COMMAND

ELEVATION SIGHT/GUN POSITIONAL COMMAND

RANGE COMMAND (TO RANGEFINDER)

PALM SWITCH

GUNNER'S HANDLE'S

GUN TURRET RATE COMMANDS

GUN TURRET DRIVE

GUN/TURRET RATE COMMANDS (OVER-RIDE)

COMMANDER'S HANDLES

TURRET POSITION ERROR

CONTROL FORCES

FIRE COMMAND

GUN TURRET

FIRE COMMAND

GUN RATE COMMANDS

Components illustration (right):

WIND SENSOR

CANT SENSOR

COMMANDER'S GPS EXTENSION

COMMANDER'S WEAPON STATION SIGHT

GUNNER'S AUXILIARY SIGHT

GUNNER'S PRIMARY SIGHT

MAIN GUN RESOLVER

TURRET ELEVATION RATE GYRO

GUN GYRO

GUN ELEVATION ACTUATOR ASSY

COMMANDER'S CONTROL HANDLE

COMPUTER CONTROL PANEL, COVER OPEN

MANUAL ELEVATION HANDLE

GUNNER'S CONTROL HANDLES

TURRET AZIMUTH DRIVE ASSY

TURRET NETWORKS BOX

ELECTRONIC UNITS

Below, a view of the gunner's auxiliary sight is at the left and the sketch at the right shows the tank commander's use of the gunner's primary sight, the unity sight, and the weapon station sight.

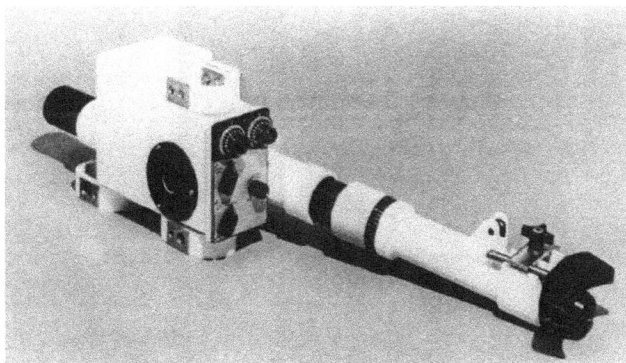

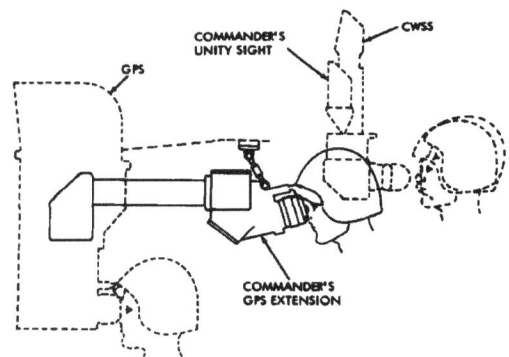

Sketch (right) labels:

GPS

COMMANDER'S UNITY SIGHT

CWSS

COMMANDER'S GPS EXTENSION

REDUCTION GEARING — RECUPERATOR

COMBUSTOR

HIGH PRESSURE COMPRESSOR

LOW PRESSURE COMPRESSOR

POWER TURBINE

LOW PRESSURE COMPRESSOR TURBINE

HIGH PRESSURE COMPRESSOR TURBINE

FINAL DRIVE

RANGE PACK

HYDROSTATIC STEER UNIT

BRAKE PACK

FINAL DRIVE

TORQUE CONVERTER

FRONT PTO's

PUMPS

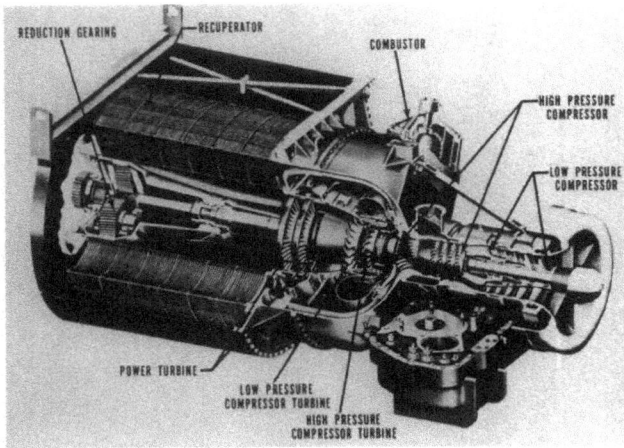

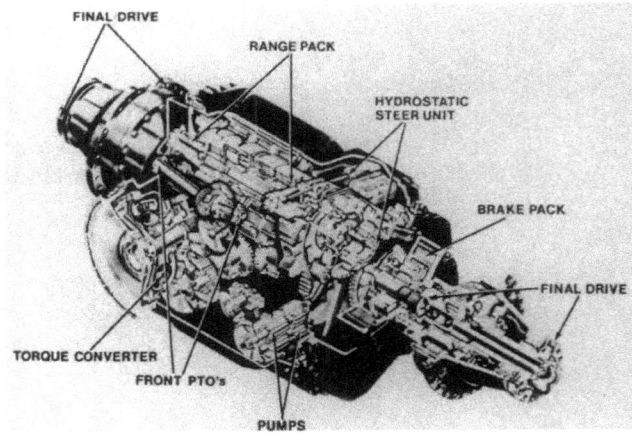

Above, cutaway drawings of the AGT-1500 gas turbine and the X-1100-3B transmission are at the left and right respectively. Below, the power pack is installed in the tank at the left and a schematic diagram of the X-1100-3B transmission is at the right.

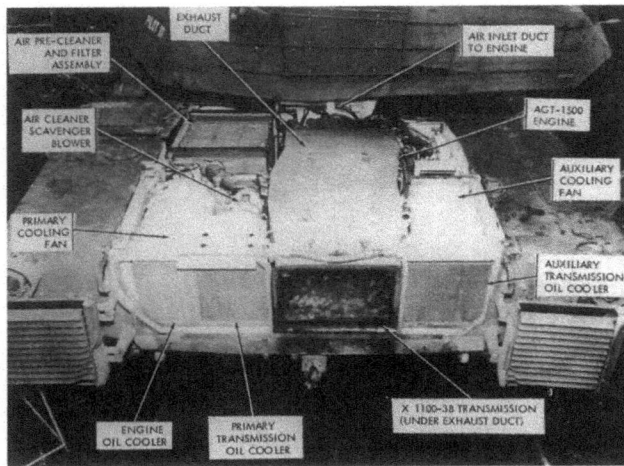

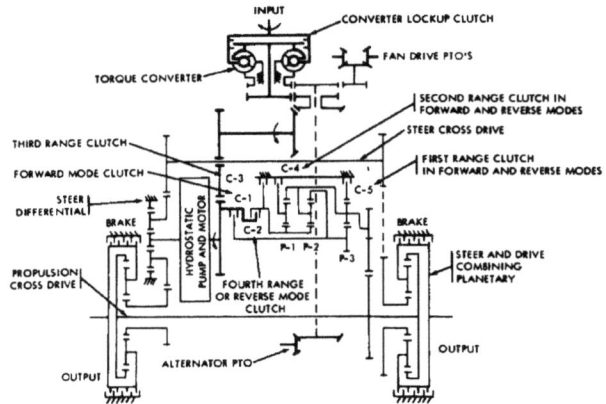

AIR PRE-CLEANER AND FILTER ASSEMBLY

EXHAUST DUCT

AIR INLET DUCT TO ENGINE

AIR CLEANER SCAVENGER BLOWER

AGT-1500 ENGINE

AUXILIARY COOLING FAN

PRIMARY COOLING FAN

AUXILIARY TRANSMISSION OIL COOLER

ENGINE OIL COOLER

PRIMARY TRANSMISSION OIL COOLER

X 1100-3B TRANSMISSION (UNDER EXHAUST DUCT)

INPUT

CONVERTER LOCKUP CLUTCH

TORQUE CONVERTER

FAN DRIVE PTO'S

SECOND RANGE CLUTCH IN FORWARD AND REVERSE MODES

STEER CROSS DRIVE

THIRD RANGE CLUTCH

FORWARD MODE CLUTCH

FIRST RANGE CLUTCH IN FORWARD AND REVERSE MODES

STEER DIFFERENTIAL

C-3

C-4

C-1

C-5

BRAKE

C-2

P-1 P-2

P-3

BRAKE

PROPULSION CROSS DRIVE

HYDROSTATIC PUMP AND MOTOR

STEER AND DRIVE COMBINING PLANETARY

FOURTH RANGE OR REVERSE MODE CLUTCH

OUTPUT

ALTERNATOR PTO

OUTPUT

The powerpack consisted of the AGT-1500 turbine engine, the X-1100-3B transmission, final drives, air cleaner, scavenging blower, and cooling system. The powerpack weighed about 8500 pounds and could be easily removed as a unit using a five ton wrecker. The sprockets at the rear of the vehicle drove the 25 inch wide tracks and the tank rode on seven dual road wheels per side. The 25 inch diameter compensating idler installed at the front of each track was interchangeable with the road wheels. It was mounted on a pivoting idler arm connected to the number one road wheel arm by a track adjusting link. Adjustment of the static track tension was by hand pumping grease into the link. When the proper tension was reached, a pressure relief valve opened releasing any additional grease.

A rear view of the cooling system for the power pack installed in the M1 tank appears at the right.

The vehicle was sprung by high strength steel torsion bars accessible from the outside on both sides of the tank. Thus, if a torsion bar fractured, both pieces could be more easily extracted. There were only two track support rollers per track compared to three on the validation phase prototype. The production tank was governed for a maximum level speed of 45 miles per hour and it could operate cross country at speeds up to 30 miles per hour.

PRIMARY TRANSMISSION OIL COOLER

ENGINE EXHAUST DUCT

ENGINE OIL COOLER

AUXILIARY TRANSMISSION OIL COOLER

SUPPORT ROLLER
SPROCKET
RETAINING RING
TRACK
MUD SCRAPER
ROADWHEEL
ROAD ARM
ROTARY SHOCK ABSORBER
TRACK ADJUSTING LINK ASSEMBLY
COMPENSATING IDLER WHEEL

ROADARM HOUSING
THREADED HOLE FOR INSTALLATION AND REMOVAL TOOL (BOTH ENDS)
ALUMINUM COVER TUBE
FACE SEALS
ROADARM/SHOCK ABSORBER HOUSING
BAR ACCESS (BOTH ENDS)
COVER TUBE RETAINER NUT
TORSION BAR (NO WRAPPING)

INTEGRAL PAD TRACK

Details of the M1 suspension are visible in the photograph at the left above with the side skirts removed. A closeup view of the track adjusting link assembly is at the top right and in the sketch below. The torsion bar and rotary shock absorber suspension unit is sketched at the left with a link of the T156 integral pad track at the lower left.

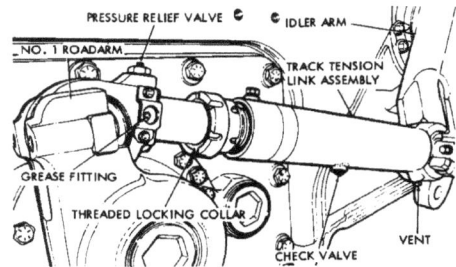

PRESSURE RELIEF VALVE
IDLER ARM
NO. 1 ROADARM
TRACK TENSION LINK ASSEMBLY
GREASE FITTING
THREADED LOCKING COLLAR
CHECK VALVE
VENT

#1 & #2 ROADARM POSITIONS DESTROYED
BUMP STOP #2 REMOVABLE
TRACK
ASSUMED MINE DAMAGE (LEFT SIDE)

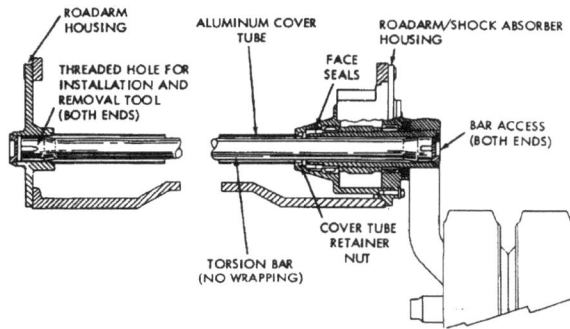

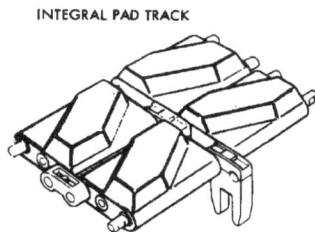

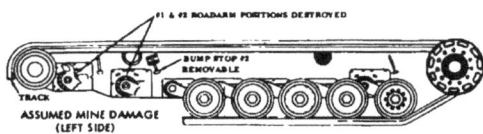

1. REMOVE FIRST TWO SKIRT PANELS (NOT SHOWN)
2. REMOVE BUMP STOP NO. 2 BY TAKING OUT FOUR BOLTS
3. A. REMOVE DAMAGED COMPONENTS (WHEELS, HUBS, ETC) FROM NUMBER 1 ARM.
   B. DISENGAGE NUMBER 1 TORSION BAR AND MOVE ROADARM UP AGAINST BUMP STOP. RE-INSTALL TORSION BAR.
   C. REMOVE NUMBER 2 TORSION BAR AND ROADARM COMPLETELY.

The method of short tracking the tank to enable it to move after mine damage is illustrated in these sketches.

4. REVERSE NO. 3 ROADARM
   A. REMOVE TORSION BAR.
   B. REVERSE ROADARM NO. 3 (MAKING IT A LEADING ARM) BY MOVING THE VEHICLE OVER A DEPRESSION IN THE GROUND (EITHER FROM MINE BLAST OR DUG BY HAND) AS SHOWN ABOVE.
   C. WITH ROADARM AT 45°, RE-INSTALL TORSION BAR.

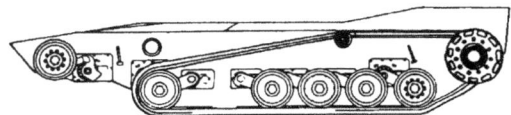

5. WITH NO. 3 ROADARM IN LEADING POSITION WRAP TRACK AROUND ROADWHEELS AND NO. 2 SUPPORT ROLLER.
   NOTE: TRACK IS TO BE SHORTENED TO LENGTH REQUIRED TO FIT AS SHOWN.
6. USING TRACK JACKS CONNECT TRACK TOGETHER BY INSTALLING CENTER GUIDES AND END CONNECTORS.
7. SHORT TRACKING IS COMPLETE MOVE TANK AS REQUIRED.

Above, the M1 appears at the left under the name of its new manufacturer and the Abrams at the right displays markings added by the troops.

On 19 February 1982, it was announced that a preliminary agreement had been reached between Chrysler and General Dynamics Corporation for the latter to purchase the Chrysler Defense Division. The deal was completed in March and the organization was renamed as the General Dynamics Land Systems Division. The personnel remained the same with Dr. Philip W. Lett, Jr. becoming Vice President of Research and Engineering. Thus Chrysler severed its connection with the development and production of tanks which had existed since before World War II.

The machine gun mounts for the tank commander and the loader can be seen in the top view below.

Details of the turret top and the rear deck of the M1 are clearly visible in these photographs taken during the troop tests of the tank.

The first M1 production tanks built at Detroit are shown here at the acceptance ceremonies on 31 March 1982. The machine guns have not been mounted on these vehicles.

It had always been intended to produce the M1 at both the Lima and the Detroit tank plants. In September 1981, the production of 60 tanks per month was authorized with 30 at Lima and 30 at Detroit. This was based on a single work shift at both plants. A total of 150 tanks per month was possible with multi-shift operation. Production of the M1 began at Detroit alongside the M60A3 during March 1982 and the first tank was accepted by the Army at a ceremony on 31 March. Production of the basic M1 continued at both plants for a total run of 2374 tanks. The last basic M1 was completed in January 1985.

Below, an Abrams is loaded aboard an Air Force C5 transport at the left. At the right is another view of the first production M1 at Detroit on 31 March 1982.

This Abrams, tank number A11, was assigned to A Company, 1st Battalion, 64th Armor, 3rd Infantry Division in Germany during late 1982. These photographs and those on the next two pages were taken by Russell P. Vaughan.

These additional views of tank A11 show an appearance of hard service. Note the battered fenders and one of the external grills from the rear doors to the engine compartment can be seen in the turret stowage rack in the upper photograph.

These two closeup views of tank A11 show details of the driver's hatch and the front of the 105mm gun mount.

The photographs on this and the following page were taken by Russell P. Vaughan in Germany during Reforger 83 in September 1983. They show M1 tanks of the 11th Armored Cavalry Regiment. Note the spare road wheel stowed on the turret roof.

The rear side skirt sections on these M1s have been cut away to prevent mud buildup. The track retaining ring on the sprocket is clearly exposed in these photographs. These rings were referred to by the troops as training wheels.

These M1s from the 1st Battalion, 64th Armor were photographed in Germany by Michael Green in May 1985. The tanks retain the original rear side skirt sections covering the track sprocket.

Above is another Michael Green photograph of an M1 from the 1st Battalion, 64th Armor. Note the spare road wheel on the turret stowage rack. Below, an Abrams is firing its cannon during a training exercise at Fort Hood, Texas on 27 January 1986.

Above are two views of the XM256 120mm smooth bore gun before its installation in the M1E1 tank. At the bottom of the page, the XM256 is firing on the test stand. Note the rounded shape of the reinforced plastic bore evacuator.

UPGRADING THE ABRAMS

The German Rheinmetall 120mm smooth bore gun was selected to replace the 105mm gun M68E1 in the Abrams and a license was obtained for its manufacture at Watervliet Arsenal, New York. During this same period, a product improvement program was underway to upgrade the basic M1 tank. On 18 September 1981, the Army Vice Chief of Staff directed that the project to install the new cannon, now designated as the 120mm gun XM256, be combined with the M1 product improvement program. In addition to the improved range and penetration performance, the 120mm gun had other advantages. Its fixed ammunition was slightly shorter than the 105mm rounds and they used a partially combustible cartridge case. Only a small stub case remained after firing and it was ejected into a container. Thus the problem of hot cartridge cases on the turret basket floor was eliminated. The larger diameter of the 120mm ammunition was, of course, a disadvantage since it reduced the number of rounds that could be stowed. A total of 40 120mm rounds were carried with 34 in the turret bustle and six in the hull. All of these were in separate compartments with blowout panels. The ready rounds on the turret basket floor, as on the original M1, were eliminated.

All of the changes were designated as Block I improvements. They included, in addition to the 120mm gun with its modified fire control equipment and ammunition stowage, better armor protection and an overpressure nuclear, biological, chemical (NBC) protection system with a microclimate controlled cooling system for the crew. The latter supplied cooling air to vests worn by the individual crew members as well as to the fighting compartment itself. An M43A1 detector and an AN/VDR-2 radiac were used in the compartment to detect low level concentrations of chemical and nuclear agents. The radiac was mounted on the hull floor at the driver's station. If the NBC system malfunctioned, a backup M13 system provided filtered air to each crew member's ventilated face mask as on the original M1. The new armor and main weapon increased the weight of the tank requiring improvements in the suspension, transmission, and final drives. The Abrams with these modifications was designated as the 120mm gun full tracked combat tank M1E1. The first two of 14 M1E1 prototypes had been delivered in March 1981, prior to the integration of the M1E1 with the product improvement program. The first prototype went to Aberdeen Proving Ground and the second to Chrysler for test purposes.

BREECH MOUNT · ROTOR · KING NUT · BORE EVACUATOR · MRS · TRUNNION BEARING · THERMAL SHROUDS · GUN TUBE

COAX READY AMMO BOX · MAIN WEAPON · FEED CHUTE · SMOKE TUBE · EJECTION CHUTE · SPENT CASE AND LINK CONNECTION BOX

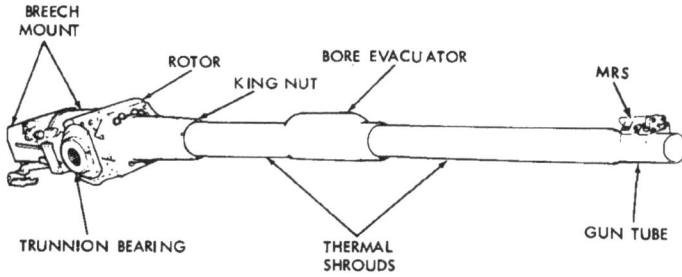

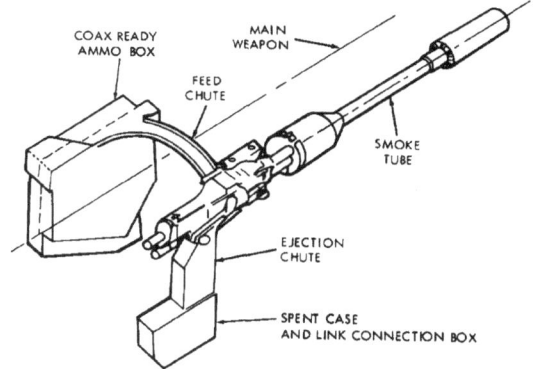

The sketches above and at the right show details of the XM256 120mm gun and the coaxial machine gun mount. Below and at the right are views of the loader's station and the doors covering the hull stowage of the 120mm ammunition.

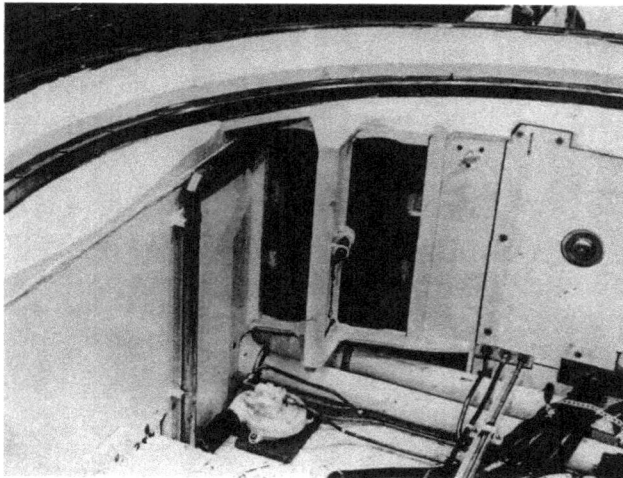

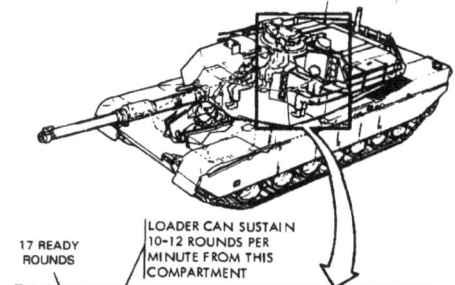

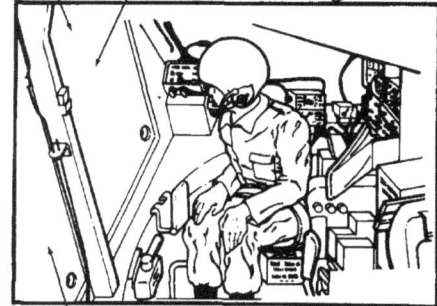

17 READY ROUNDS · LOADER CAN SUSTAIN 10-12 ROUNDS PER MINUTE FROM THIS COMPARTMENT · 17 SEMI-READY ROUNDS

Below are the changes evaluated on the M1E1 for application to the M1A1 production tank.

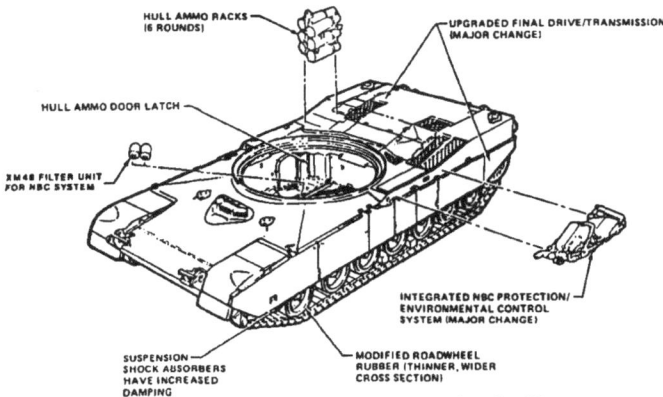

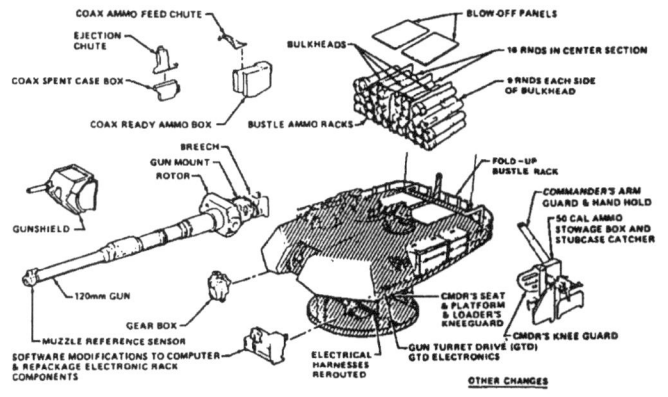

HULL AMMO RACKS (6 ROUNDS) · UPGRADED FINAL DRIVE/TRANSMISSION (MAJOR CHANGE) · HULL AMMO DOOR LATCH · XM48 FILTER UNIT FOR NBC SYSTEM · SUSPENSION SHOCK ABSORBERS HAVE INCREASED DAMPING · MODIFIED ROADWHEEL RUBBER (THINNER, WIDER CROSS SECTION) · INTEGRATED NBC PROTECTION/ENVIRONMENTAL CONTROL SYSTEM (MAJOR CHANGE)

OTHER CHANGES
- OEM STOWAGE REARRANGED
- DUAL AIR HEATER
- HULL NETWORK BOX
- ELECTRICAL HARNESSES

COAX AMMO FEED CHUTE · EJECTION CHUTE · COAX SPENT CASE BOX · COAX READY AMMO BOX · BREECH · GUN MOUNT ROTOR · GUNSHIELD · 120mm GUN · GEAR BOX · MUZZLE REFERENCE SENSOR · SOFTWARE MODIFICATIONS TO COMPUTER & REPACKAGE ELECTRONIC RACK COMPONENTS · ELECTRICAL HARNESSES REROUTED · BLOW-OFF PANELS · BULKHEADS · 16 RNDS IN CENTER SECTION · 9 RNDS EACH SIDE OF BULKHEAD · BUSTLE AMMO RACKS · FOLD-UP BUSTLE RACK · COMMANDER'S ARM GUARD & HAND HOLD · 50 CAL AMMO STOWAGE BOX AND STUBCASE CATCHER · CMDR'S SEAT & PLATFORM & LOADER'S KNEEGUARD · CMDR'S KNEE GUARD · GUN TURRET DRIVE (GTD) GTD ELECTRONICS

NOTE: NO MAJOR CHANGE FROM M1 (105MM) DESIGN

OTHER CHANGES
- COMBUSTIBLE CASED AMMUNITION WITH METAL STUB
- REDESIGNED LOADER'S SEAT
- REDESIGNED STOWAGE UNDER LOADER'S SEAT
- NEW LOADER'S SHOULDER GUARD
- TURRET NETWORKS BOX
- TANK COMMANDER'S PANEL

225

Above, left to right, are the 120mm XM827 APFSDS-T and the XM865 TPCSDS-T rounds. Below, left to right, are the 120mm XM830 HEAT-MP-T and the XM831 TP-T rounds. Various components on the M1E1 tank are identified on the sketch at the right.

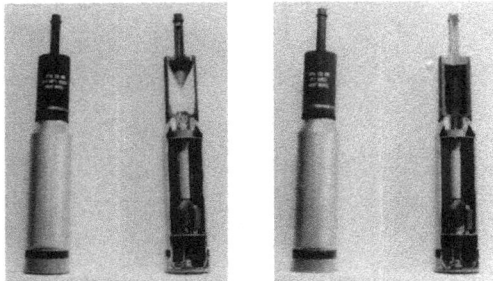

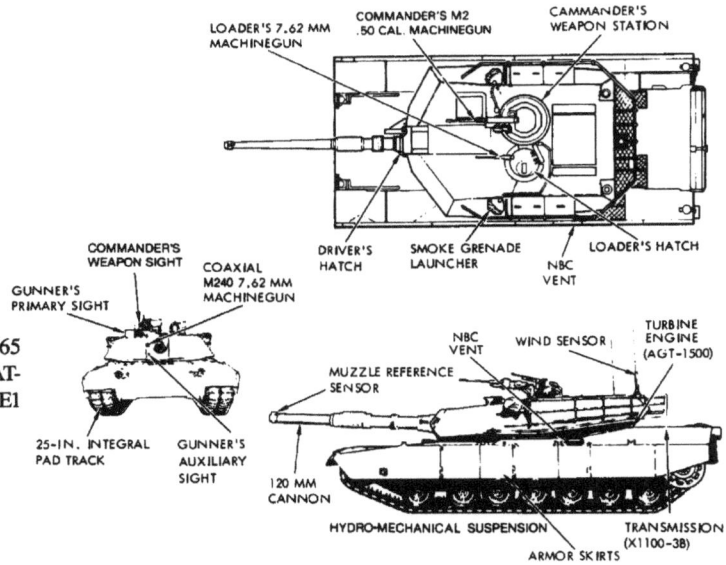

The nuclear, biological, chemical (NBC) system on the M1E1 is sketched at the right and the location of the NBC system components can be seen in the drawing below.

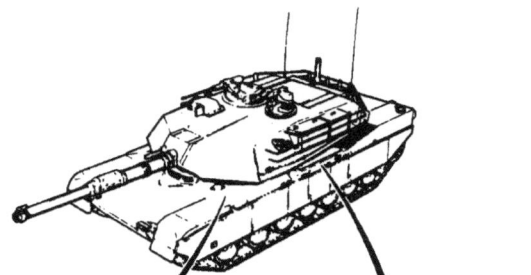

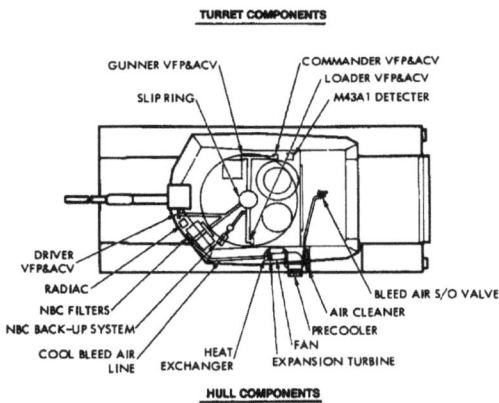

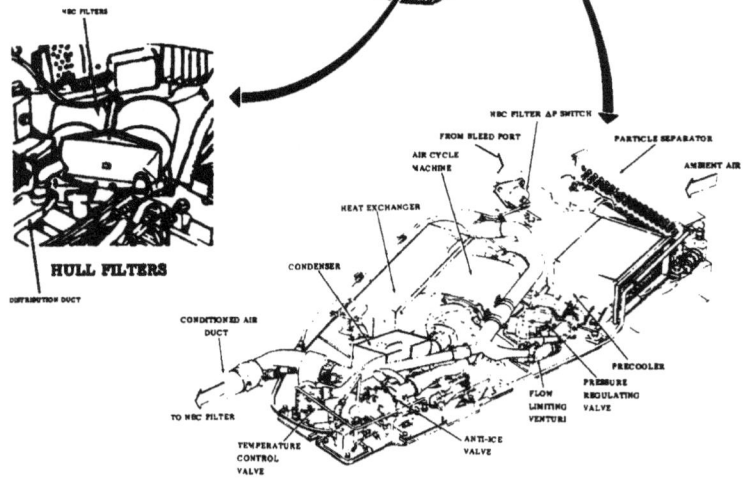

The tank commander's .50 caliber M2HB machine gun appears below and the loader's 7.62mm M240 machine gun is at the right.

One of the 120mm gun tank M1E1 pilots is shown in these two photographs. The steel plates were welded to the turret front to simulate the weight of the heavier armor planned for the production version of the tank. Note that this tank retains the original rear side skirt sections.

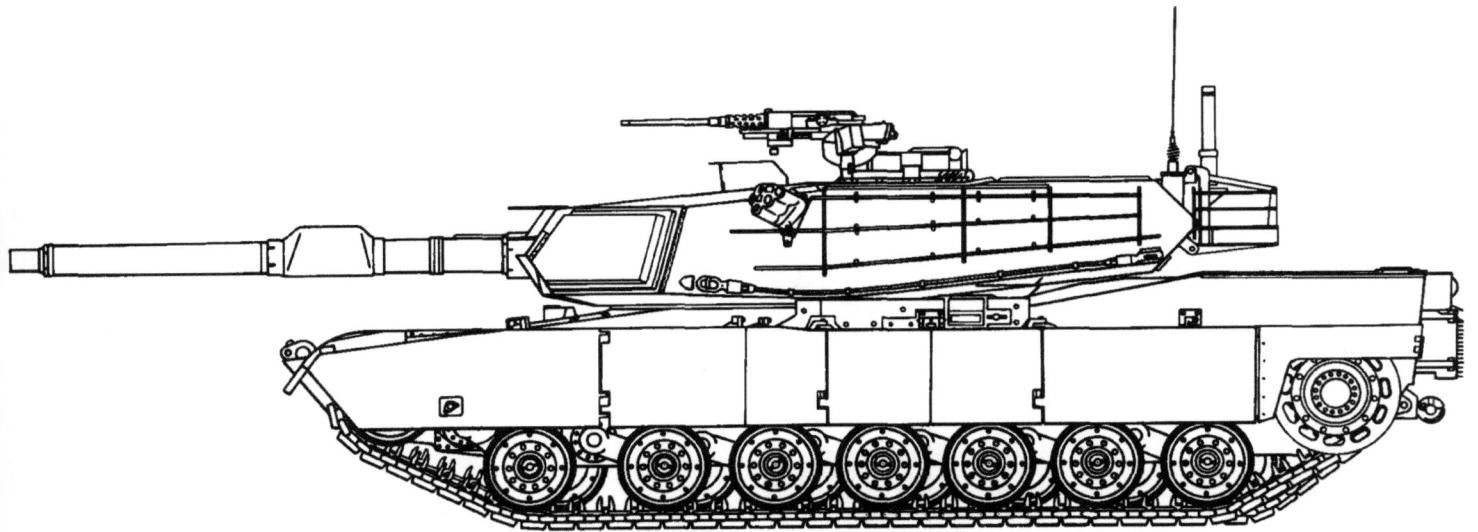

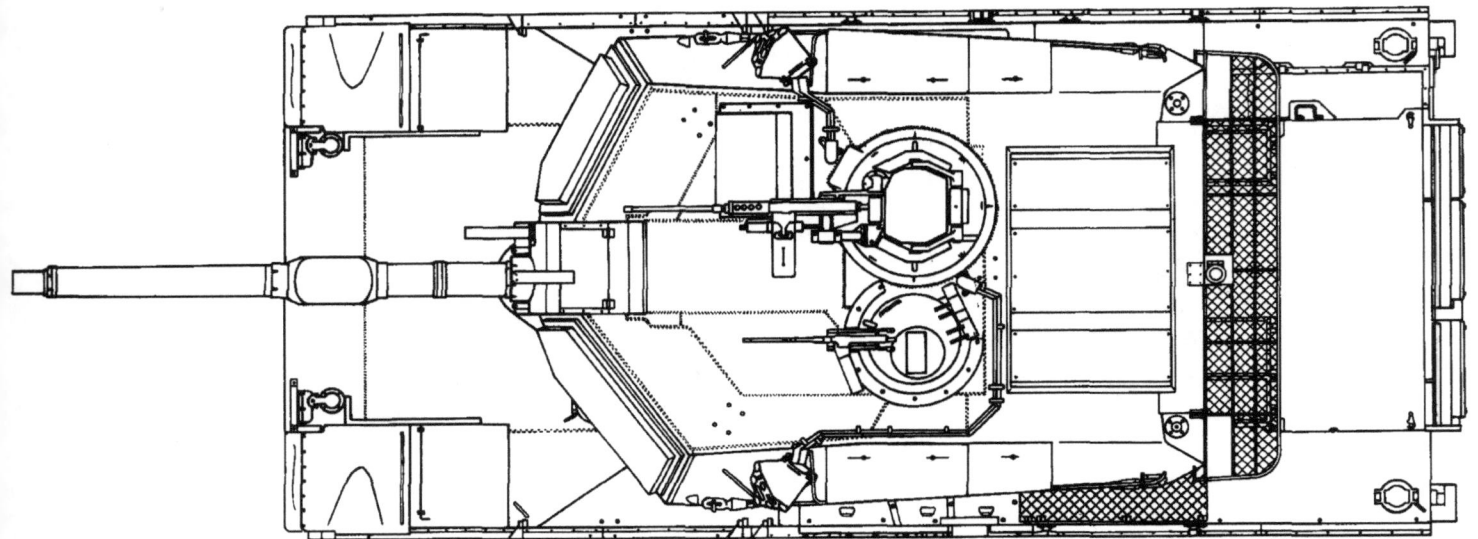

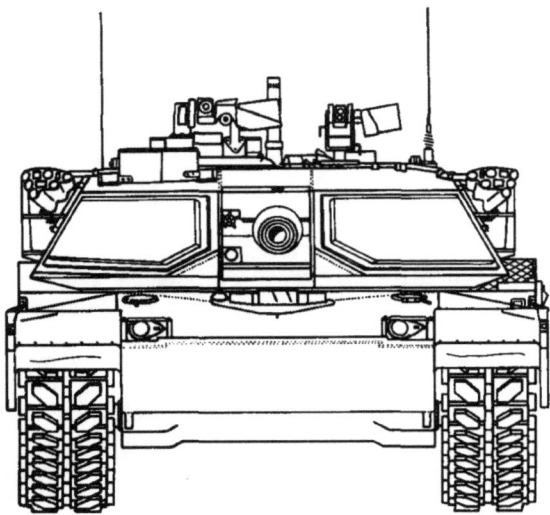

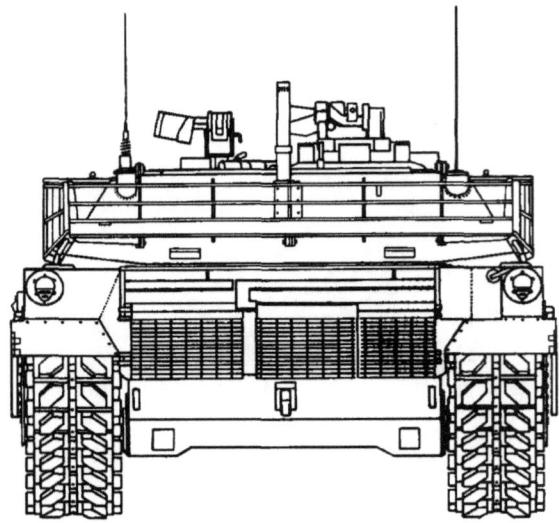

120mm Gun Tank M1E1

Another M1E1 pilot appears in these views. Note the new rear side skirt sections that are cut away to prevent mud buildup around the sprocket. The M1E1s also still have the track retaining rings on the sprocket.

These photographs of the IPM1 taken at Aberdeen Proving Ground by Robert Lessels show the final type of turret stowage rack in great detail. The machine guns and smoke grenade launchers have not been installed on this vehicle.

Concurrent with the development of the 120mm gun tank, many of the Block I improvements were introduced on to the M1 production line. The resulting vehicle was designated as the improved performance M1 (IPM1). This tank, which retained the 105mm gun M68A1, featured the heavier armor protection as well as the upgraded suspension, transmission, and final drives. As on the M1E1, the final drive gear ratio was changed from 4.30:1 to 4.667:1 and the pressure in the hydraulic shock absorbers was increased from 3000 psi to 3500 psi. The changes in the final drive gear ratio reduced the maximum governed speed from 45 to 41.5 miles per hour. The new detachable stowage rack was installed on the turret bustle. The first IPM1 was delivered in October 1984 and production continued until May 1986 with a total run of 894 tanks.

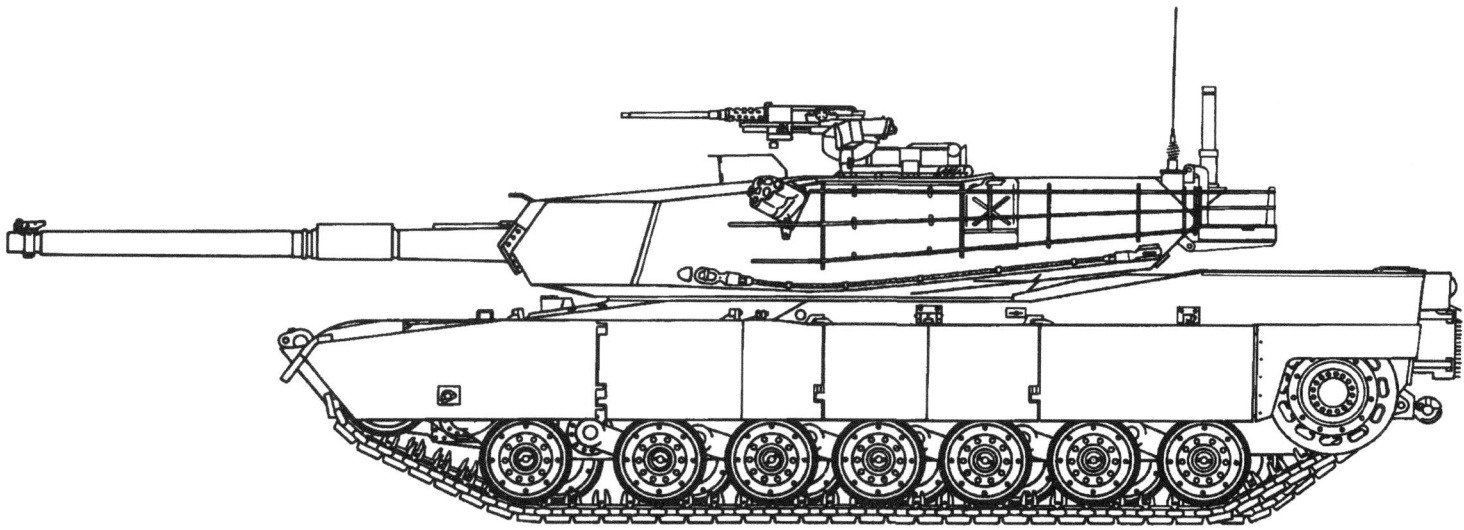

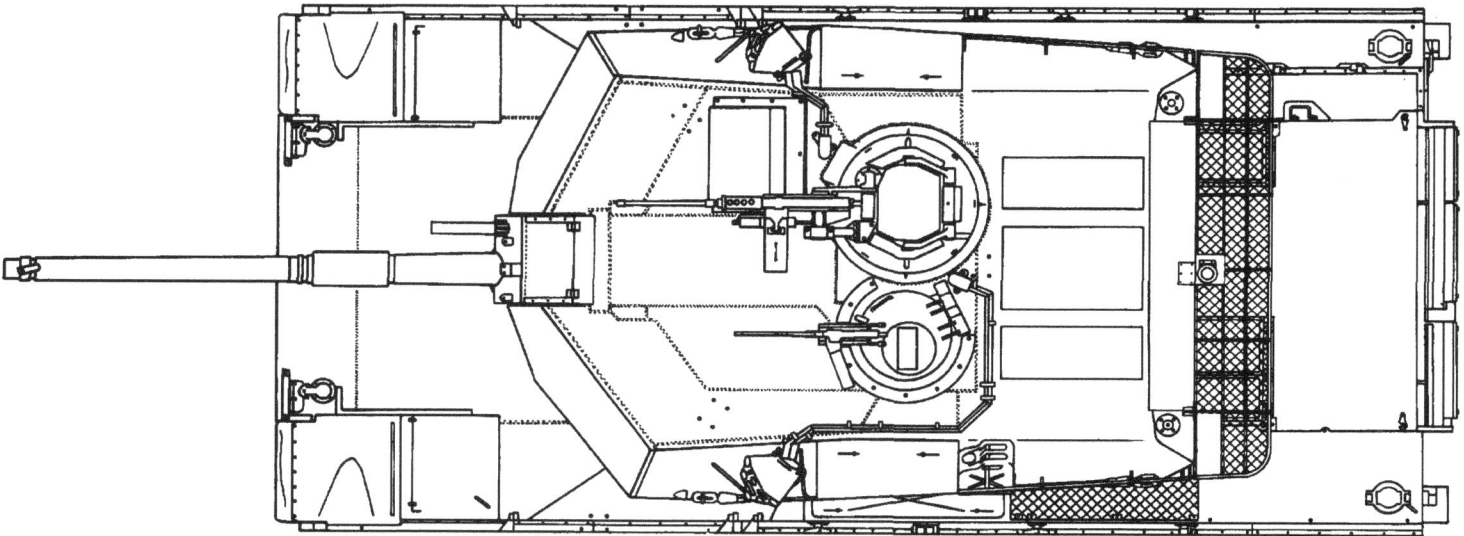

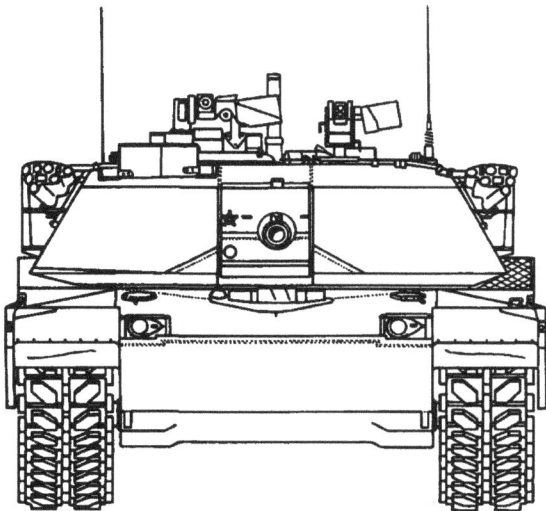

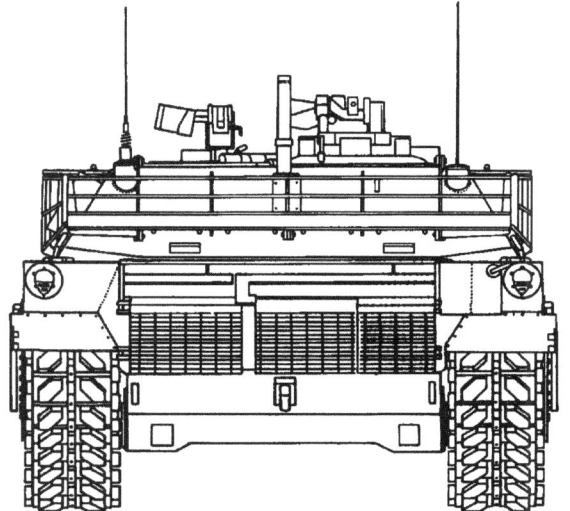

Scale 1:48

© D.P. Dyer

105mm Gun Tank IPM1

This IPM1 has the interim turret stowage rack. Brackets are welded on the lower rear of the turret bustle to mount a rear stowage rack, but it has not yet been installed. Addition of that rack provided stowage space similar to that of the final type rack in the previous photographs.

None of the secondary armament has been installed on this IPM1 photographed at Aberdeen Proving Ground.

The steel plates welded to the front of the turret and hull to simulate the weight of heavier armor can be seen on this M1E1 pilot above and below. These photographs and those at the bottom of the page were taken by Robert Lessels at Aberdeen Proving Ground in December 1984.

To simulate the weight of the heavier armor intended for the production version of the M1E1, steel plates were welded on to the turret front of the prototypes. After an extensive DT/OT evaluation program, the new version of the Abrams was standardized as the 120mm gun full tracked combat tank M1A1 on 28 August 1984. Approval for full production was received in December and the first production tanks were delivered at Detroit in August 1985. Later, production at the Lima plant also shifted to the M1A1. Production testing of the M1A1 began in fiscal year 1986. When the production M1A1 tanks became available, the Department of the Army directed that they be shipped first to the units in Europe to replace the basic M1s already in service. The latter were returned to the continental United States and reissued to active Army and selected Army National Guard units. In Europe, the 1st Armored and 3rd Infantry Divisions as well as the 2nd Armored Division (Forward) were the first to receive the M1A1s during 1988. At that time, the 3rd Armored Cavalry Regiment was the only unit in the continental United States equipped with the M1A1.

In March 1983, a program was initiated to develop an enhanced version of the 105mm gun. The gun tube was extended by 1.5 meters compared to the M68A1 and it could tolerate a higher chamber pressure. Designed to replace the 105mm gun M68A1 in the M1 and the IPM1, it was

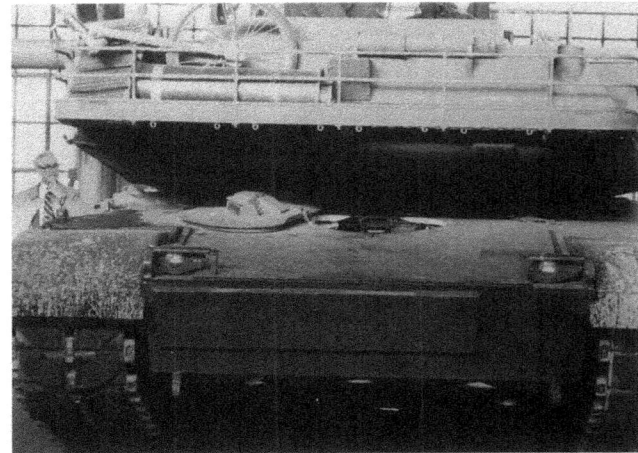

expected to have improved penetration performance, particularly with new ammunition. At that time, it was expected that the installation of the enhanced 105mm gun would be a less costly method of upgrading the tank's firepower compared to retrofitting the M1 and IPM1 with the 120mm gun M256. Later, the enhanced 105mm gun was abandoned and the 120mm gun was selected for retrofitting in the earlier models of the Abrams.

Details of the side skirts are visible in these views of the M1E1. The port for the NBC system on the left side of the tank appears at the right below.

ENGINE
AIR INLET
TOWERS

CAP

ENGINE
EXHAUST
TOWER

TOWERS REMOVED BY
TURNING TURRET

2 METERS

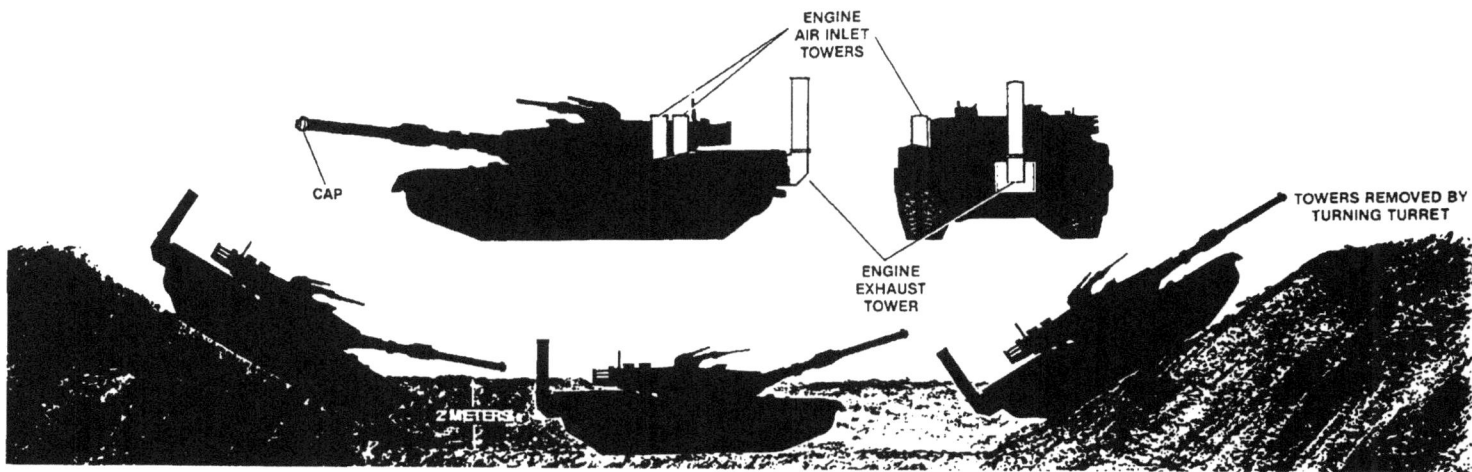

The sketch above illustrates the fording sequence of the Abrams when equipped with the deep water fording kit.

The United States Marine Corps also expressed interest in obtaining the M1A1. This resulted in the development of a deep water fording kit for use on the M1A1 with the cannon in the forward position. A prototype of this kit was demonstrated in October 1984.

Below and at the right, the M1E1 is operating with the deep water fording kit in October 1984.

At the right, the M1E1 climbs out onto the beach. The inlet and exhaust towers can be seen on the rear of the tank.

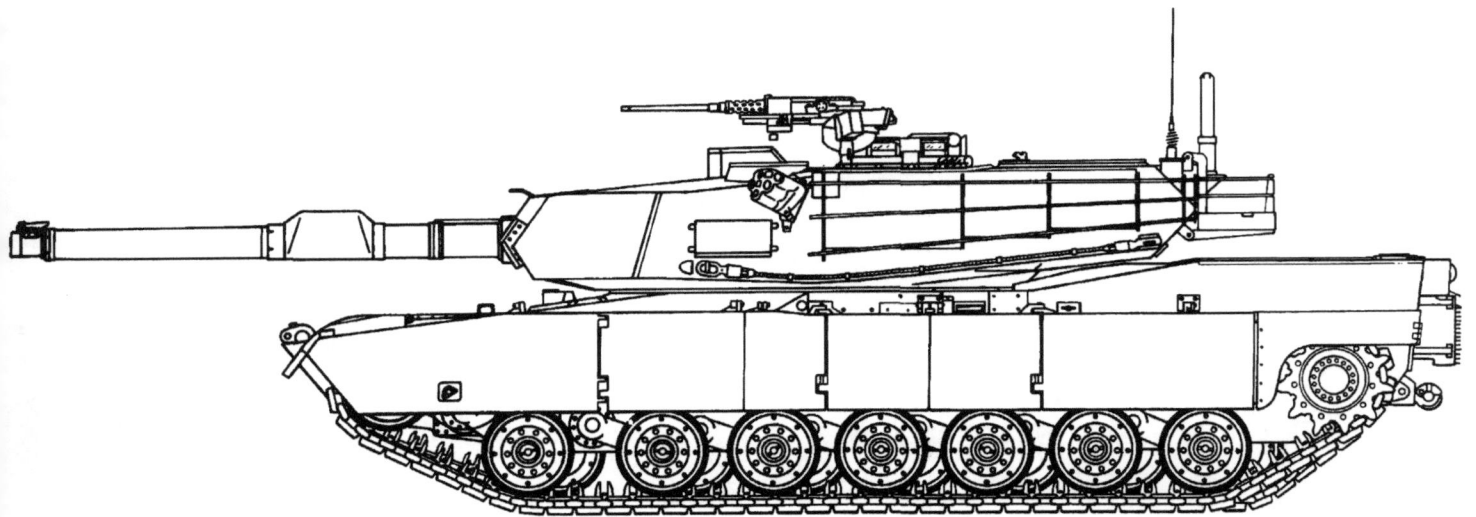

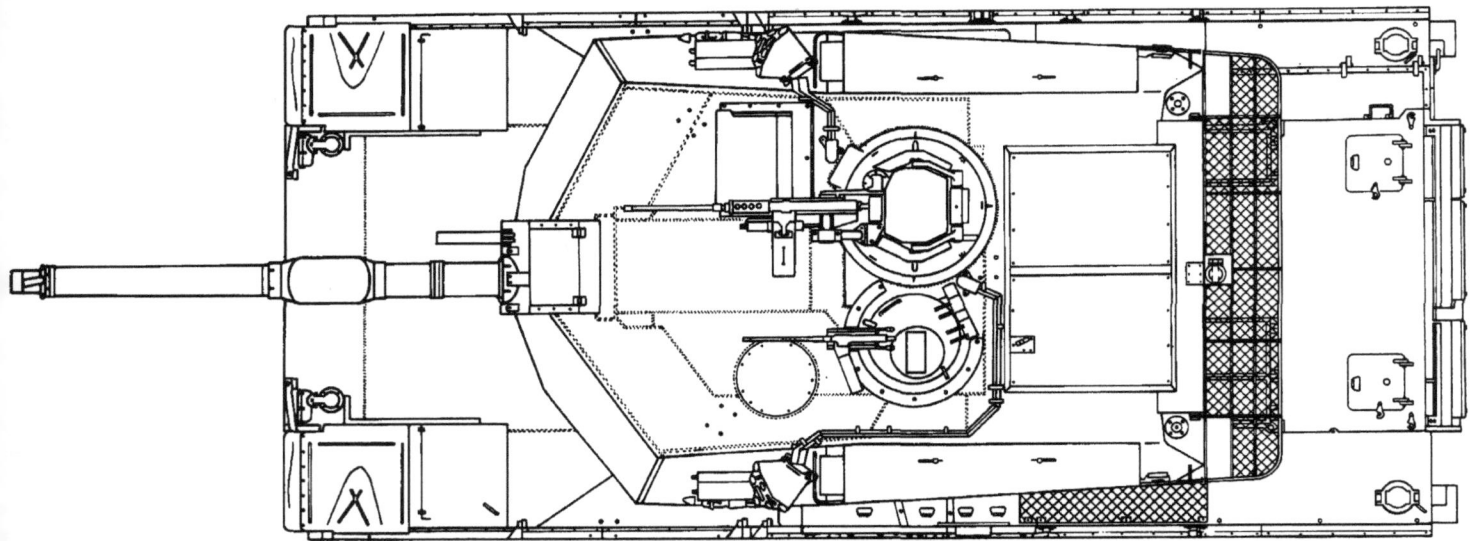

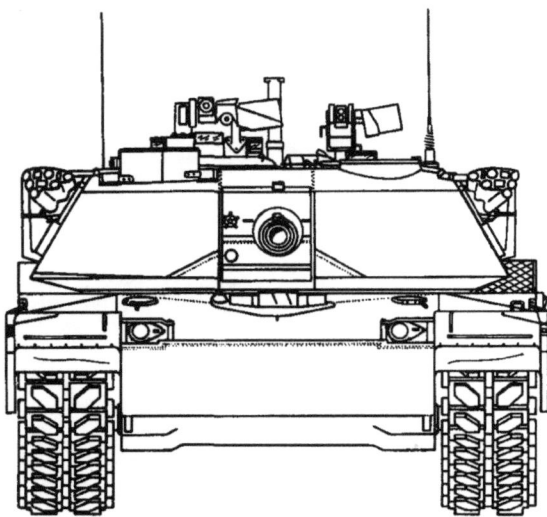

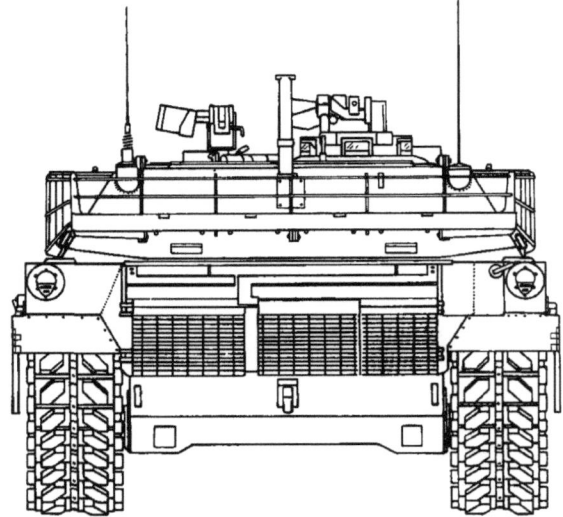

120mm Gun Tank M1A1

The photographs above and at the left below show a late production M1A1 tank with the blanked off port in the left front turret roof for the future installation of the commander's independent thermal viewer. An earlier production M1A1 appears below at the right.

Two late production M1A1 tanks from the 3rd Armored Cavalry Regiment are shown here operating at Fort Bliss, Texas.

The band around the bottom of the turret on these late production M1A1 tanks is a strip used to attach sensors for the MILES training equipment.

As indicated by its markings, the M1A1 above and below is from A Company, 4th Battalion, 66th Armor of the 3rd Infantry Division. These photographs and those on the opposite page were taken by Michael Green near Aschaffenburg, Germany during Reforger 88 in September 1988.

Another tank from A Company, 4th Battalion, 66th Armor is shown here during Reforger 88. Note that this M1A1 retains the early rear side skirts, but the track retaining ring has been eliminated.

These M1A1s from A Company, 3rd Battalion, 63rd Armor show a variety of rear side skirts. Above, one tank has the rear side skirt section removed completely and others have the new design which has been cut back to expose the sprocket. However, tank A14 "Death Stalker" retains the old type rear skirt.

The M1A1s shown here are fitted with the new design rear side skirt. These photographs as well as those on the previous page were taken by Michael Green during Reforger 88.

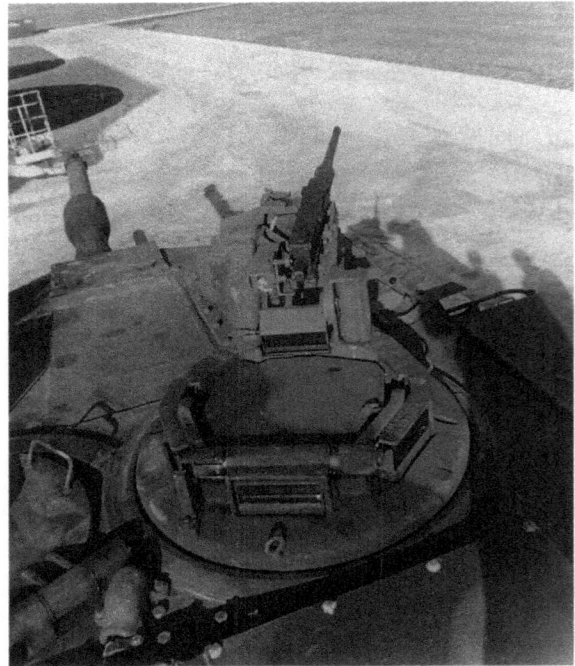

Above and at the right is the original tank commander's weapon station installed on the M1A1. Compare these photographs with those of the new design at the bottom of the page. The Block II improvements for the Abrams are listed in the sketch below.

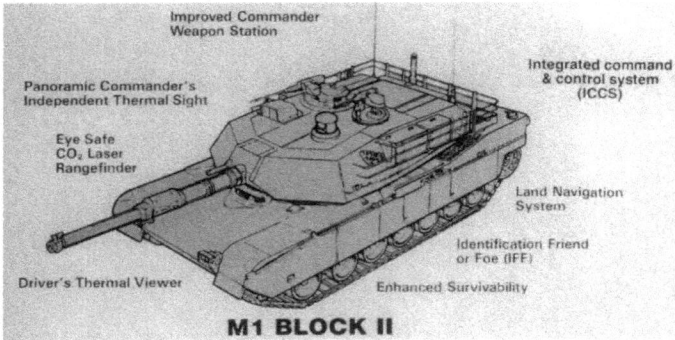

Improved Commander Weapon Station

Panoramic Commander's Independent Thermal Sight

Eye Safe CO₂ Laser Rangefinder

Driver's Thermal Viewer

Integrated command & control system (ICCS)

Land Navigation System

Identification Friend or Foe (IFF)

Enhanced Survivability

M1 BLOCK II

Below and at the right are views of the new non-rotating tank commander's weapon station. Note the increase in the size of the hatch opening and that the .50 caliber machine gun is now on a flexible mount rotating around the hatch.

After the introduction of the M1A1, some additional modifications were considered as Block II product improvements. Approved by the Army Vice Chief of Staff on 1 February 1985 and reevaluated in August of that year, some of these changes were scheduled for incorporation into the M1A1 production tanks starting in late 1988. With these new features, the tank was nicknamed the M1A1+. The modifications included an improved non-rotating commander's weapon station (CWS), a commander's independent thermal viewer (CITV), a carbon dioxide laser range finder, an intervehicular information or battle management system, survivability improvements, and a driver's thermal viewer. The new CWS featured a larger hatch with periscopes providing near panoramic vision. The ring mounted commander's machine gun was fired manually

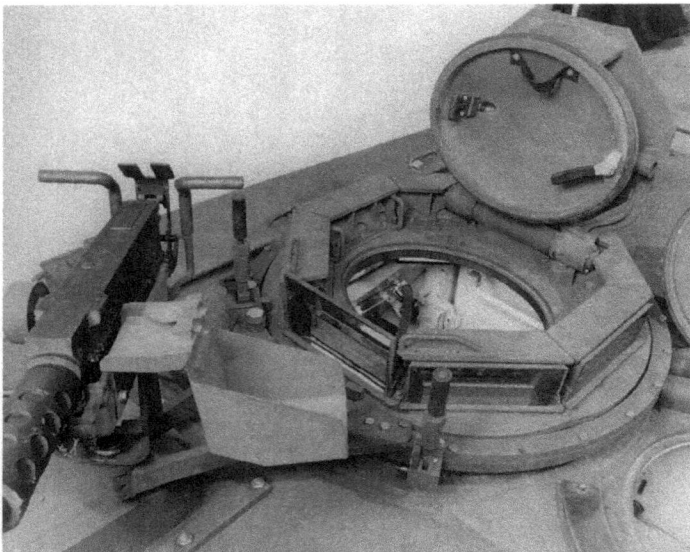

At the right is the mock-up of the commander's independent thermal viewer (CITV) installed in the tank turret roof.

with the hatch open and the internal power and manual controls for this weapon were eliminated. The CITV allowed the commander to search for new targets while the gunner was engaging a previous target. Prior to the availability of the CITV, holes were cut in the turret roof forward of the loader's hatch on late production M1A1s to permit the later installation of the new thermal viewer. The carbon dioxide laser range finder could operate under the same visibility conditions as the tank's thermal sight permitting the engagement of targets observed through the thermal sight. It also was eye safe, eliminating the need for filters during training operations. The intervehicular information or battle management system utilized a computer to process information required for command, control, or maintenance purposes. The survivability enhancements originally were intended to include improved protection against top attack weapons. However, this feature was deferred to later production tanks. Improved armor protection was obtained with a new special armor incorporating depleted uranium. This feature was introduced into the M1A1 production starting in October 1988. The new armor greatly increased the penetration resistance particularly against kinetic energy rounds. The driver's thermal viewer

allowed him to see under the same visibility conditions as the gunner and the tank commander when they were using the gunner's primary sight. Thus he could drive without assistance from them in smoke or fog. Although originally intended for the first production of the improved M1A1, both the commander's independent thermal viewer and the driver's thermal viewer were postponed until later production because of cost restrictions. Also deferred to a later date was an identification, friend or foe (IFF) device. This was intended to identify friendly tanks on the battlefield and to prevent their accidental destruction.

Below, this view of a late production M1A1 shows the blanked off mounting port for the future installation of the CITV in the turret roof.

Above is the M1A1 modified as a prototype of the 120mm gun tank M1A2 with the new commander's station and the mock-up of the CITV. The other new features of the M1A2 are listed in the photograph.

On 14 December 1988, General Dynamics Land Systems Division was awarded a contract for the full scale development of the improved Abrams. Scheduled for fielding in 1992, this vehicle, to be designated as the 120mm gun tank M1A2, included the commander's thermal viewer, the driver's thermal viewer, a position/navigation (Pos/Nav) system, and a new electronic system in addition to other new features previously described.

Further details of the prototype M1A2 can be seen below.

The M1 in the photographs on this page is equipped with an externally mounted auxiliary power unit (APU).

In an effort to reduce fuel consumption and engine wear, it was proposed that an auxiliary power unit (APU) be installed in the Abrams. This unit would provide electrical power when the tank was not moving, thus reducing the engine idling time. The limited space inside the vehicle did not provide a practical location for such an APU. On 30 October 1983, the Army Vice Chief of Staff directed that an external APU be developed for the Abrams. Four such units were procured and two were tested at Fort Knox and two at Fort Hood. The unit consisted of a 5.6 kw diesel powered generator mounted on the right rear fender of the tank.

In the view below, the external APU can be seen installed at the right rear of the tank.

Above are two photographs by Robert Lessels of the original T156 integral pad track installed on the M1. Note the worn condition of the track in the left view. All of the photographs on this page were taken at Aberdeen Proving Ground during June 1985.

One problem that continued to plague the Abrams was the short life of the T156 track. This double pin track with integral rubber pads required replacement after about 700 to 800 miles under normal operation. This was far short of the design goal of 2000 miles. Efforts were made to increase the track life of the T156 and a program at Aberdeen evaluated other track designs against it. The T156 was similar in design to the T97 track used on the M60 series tanks. It was selected originally because it was lighter in weight than a replaceable pad track and the time required to change a complete track with integral pads was much less than that necessary to replace all of the individual pads. Two new tracks tested in comparison with the T156 were the German Diehl N570 and the T158 developed by FMC Steel Products and the Goodyear Tire and Rubber Company.

Both of the new tracks used replaceable rubber pads and the Diehl N570 was the same type used on the German Leopard II tank. The wear life was greatly increased with both tracks, but they added about 3000 pounds to the weight of the tank. An advantage of the T158 track was that it could directly replace the T156 without any changes. However, the German Diehl N570 required the installation of new sprockets and road wheels. After further modification and additional tests, FMC was awarded a production contract for the T158 track. Under this contract, FMC guaranteed the Army a track life of 2100 miles.

Below and at the right are views of the Diehl replaceable pad track and its installation on the M1. Note the 12 tooth sprocket used with this track.

The replaceable pad T158 track manufactured by FMC is illustrated in the sketches at the right and the photograph below.

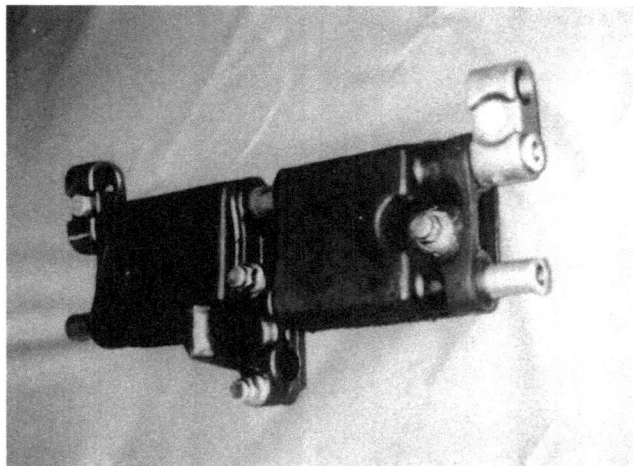

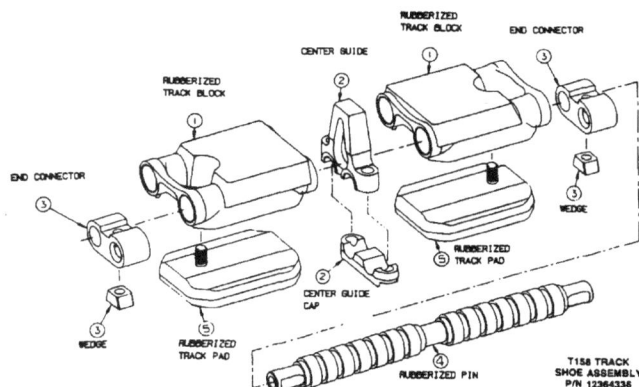

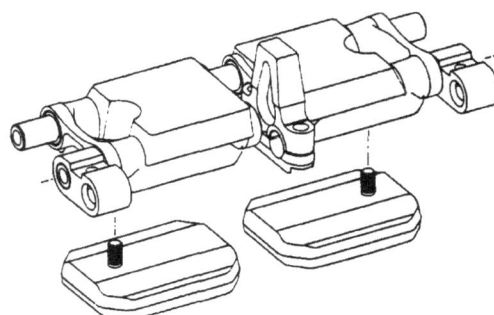

As issued to the troops, the M1 tank was equipped with protective systems to prevent excessive damage to the power train in the event of an engine malfunction or a component failure. One of these was incorporated in the electronic system that controlled the engine throttle response. If this unit lost electrical power or sensed a fuel control malfunction, it put the engine into idle and disconnected it from further control. In this condition, the tank could not move unless the electronic control unit could be reset. Also, if the signal to the electronic transmission control was lost, the driver had no way to engage the transmission from inside the tank. Although such protective systems were useful in preventing damage during peacetime operations, they could immobilize and endanger the tank under combat conditions.

In July 1986, The Science Advisor to the Commander-in-Chief U.S. Army Europe and Seventh Army requested an investigation to determine if a simple mechanical override for these electronic systems could be devised which would be used only under combat conditions. This study

The SHAFTS installation in the driver's compartment of the M1 appears at the right. The two lever control box can be seen to the left of the driver's seat.

produced the shift hand actuated fuel transmission system (SHAFTS) for installation in the M1. It consisted of a mechanical fuel metering valve and a mechanical transmission shifter located in the engine compartment operated by a two lever control box mounted at the left side of the driver's station. In an emergency combat situation, the SHAFTS allowed the driver to bypass the protective systems and operate the turbine engine and transmission up to a maximum output of about 400 horsepower. Thus the tank could extract itself from a dangerous position when the risk of damage to the power train was justified.

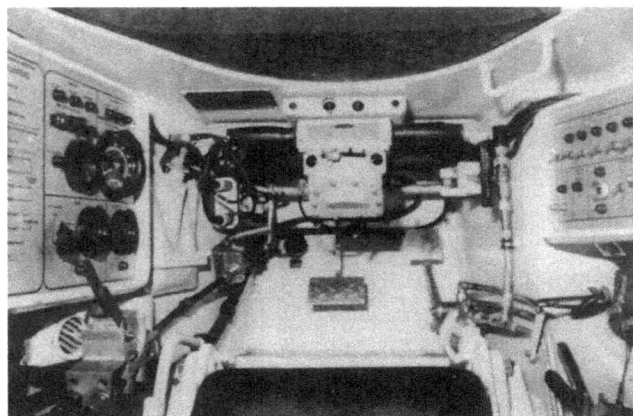

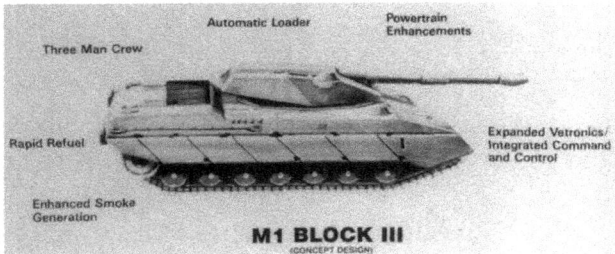

Two artist's design concepts for a future Block III Abrams tank are shown above.

Long range modifications of the Abrams were considered as Block III product improvements. Concept studies of this version of the tank featured improved armament in a low silhouette turret or externally mounted. An automatic loader was provided and the crew was reduced to three men. New armament under consideration included lightweight guns ranging from 120mm to larger calibers. A more compact arrangement of the propulsion system mounted the AGT-1500 gas turbine in the transverse direction. Referred to as the transverse mounted engine propulsion system (TMEPS), it provided approximately 46 cubic feet of additional free space inside the engine compartment which could be used for extra ammunition, fuel, or an auxiliary power unit mounted under armor. Rapid refueling and rearming equipment also was proposed as well as improved armor protection and smoke generators. The application of very high speed integrated circuit technology would improve the fire control as well as other command and control systems. In addition to the lower silhouette, reduction of the thermal, magnetic, electromagnetic, and radar signatures also would improve the survivability of the tank. Improved suspension systems also were considered.

The mock-up of the transverse mounted engine propulsion system (TMEPS) appears above at the right. The sketches below compare the space requirements for the current propulsion system with that for the TMEPS and show ways that the additional free space with the latter could be utilized.

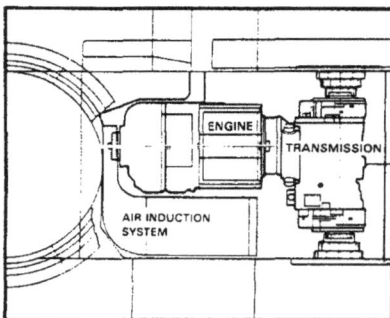

CURRENT PROPULSION SYSTEM COMPARTMENT

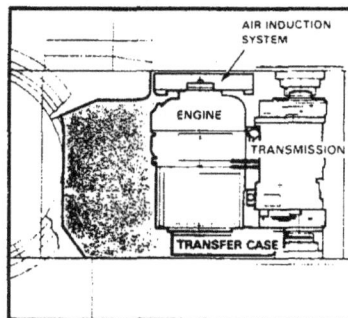

TRANSVERSE MOUNTED ENGINE PROPULSION SYSTEM

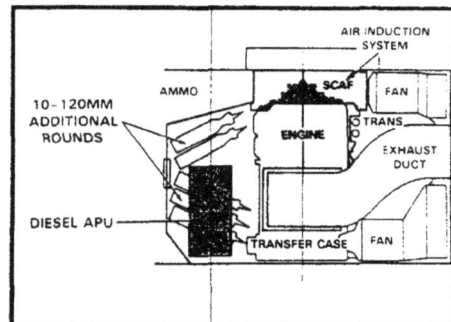

PART IV

VEHICLES BASED ON THE MAIN BATTLE TANK AND FUTURE DEVELOPMENT

Above, the track width mine clearing plow and the track width mine clearing roller are mounted on the M1 tank at the left and right respectively. The plow is shown in the travel position.

COMBAT ENGINEER VEHICLES

Several kits to permit special operations were developed for use on the Abrams. One of these installed the tank mounted mine clearing roller (TMMCR) on the front of the vehicle. This was the same basic roller assembly used with the M48 and M60 series tanks. It consisted of two roller banks of five discs each, with one located in front of each track. A chain carrying a dog-bone weight connected the two roller banks. This was intended to detonate any tilt rod mines in the space between the two roller banks. Weighing about ten tons, the TMMCR was patterned after Soviet equipment previously evaluated at Fort Knox. Designed to withstand two detonations from mines containing 22 pounds of explosive, the roller banks could be jettisoned by the driver from inside the tank using a quick disconnect system. The mine clearing roller also was intended for installation on the ROBAT.

Another device adapted for use with the M1 tank was the track width mine clearing plow. The plow assembly could be installed on the tank by four men in less than one hour using a suitable lifting device. The latter could be an A frame, a combat engineer vehicle, or a tank recovery vehicle. The plow consisted of two units, one in front of each track, with skids to control the depth of cut and mold boards to cast the mines and debris to the side. The plow could clear mines buried to a depth of two to four inches in front of each track. Like on the track width mine clearing roller, a chain and dog-bone weight assembly was used to clear tilt rod mines between the two plow units. The plow could be raised for travel or lowered for use from within the tank.

At the right, the bulldozer kit is installed on the M1 during the test program.

A less successful project was the original attempt to develop a tank mounted bulldozer. Such bulldozers had been provided for use with all of the standard U.S. tanks since World War II. The M8A3 bulldozer was employed with the M48A3 tank in Vietnam and the M9 was standardized for use with the M60 series tanks. The revised Materiel Need document for the Abrams included a requirement for a bulldozer kit which could be mounted on any M1 by battalion maintenance personnel, but no funds were available for such a development. However, arrangements were made to test a privately developed American bulldozer kit as well as one of Israeli manufacture during the Summer of 1985. Unfortunately, mechanical problems with the American design prevented it from participating in the test program. As a result, only the Israeli bulldozer kit was evaluated. The tests confirmed the value of a tank mounted bulldozer for clearing debris and preparing defensive earthworks. However, during the deep digging required for preparing tank firing positions, serious overheating occurred in the tank transmission. Other problems included the restriction of the driver's field of vision by the dozer blade and some interference with the movement of the tank's main gun. Further study indicated that solving the transmission overheating problem would

253

Above is the mock-up of the combat engineer vehicle T118 with the original long boom. The sketches at the bottom of the page show the dimensions of the CEV T118 with the original long boom (left) and the final shortened version (right).

require redesign of the tank's power train. This was not considered practical and it was recommended that the deep digging requirement be eliminated, particularly since earthworks provided very little protection against armor piercing ammunition using long rod penetrators. Without the deep digging requirement, the dozer blade could be made smaller and narrower reducing the problems with driver vision and gun interference. Such a bulldozer could still be effectively employed for debris clearance and the breaching of obstacles.

During the development cycle of the various main battle tanks, consideration was given to the use of the tank chassis as the basis for other armored vehicles. These included recovery vehicles, bridge launchers, combat engineer (CEV) and other specialized armored vehicles. Early in the T95 program, concept studies proposed the modification of the tank chassis for use as a combination recovery and combat engineer vehicle. However, the combined concept was rejected and the development of the two types of vehicles proceeded along separate lines. As a result, the

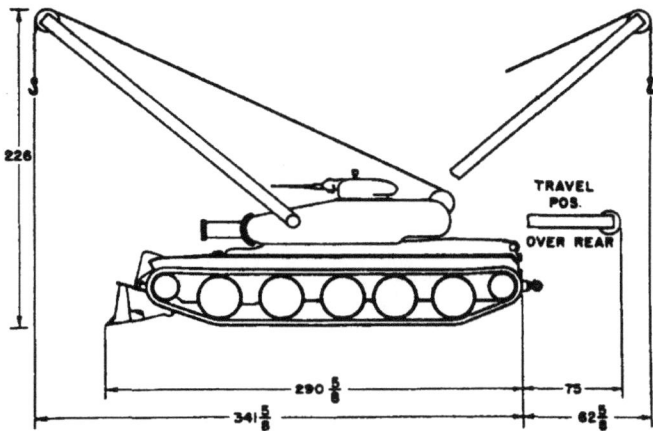

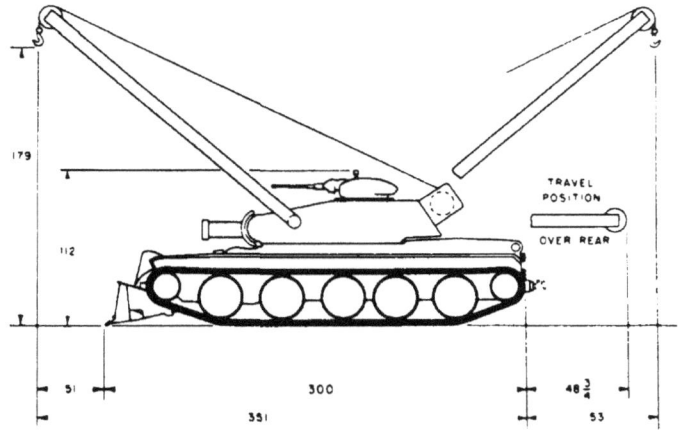

The left and right sides of the first pilot CEV T118, registration number 9B2027, can be seen here. Compare the shortened boom with the original version on the mock-up shown on the previous page.

combat engineer vehicle T118 was designed based on the T95 tank chassis. This project was initiated by OTCM 36699 dated 24 December 1957 which authorized the construction of three pilot vehicles. The 165mm M57 (T156) demolition gun was approved for installation in the three pilots despite problems with this weapon during its evaluation. Secondary armament consisted of a .30 caliber M37 coaxial machine gun in the turret and a .50 caliber M2HB machine gun in the commander's cupola.

In May 1958 a mock-up of the T118 was displayed at Detroit to representatives of CONARC and the Corps of Engineers. As a result of this meeting some modifications were made to the design. These included a shorter boom to reduce the rear overhang when it was stowed in the travel position. The three pilot vehicles were then fabricated with the first, registration number 9B2027, being delivered to Aberdeen Proving Ground in December 1959. Numbers two and three (registration numbers 9B2028 and 9B2029) followed and were delivered respectively to the Armored Board at Fort Knox and to the Engineer Test Unit at Fort Belvoir, Virginia. The pilot T118s were equipped with a front mounted hydraulically operated bulldozer, an A frame boom, and a 60,000 pound capacity engine driven winch using one inch diameter wire rope. The boom was raised or lowered hydraulically and the modification had reduced the rear overhang in the travel position from the original 75 inches on the mock-up to 48¾ inches on the three pilots.

These photographs of the first pilot CEV T118 show details of the bulldozer blade. Note the folding section in the center to permit a better forward view for the driver with the blade in the travel position.

Additional views of the first pilot CEV T118 appear above. The lightweight road wheels have been installed on this vehicle. Details of the boom and winch can be seen in the top view at the right.

After cancellation of the T95 program, the CEV project was reoriented to utilize the chassis of the M60 tank and the designation was changed to the T118E1. After further modification, it was standardized based on the M60A1 tank as the combat engineer vehicle M728. During the period of the T118 development, the hydraulically operated bulldozer also was proposed for installation on the T95 and T96 battle tanks.

A sketch and an artist's concept of the bulldozer kit proposed for the T95 and T96 tanks are shown below. This was the same bulldozer eventually mounted on the CEV T118. Obviously, the folding section necessary for the driver's forward vision was a later modification.

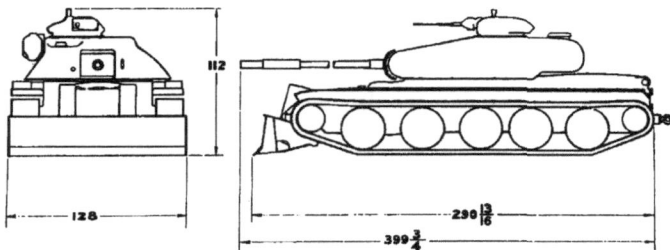

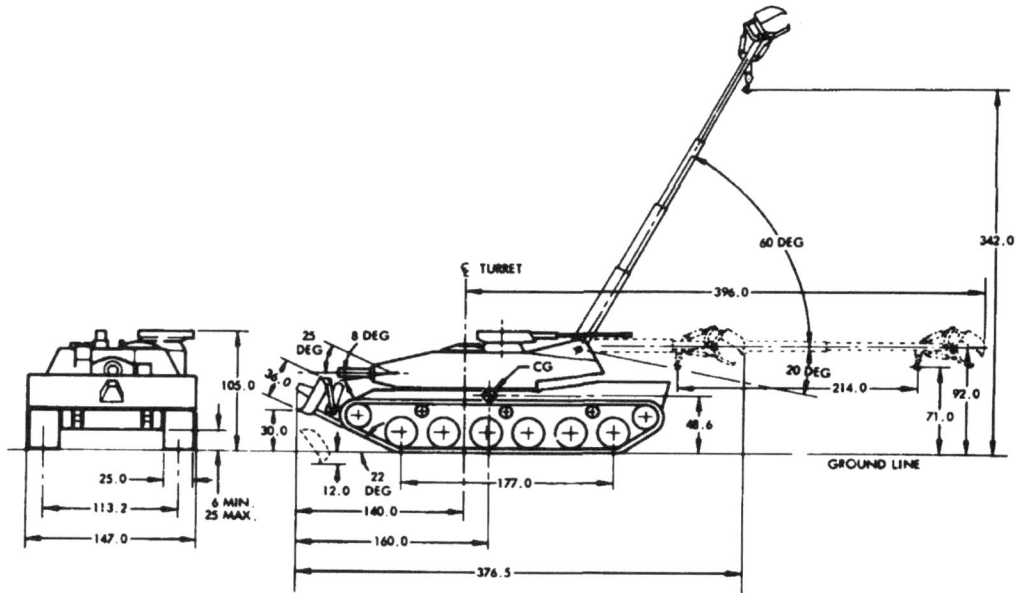

Above is a dimensional sketch of the proposed combat engineer vehicle XM745 based on the chassis of the MBT70 tank. The long reach of the back hoe bucket and gripper is obvious.

During the MBT70 development program, several companion vehicles were proposed based on the same chassis. One of these was designated as the CEV XM745. A concept drawing, layout number LK 1041 dated 28 August 1967, depicted a vehicle on the MBT70 chassis armed with the 165mm demolition gun mounted in a four man turret. However, some reports refer to the installation of a 152mm demolition gun in the XM745. The crew consisted of the driver, commander, gunner, and loader. The latter also manned a 25mm automatic gun mounted on top of his vision cupola. This weapon had a 360 degree traverse and an elevation range of +70 to –18 degrees. Other secondary armament consisted of a 7.62mm machine gun coaxial with the demolition gun. A 60,000 pound capacity tow winch was located in the turret bustle below the mount for the 20,000 pound capacity boom assembly. The latter could be extended from 144 inches to approximately 360 inches and it was fitted with a back hoe bucket and gripper as well as a lifting hook. An hydraulically operated bulldozer fitted with a ram was installed on the front of the vehicle. The CEV was powered by the AVCR-1100 engine using a General Motors Allison XHM-1500 transmission. As an alternate CEV for use with the MBT70, it was proposed to install a three man turret with similar equipment on the chassis of the M60A1 tank. In this case, the driver remained in the hull. After the cancellation of the MBT70 and XM803 programs, the CEV role continued to be performed by the M728 based on the M60A1 tank.

The alternate CEV proposal mounting a three man version of the XM745 turret on an M60A1 chassis is sketched below. Here the driver remained in the hull.

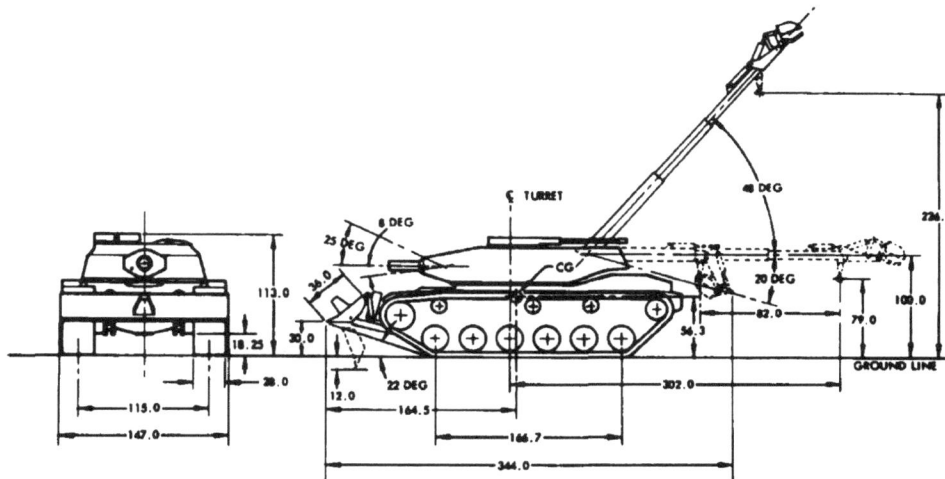

257

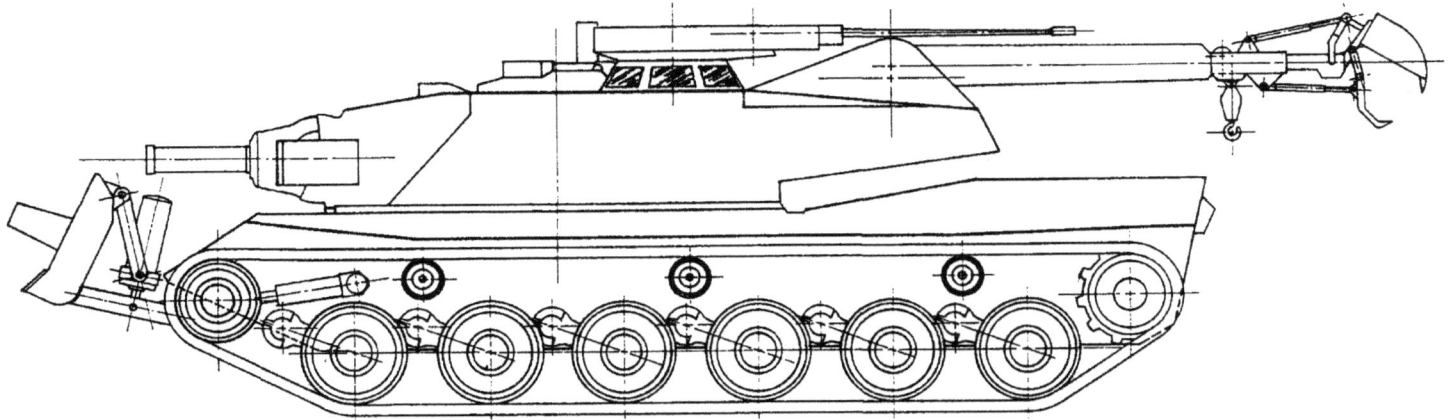

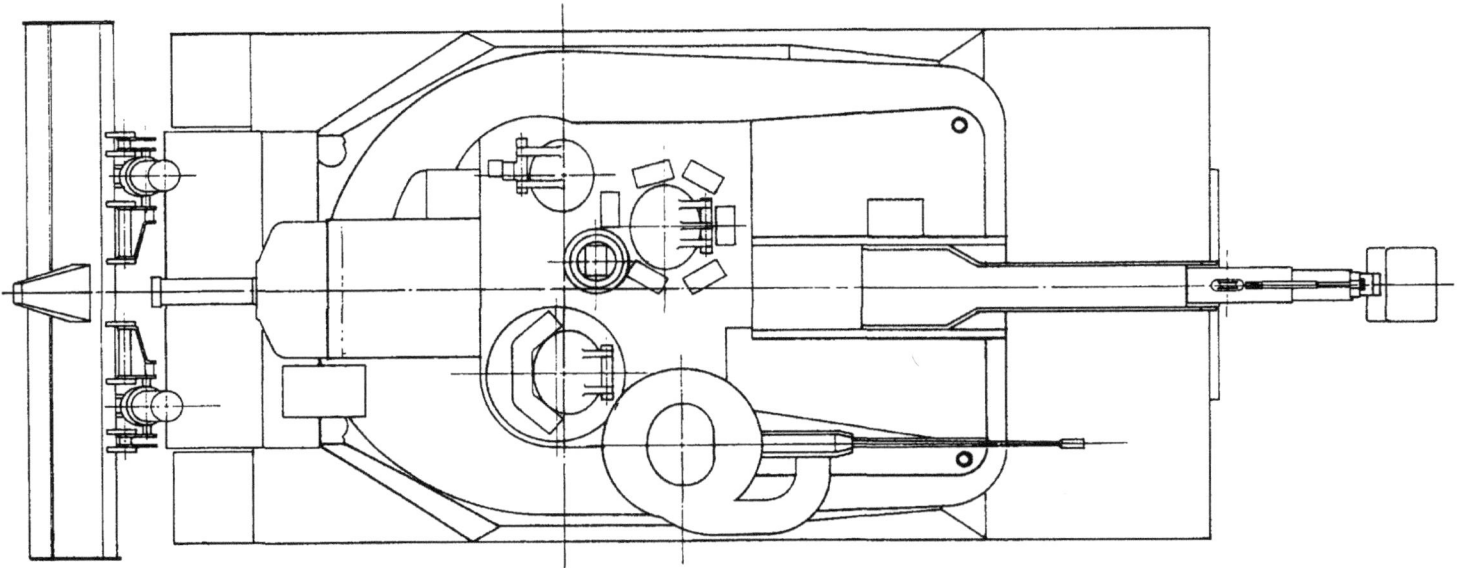

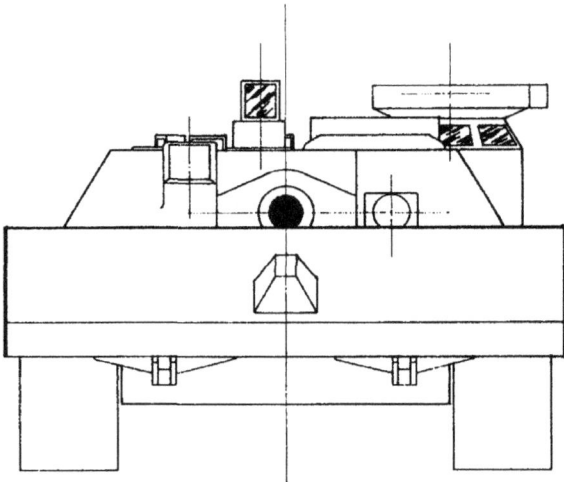

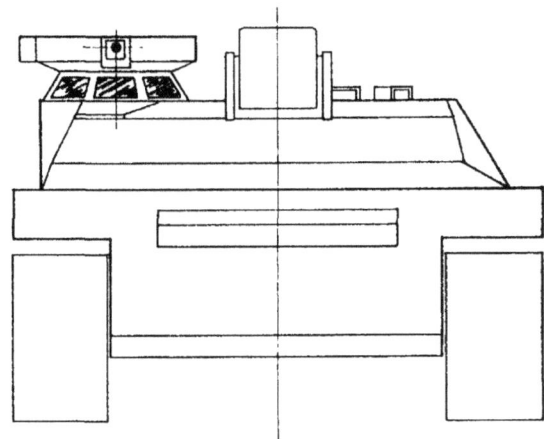

Scale 1:48

© Dan Graves

Combat Engineer Vehicle XM745

Above is the XM1060 robotic obstacle breaching assault tank (ROBAT) after completion at Detroit.

By the early 1980s, two new programs were underway to develop specialized engineer armored vehicles for use during assault operations. One of these, designated as the robotic obstacle breaching assault tank (ROBAT), was intended specifically for clearing and marking a path through a minefield. The other program covered the development of the counter obstacle vehicle (COV) for use in breaching a wide variety of obstacles as well as penetrating minefields.

A demonstrator version of the ROBAT using a turretless M60 series tank chassis was under test at Fort Knox in early 1982. The vehicle could be driven by remote control and it was fitted with explosive line charges which could be projected into a minefield. Detonation of the line charges exploded the mines clearing a path through the field. The ROBAT could then be driven by remote control along this path with its roller type mine exploder detonating any mines missed by the line charges. After these preliminary tests, two prototype ROBATs were authorized for construction at Detroit and the first arrived at Aberdeen Proving Ground on 17 March 1986. After initial inspection and some minor modification, automotive testing began in May followed

Two additional views of the ROBAT are shown below. In the left photograph, the launcher for the explosive line charge is erected in the firing position.

259

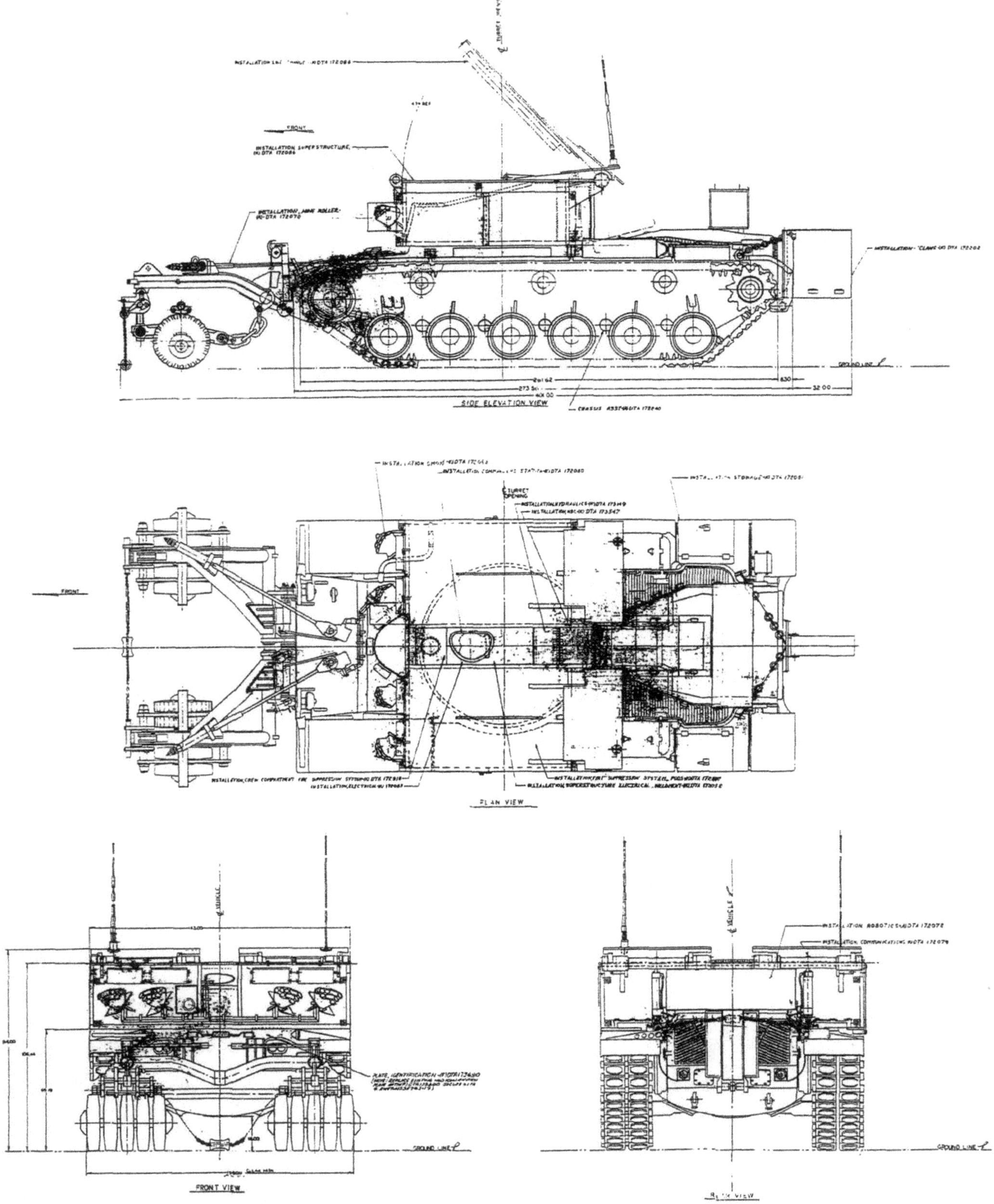

The XM1060 robotic obstacle breaching assault tank (ROBAT) is shown in these drawings.

by safety checks with inert line charges in June and July. The second prototype arrived at Aberdeen during the third week of August and the first vehicle was then shipped to Fort Knox for operational testing which began in late September 1986.

The two ROBAT prototypes were designed and fabricated at the TACOM Research, Development, and Engineering Center and they differed widely from the original demonstrator vehicle. They were based on a modified M60A3 tank chassis with the turret replaced by a heavy armor base plate. Two armored pods containing the explosive line charges were mounted on top of the base plate, one on each side. The two man crew consisted of the driver and the commander. The driver occupied the normal position in the tank hull with the commander located above and to his rear between the two line charge pods. The intended mode of operation was for the crew to drive the ROBAT close to the assault area, remove the remote control equipment to a suitable observation point, and then drive the vehicle by remote control to the edge of the minefield. They would then launch the line charges and proceed to clear and mark a path through the minefield. The ROBAT prototypes, designated as the XM1060, were

fitted with four kits consisting of a mine clearing roller, the remote control kit, a nuclear, biological, chemical (NBC) kit, and a cleared lane marking kit (CLAMS). The latter marked the cleared lane with chemiluminescent light sticks. Following the completion of the operational testing at Fort Knox in February 1987, both ROBAT prototypes were returned to TACOM for the correction of deficiencies revealed during the test program.

The counter obstacle vehicle (COV) was a project of the Engineer Support Laboratory at the Belvoir Research and Development Center. To defeat all types of obstacles in an hostile environment, it was equipped to breach minefields and was fitted with a combination of excavating, dozing, lifting, and hauling equipment. It also could be used to construct roads in a combat zone. A contract was awarded to the BMY Division of Harsco Corporation to perform the detailed design work and to fabricate two test bed vehicles. The design work began at BMY's York, Pennsylvania plant in June 1982 and continued for one year. Fabrication of the test beds started in July 1983 and extended through December 1984. Tests at BMY ran from January through April 1985. Shipped to the Engineer Proving Ground at Fort Belvoir, the COV test bed was evaluated from October 1985 through February 1986.

The original concept for the COV was based on utilizing the chassis of the M88A1 medium recovery vehicle also manufactured by BMY. However, by the time the actual vehicle was complete, the basic design was drastically modified. In the COV, the three man crew was seated in tandem along the vehicle center line with the driver in front followed by the equipment operator and the vehicle commander. Two hydraulically operated rotatable telescopic arms were mounted at the front of the vehicle, one on each side. In the stowed position, they extended to the rear on each side of the crew compartment. These arms were fitted with 1.3 cubic yard excavator buckets or other attachments

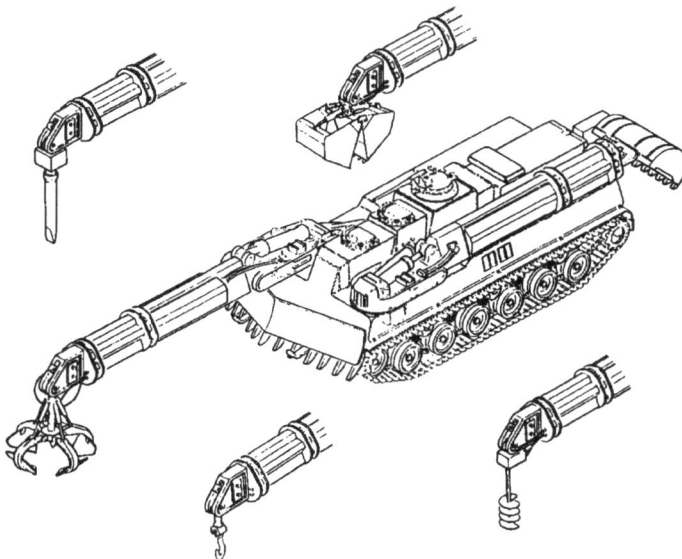

The various attachments for the counter obstacle vehicle (COV) are sketched at the left. Below, an early COV concept drawing appears at the left and the vehicle is being assembled at the right.

Above is the counter obstacle vehicle (COV) test bed photographed in January 1985. The outboard extensions have not been installed on the mine plow which is in the travel position.

including a 24 inch diameter earth auger, a four tine 4500 pound capacity grapple, a lifting hook, or an hydraulic hammer. The COV test bed was powered by a 908 gross horsepower AVDS-1790-5 diesel engine with an XT-1410-5X transmission. The running gear was similar to that on the M88A1, but it featured higher strength torsion bars and an hydraulically actuated lockout system on the first, second, and sixth road wheel stations to provide stability during digging or lifting operations. A full width mine plow developed in Israel was installed on the front of the vehicle. The teeth of the mine plow could be fitted with covers and the two end sections removed permitting it to operate as a conventional dozer blade. For mine clearing operations, it was intended to mount a mine roller or a mine clearing line charge (MICLIC) as well as the cleared lane marking system (CLAMS).

The fully loaded weight of the COV test bed was over 75 tons. This was considered excessive, particularly since the armor protection on the test bed vehicle was inadequate for its proposed use. Further work was initiated in an effort to reduce the weight and increase the level of protection by modifying the design and utilizing new materials.

Below, the COV is operating the excavator buckets (left) and removing a wrecked automobile with the grapnel (right).

Above, the COV is being operated by remote control and at the right is a closeup view of the earth auger installed on one of the telescopic arms.

Above, the COV is climbing a vertical wall with the help of the hydraulically operated telescopic arms. The drawings at the right and below show the mine plow with its outboard extensions installed on the COV.

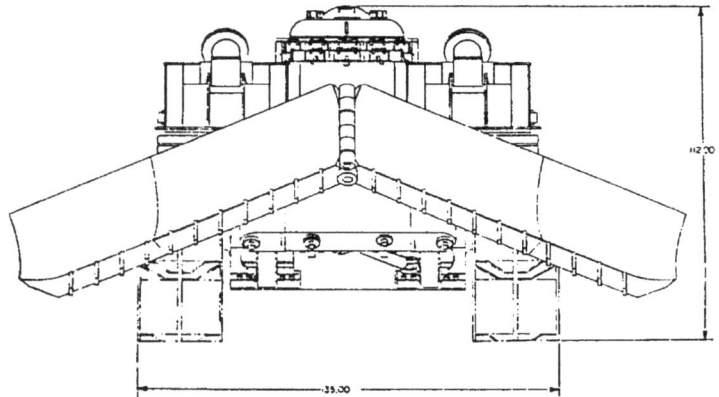

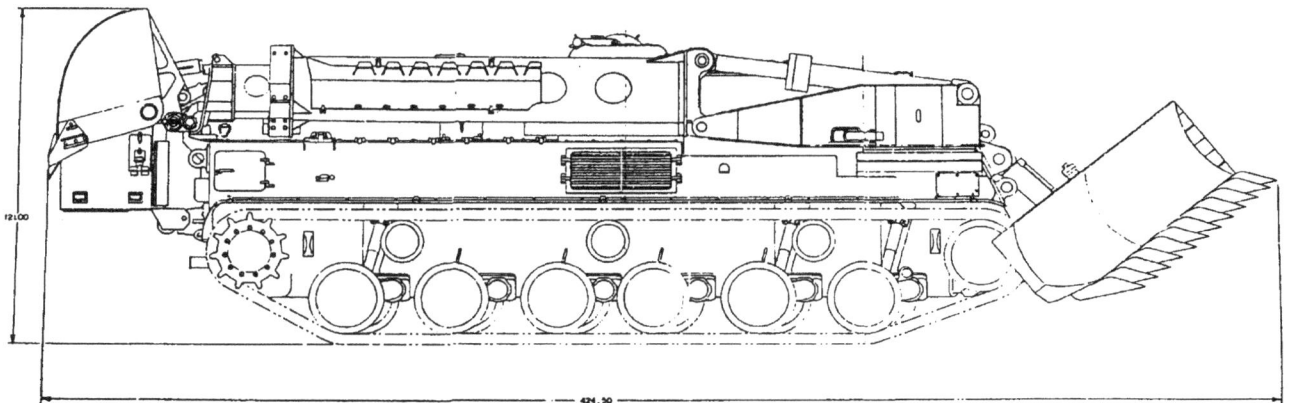

263

The drawing above shows the proposed armored vehicle launched bridge (AVLB) based upon the T95 tank chassis.

BRIDGING EQUIPMENT

During the development of the CEV T118 based on the T95 tank, an additional chassis was authorized for use with an armored vehicle launched bridge (AVLB). The T95 tank chassis for the AVLB development was shipped to the Corps of Engineers at Fort Belvoir during January 1959. This turretless vehicle was to be fitted with a 45 ton capacity folding bridge capable of spanning a 60 foot gap. Manned by a crew of two, it was 408 inches long and 144 inches wide with the bridge in the travel position. Overhang of the bridge and launching mechanism was 63 inches in front and 73 inches in the rear. The estimated weight was 44 tons for the vehicle and 11 tons for the bridge giving a total weight of 55 tons. After the cancellation of the T95 program, the project was dropped and the requirement for AVLBs was met by 60 ton capacity folding aluminum bridges installed on M48A2 and M60A1 tank chassis. Bridges capable of spanning 40 or 60 foot gaps were used.

The dimensions of the 60 ton capacity AVLB on the M60A1 tank chassis can be seen at the right.

The companion vehicles for the MBT70 also included an AVLB. The proposed vehicle based on the MBT70 chassis was designated as the XM743 and it carried the 60 ton capacity XM744 double folding bridge which could span a gap 100 feet wide. Research had indicated that a 100 foot bridge would be far more effective than the 60 foot model mounted on the M60A1 AVLB. Alternate versions of the XM743/744 combination were considered which launched the bridge in either the forward or the aft direction. The study indicated that the installation for aft launching provided much better visibility for the driver. The XM744 bridge also was considered for use on the M60A1 chassis using the aft launch design.

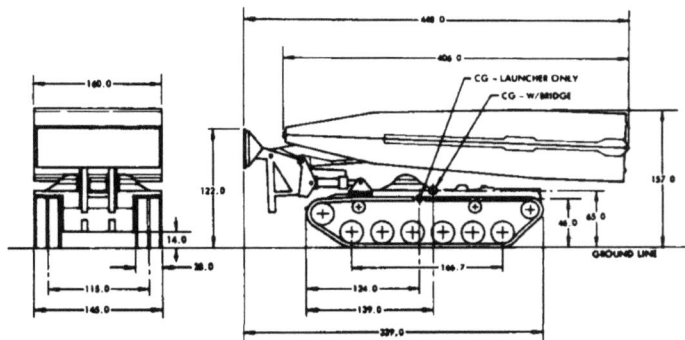

264

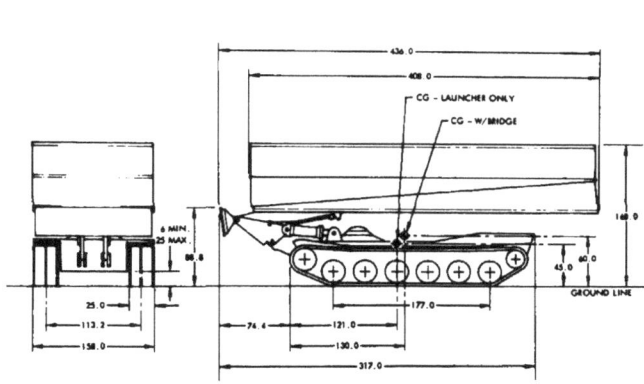

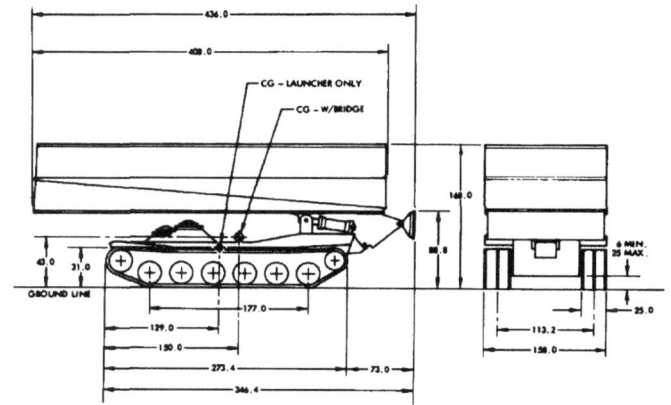

The sketches above show the dimensions of the AVLB XM743/744 combination based upon the MBT70 chassis. This unit was designed to launch the bridge in either the forward (left) or aft (right) direction. At the right, the XM744 bridge is mounted on the M60A1 chassis and it is arranged to launch in the aft direction.

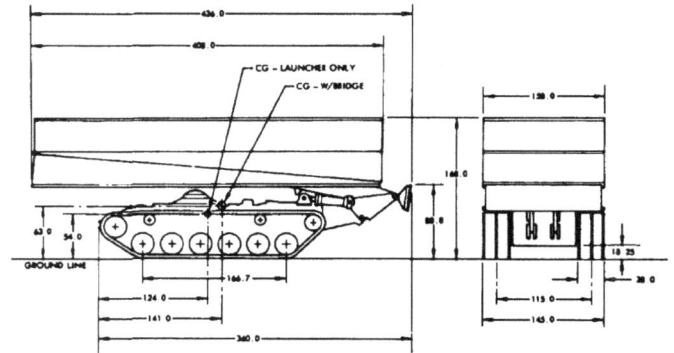

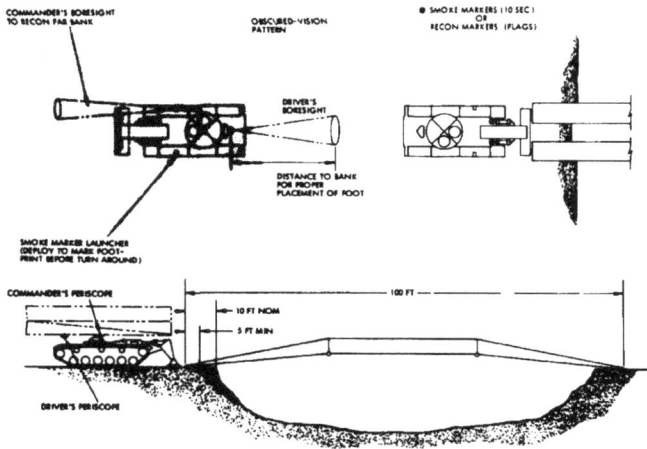

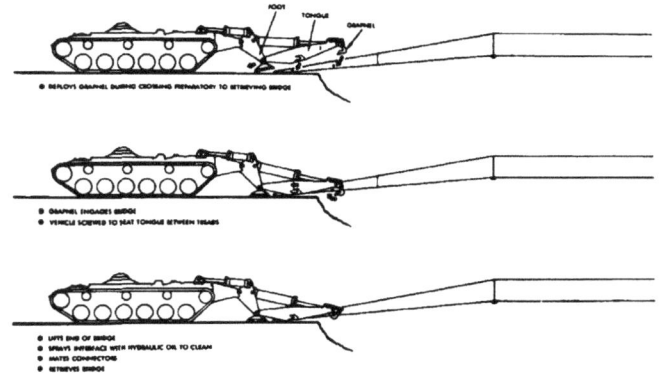

Above, the aft launch (left) and retrieval (right) procedures are shown for the XM744 double folding bridge. Below is a drawing of the bridge itself fully extended for use.

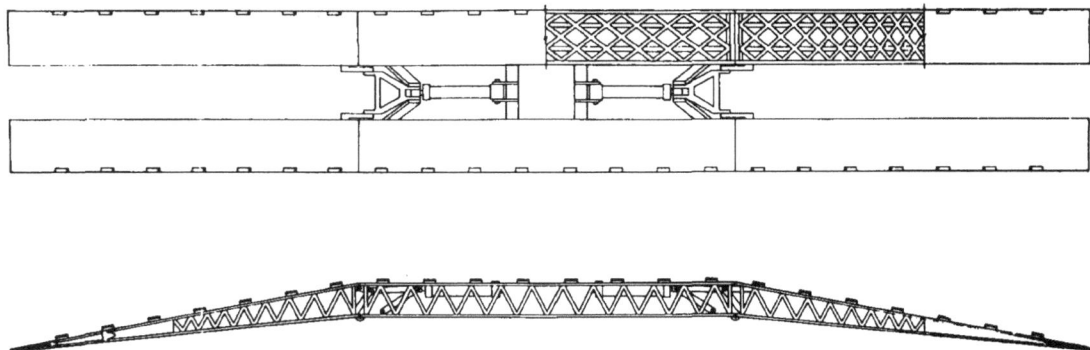

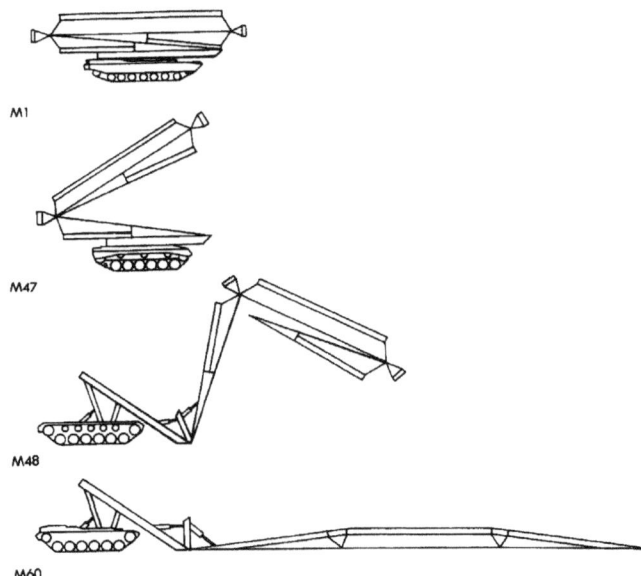

Above from left to right are the HAB, the TLB, and the 60 ton AVLB being launched from the TLB trailer. The sketch at the right shows the HAB installed on the M1, M47, M48, and the M60 tank chassis.

With the appearance of the heavier M1 tank series, the 60 ton capacity AVLBs were no longer adequate. Also, research had shown that it would be highly desirable to have assault bridges capable of spanning gaps greater than 60 feet, preferably up to 100 feet. In addition, the AVLB on the M60A1 tank chassis lacked the speed and mobility required to operate with the new tanks.

After preliminary concept studies, the Army issued a request for proposals from industry for the design of a heavy assault bridge (HAB) with a 70 ton capacity. The longest possible bridge consistent with mobility considerations was required. It also was to be capable of installation on either the M60A1 or the M1 turretless tank chassis. After review of the various proposals, a contract was awarded to the BMY Division of Harsco with Israel Military Industries (IMI) as a major subcontractor. The latter was to design and produce three prototype bridges from high strength aluminum alloys and carbon-epoxy composite materials. BMY was to develop the hydraulic launch mechanism and install the bridge and launcher assembly on two M1 tank chassis and on one M60 series tank chassis. The first prototype HAB began testing at BMY early in 1986. These tests revealed the necessity for strengthening some areas to prevent crack formation. The first HAB was modified to incorporate these changes which also were applied to the remaining prototypes. The 32 meter (106 feet) double fold HAB was capable of spanning a 30 meter (100 feet) gap. Installed on the M1 tank chassis, the prototype HAB was 538 inches long and 158 inches wide with a weight of 123,000 pounds. The same installation could carry and launch the standard 60 ton capacity AVLB and the 70 ton capacity trailer launched bridge of the United States Marine Corps.

At the right is an artist's concept of the TLB in operation.

With the adoption of the M1A1 tank, the Marine Corps also needed a new assault bridge. However, unlike the Army, the Marine Corps required the assault bridge to be transportable in a C130 aircraft. The HAB could only be handled in the much larger C5A. On 20 April 1984, Israel Military Industries (IMI) received a contract to design and fabricate two prototypes of a trailer launched bridge (TLB) with a 70 ton capacity to meet this requirement. This time, BMY was the subcontractor to develop the launcher and trailer assembly. The first prototype TLB began company testing in mid 1986.

The TLB was 24 meters (approximately 79 feet) in length and it was designed to span a gap 22 meters (approximately 72 feet) wide. Like the HAB, the TLB was a three section double folding bridge. It was launched from a tilting frame mounted on a two axle trailer. To fit within the 117 inch wide space available in the C130 aircraft, the TLB's normal width of 150 inches was reduced to 115 inches by

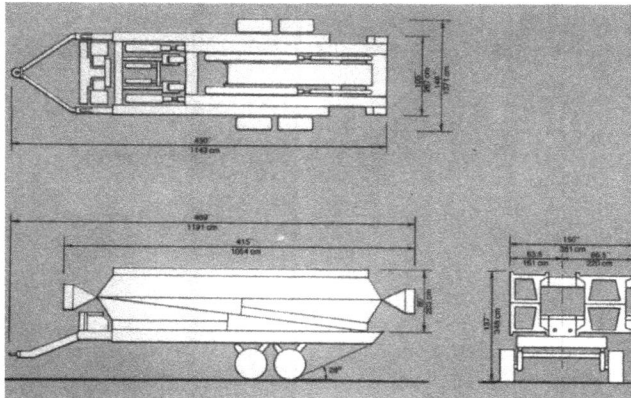

The TLB can be towed and launched by any of these vehicles:

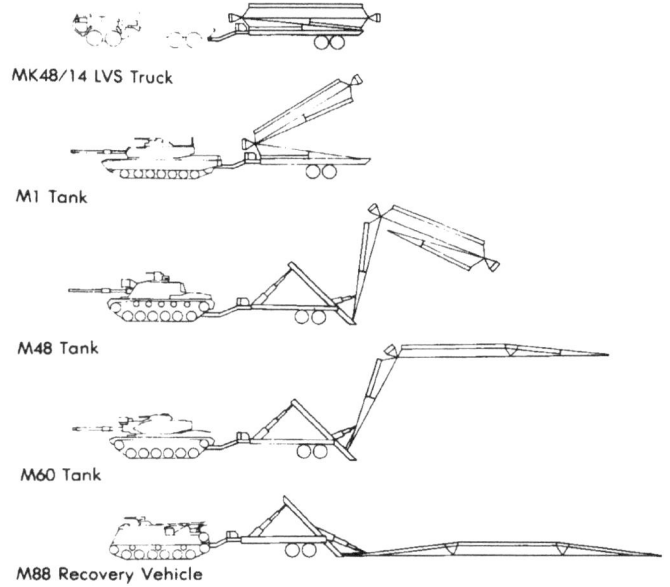

MK48/14 LVS Truck

M1 Tank

M48 Tank

M60 Tank

M88 Recovery Vehicle

The drawing at the upper left shows the dimensions of the TLB stowed on its trailer and the width reduction when the central beam unit is telescoped can be seen in the lower drawing at the left. At the right is the launch sequence for the TLB.

telescoping the central beam unit. The trailer suspension and wheels were removed when stowed in the aircraft. Both the trailer and the bridge were fabricated from 7005 and 7075 high strength aluminum alloys. To accommodate wide vehicles such as the M1A1 tank, one bridge trackway was designed much wider than the other with no outer curb. This allowed the vehicle to overhang on that side. Even in this condition, the load was carried on the major structure. The TLB was deployed by three sets of hydraulic cylinders powered by two diesel engines. Increasing the hydraulic pressure could compensate for the loss of one cylinder in each set and operations could continue with only one diesel engine available. On its trailer, the TLB

had an overall length of 469 inches and weighed about 31,000 pounds. The TLB's trailer launcher also could be fitted with the HAB or the standard 60 ton capacity AVLB.

Another towed assault bridge (TAB) developed by IMI and evaluated at Aberdeen Proving Ground also had a 70 ton capacity and was capable of spanning a 10 meter (approximately 33 feet) gap. Weighing about 11 tons, the TAB could be divided into two sections along its longitudinal axis for stowage in a C130 aircraft. The load carrying body of the TAB was 12 meters (approximately 39 feet) in length. The overall length of 19 meters (approximately 62 feet) included an extendable front section and the coupling hook.

Below the TAB is being pushed into position (left) and crossed (right) by an M60A1 tank.

Above, the standard M88A1 recovery vehicle appears at the left. It could be identified from its predecessor, the M88, by the grill door for the auxiliary power unit on the right side of the cab. The door for this unit on the M88 is marked by the arrow B in the right-hand photograph.

RECOVERY VEHICLES

After the concept of a combined tank recovery and combat engineer vehicle was dropped, work continued on the development of a dedicated recovery vehicle. The T88 developed by BMY was standardized in February 1959 as the medium recovery vehicle M88. It remained in first line service until replaced by the M88A1 which entered production in June 1975. The primary difference between the M88 and the M88A1 was the replacement of the gasoline engine in the earlier vehicle with the AVDS-1790-2DR diesel in the M88A1.

During the MBT70 program, a design was prepared for a recovery vehicle based upon that chassis. Designated as the RV XM742, it was fitted with a turret mounted telescoping boom which projected forward when stowed in the travel position. The four man crew was housed in a fully rotating turret and the vehicle was equipped with the same dozer kit as the proposed CEV XM745. As an alternate recovery vehicle for use with the MBT70, the same telescoping boom arrangement in a low silhouette turret was

proposed for installation on the M60A1 chassis. With this design, the driver was located in the hull and the three remaining crew members rode in the turret basket.

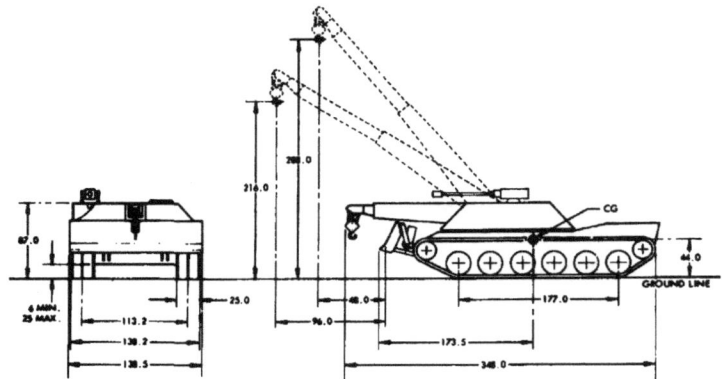

The dimensions of the XM742 recovery vehicle are shown above and they can be compared below with those for the M88A1 (left) and the proposed recovery vehicle mounting a low silhouette version of the XM742 turret on the M60A1 tank chassis (right).

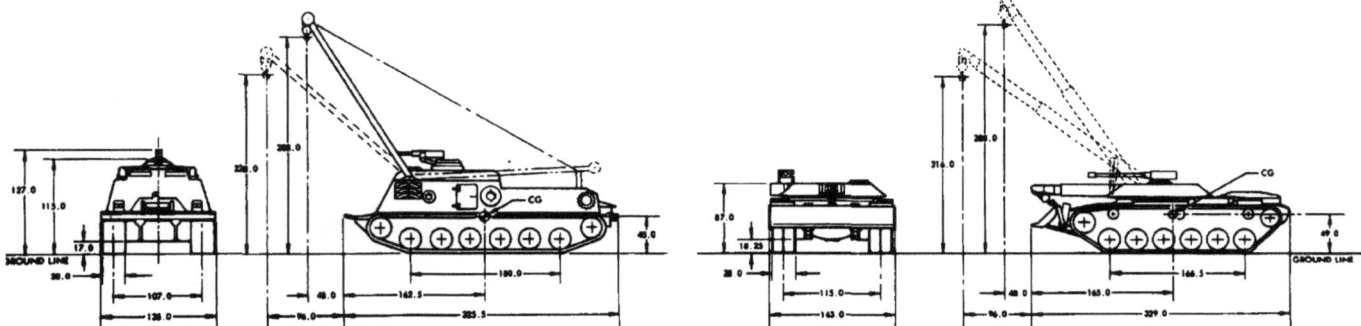

268

The sketch at the right lists the improvements over the M88A1 proposed for the M88AX.

Although the M88A1 was satisfactory for use with the M60 series of tanks, the introduction of the 60 ton M1 and the even heavier M1A1 created a serious problem. When a tank being towed is heavier than the towing vehicle, it becomes very difficult to handle. On a slope exceeding 10 per cent, it is extremely dangerous and standard practice required two M88A1s to safely tow a single M1 tank. Recognizing this problem, BMY initiated an independent development program in 1984 to upgrade the M88A1. The result was a test bed vehicle referred to as the M88AX. The 750 horsepower engine in the M88A1 was replaced by the 1050 horsepower AVDS-1790-8AR diesel with a modified transmission. Ballast weight was added to increase the vehicle weight to 65 tons. The M88AX was submitted to the Army for evaluation at Aberdeen Proving Ground in April 1985. These tests revealed that the M88AX could tow an M1 tank more rapidly than the M88A1 could move the lighter M60A1. Some problems occurred with the brakes during the test, but modifications were made and satisfactory performance was demonstrated by December 1985. For comparison purposes, an M1A1 tank also was tested as a towing vehicle to evaluate its power train for use in a recovery vehicle. Some concern had been expressed that it would not be suitable for such service as it had been optimized for high speed and agility for use in the main battle tank. However, with its 1500 horsepower gas turbine, the M1A1 turned in an excellent performance exceeding that of the M88AX.

Further analysis of the M88AX revealed that it would be expected to have greater mobility handling the M1 in European terrain than the M88A1 with the M60A1. Thus it was considered suitable for such service. As a result of the evaluation, the Army awarded a contract in January 1987 to BMY for the procurement of five prototype vehicles now designated as the M88A1E1. With the heavier M1A1 scheduled for early fielding in Europe, the need for an improved recovery vehicle was becoming urgent and the Army intended to procure the M88A1E1 as the M88A2, if it proved to be satisfactory. The combat loaded weight

of the M88A1E1 was 67 tons compared to 56 tons for the M88A1. Six tons of armor added as an overlay on the cab and as track skirts increased the level of ballistic protection. With the new power plant, the M88A1's maximum speed of 26 miles per hour without a towed load was

The M88AX is towing the Abrams tank in these photographs taken during the test program. Steel slabs have been mounted on the vehicle to equal the weight of the proposed heavier armor.

In the view below, the M88AX is towing the M1A1 tank and the sketch at the right shows the configuration of the M88A1E1 recovery vehicle.

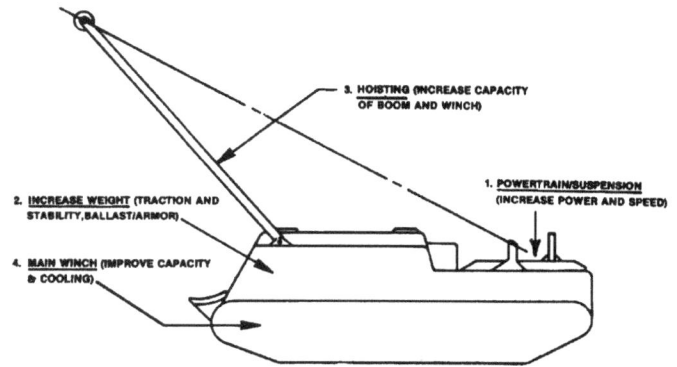

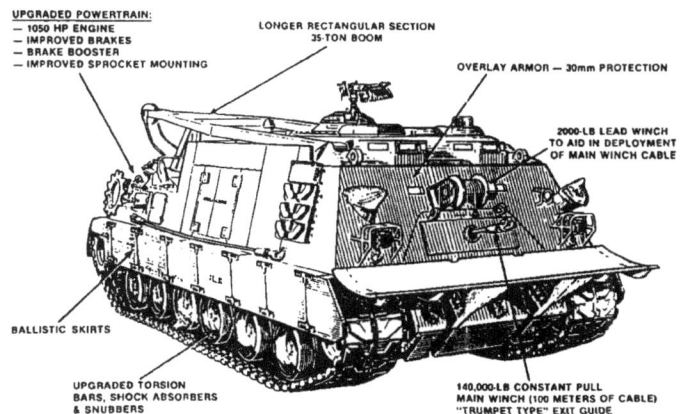

The photographs on this page show the M88A1E1 towing the M1A1 tank during the evaluation program. Note the blocky appearance resulting from the overlay armor and the side skirts.

retained. Fuel capacity was increased from 400 to 450 gallons to maintain the same range of 300 miles. The M88A1E1 utilized the same A frame boom arrangement as the M88A1, but the boom capacity and the hoist winch pull was increased from 50,000 to 60,000 pounds. The main winch pull was increased from 90,000 to 140,000 pounds.

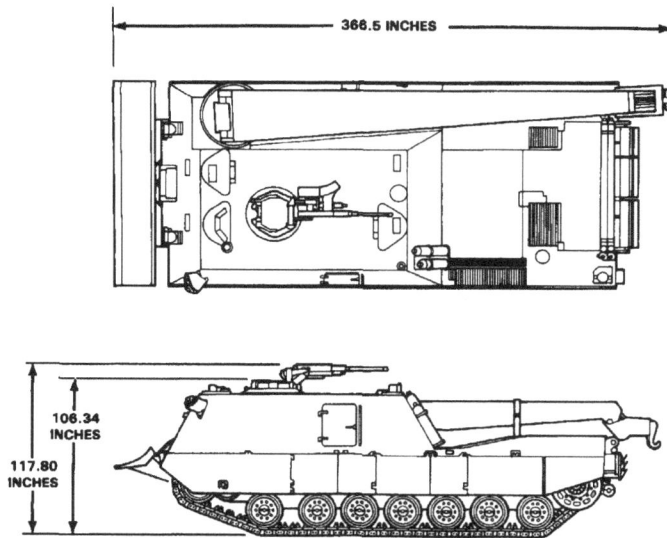

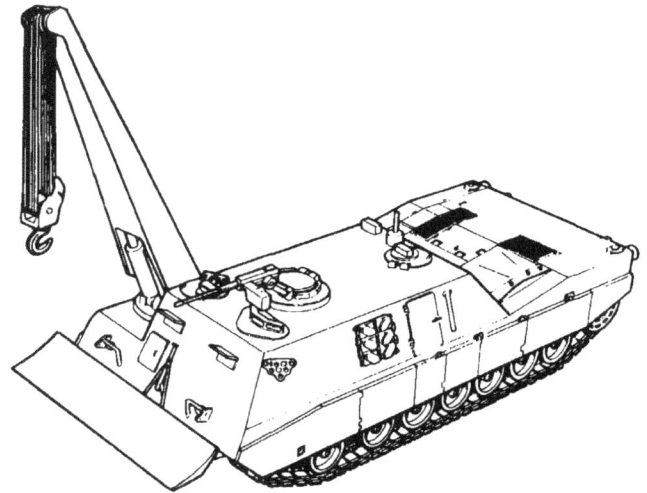

The drawings above depict the early configuration of the recovery vehicle based upon the Abrams tank with the crane mounted on the right side of the hull.

TACOM also had anticipated the requirement for a new recovery vehicle and had developed preliminary drawings of such a vehicle based on the M1A1 tank. In mid 1985, the Army drafted a required operational capability (ROC) document for a new recovery vehicle, designated as the RV90, incorporating performance requirements based on the Abrams. General Dynamics Land Systems further developed the concept utilizing the M1A1 tank chassis and the 35 ton capacity rotating crane from a new German recovery vehicle. Based on the Abrams with its 1500 horsepower turbine, the performance of the RV90 obviously would exceed that of the M88A1E1 powered by the 1050

horsepower diesel. This already had been observed at Aberdeen when comparing the towing ability of the M88AX and the M1A1 tank. In view of the Army's intention to procure the M88A1E1 at the earliest possible date, the ROC document had been revised to correspond with its reduced performance level. However, in September 1986, General Dynamics offered to build a recovery vehicle based on the Abrams using their own funds, if the Army would agree to test it in competition with the M88A1E1. The Army declined the offer, not wishing to disrupt its procurement plans for the M88A1E1 which it believed would be adequate for the job, less costly, and available at an earlier date. At this point, Congress entered the picture with the House Armed Services Committee requiring that the Army conduct a fair competition between the two vehicles and that they be tested against the original ROC document and not the revised version.

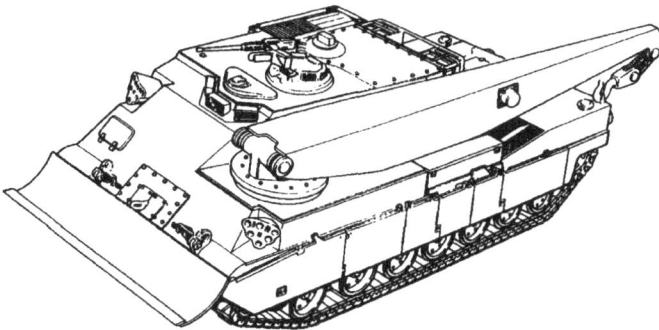

The final design of the Abrams recovery vehicle (ARV) is sketched at the left. Note that the crane has been relocated to the left side of the vehicle. Below, a model of the ARV appears at the left and the vehicle itself is at the right.

Above is a photograph of the Abrams recovery vehicle after completion at the Detroit Army Tank Plant in February 1988.

The new Abrams recovery vehicle (ARV) was rolled out at the Detroit Army Tank Plant in February 1988. One obvious difference from the earlier concept studies was the relocation of the 35 ton capacity rotating crane from the right to the left side of the hull. Manned by a crew of three, the 67 ton vehicle had space for the four man crew of a disabled tank. Its maximum speed was a little over 40 miles per hour and the maximum range was about 300 miles. Like the M1A1 tank, it was equipped with an NBC overpressure system and microclimate cooling vests for the crew. It also included an automatic Halon fire extinguisher system.

Below, the new Abrams recovery vehicle is towing a late production M1A1 tank.

The photographs above and below show the M88A1E1 recovery vehicle towing an M1A1 tank during test operations.

The comparative tests between the Abrams recovery vehicle and the M88A1E1 were carried out during May and June 1988. After review, the Army selected the M88A1E1 as the winner and put additional funds on the full scale development contract already awarded to BMY. At that time, the Army intended to procure 849 of the new vehicles as the M88A2 through Fiscal Year 1994. However, this decision was reversed in April 1989 when the Secretary of Defense cancelled the contract and allocated the funds to higher priority programs.

Below is another view of the M88A1E1 hauling the M1A1 over rough terrain.

Above is the full size mock-up of a new tank design concept with the main gun in an external mount.

THE FUTURE

The Abrams tank obviously will play a major role in the United States armed forces well into the next century. Like its predecessors, it will be modified and improved to counter changes in the threat that it will be called upon to meet. The Abrams itself was the last of a long line of research projects and experimental vehicles intended to replace the Patton tank and its product improved version, the M60 series. As with the earlier tanks, even before the Abrams was standardized, research projects were studying possible future replacements. The future close combat vehicle studies at TACOM investigated a wide variety of configurations. Frequently, the results of such research programs are applied as product improvements to the tank in service rather than replacing it with a completely new vehicle. This approach already has been applied to the

Abrams with the Block I and II improvements resulting in the M1A1 and the M1A2. The Block III requirements will eventually produce an M1A3 or a completely new tank.

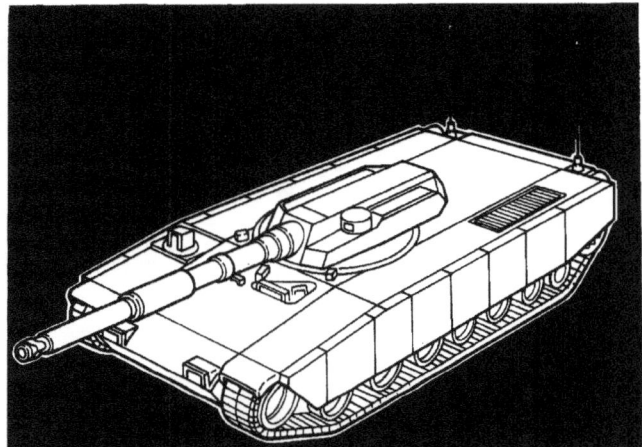

The drawing at the right illustrates the external gun tank concept.

274

The tank test bed (TTB) based upon the Abrams tank chassis appears above. The bore evacuator has been omitted from the 120mm gun in this installation.

Many of the new research projects utilize the Abrams tank chassis in their investigations. For example, the tank test bed (TTB) program was initiated in 1980 to explore the application of new technology and advanced vehicle designs for future tank use in the 1988-1996 time period. Using a modified M1 tank chassis, the three man crew of the TTB was seated side by side in the front hull. An M256 120mm gun was installed in a small unmanned turret and it was fed automatically from a 44 round magazine in the hull. Three externally mounted electro-optical sight groups were installed with two on the sponsons and one on the turret.

Below, the stations for the three man crew of the tank test bed are at the left and the automatic loader for the 120mm gun can be seen at the right.

TANK TEST BED (TTB) CREW PANELS

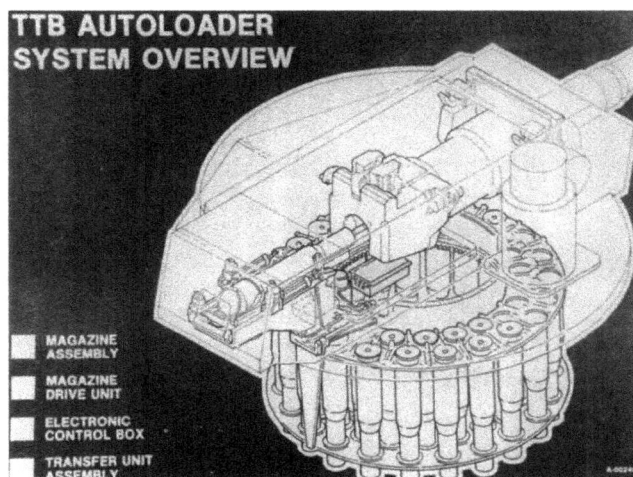

TTB AUTOLOADER SYSTEM OVERVIEW

MAGAZINE ASSEMBLY

MAGAZINE DRIVE UNIT

ELECTRONIC CONTROL BOX

TRANSFER UNIT ASSEMBLY

Details of the original surrogate research vehicle (SRV) are visible in the two top photographs. At the right is a later view of the SRV after some modification.

In November 1980, the surrogate research vehicle (SRV) project was launched, also using the modified Abrams tank chassis. As on the TTB, the forward fuel cells were removed to provide space for three crew stations side by side in the front hull. Two additional crew positions were located in the rotating basket suspended from the turret ring. A major objective of this program was to evaluate different crew arrangements, such as all three men side by side in the front hull versus the driver in the hull with the tank commander and the gunner in the turret. On the SRV, the tank turret was replaced by a low silhouette structure fitted with remote electro-optical vision equipment and a laser simulated weapon system. Phase I of the SRV testing began at Fort Knox in 1983 to provide information on the effect of crew arrangement to the tank test bed program. Phase II tests of the SRV investigated surveillance problems during 1984/85. This was followed by Phase III in 1986/87 devoted to command and control studies.

The configuration of the surrogate research vehicle is illustrated in the drawings below and at the right.

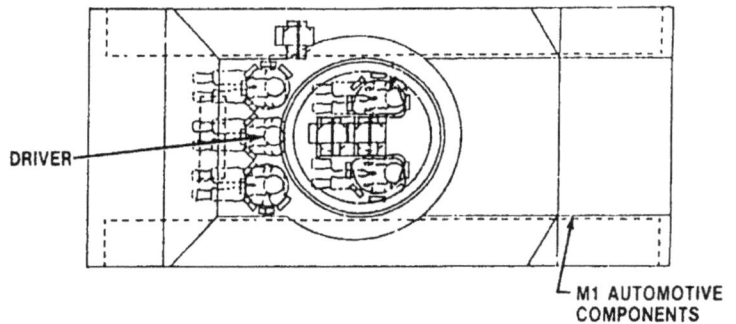

DRIVER

M1 AUTOMOTIVE COMPONENTS

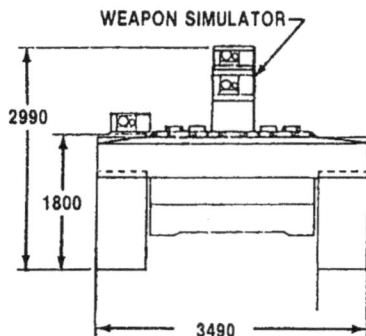

WEAPON SIMULATOR

2990

1800

3490

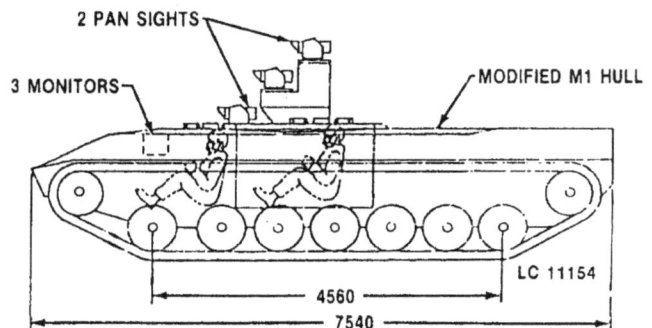

2 PAN SIGHTS

3 MONITORS

MODIFIED M1 HULL

4560

7540

LC 11154

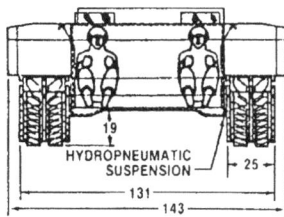

The automotive technology demonstrator (ATD) is sketched above and at the right.

A third contributor to the tank test bed program was the automotive technology demonstrator (ATD). The original TTB and SRV utilized the Abrams chassis, power train, and suspension. The objective of the ATD was to evaluate new components in all of these areas. Designed for a two man crew, its weight could be adjusted by adding ballast for a specific test.

Among the components to be evaluated are the advanced integrated propulsion systems (AIPS). In July 1984 contracts were awarded to General Electric Aircraft Engines and to the Cummins Engine Company to build demonstration systems. Thus the diesel versus turbine controversy appeared again with General Electric building a gas turbine system and Cummins producing a diesel power plant. Both units occupied approximately half the volume of the standard Abrams propulsion system and offered much greater fuel economy. Thus they were prime candidates for the Block III tank.

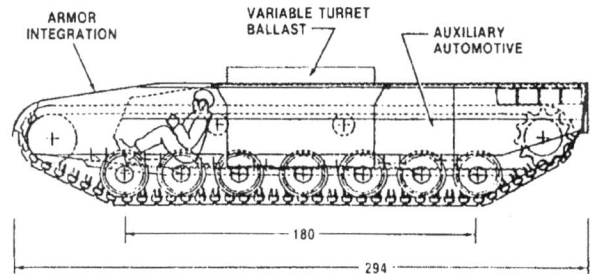

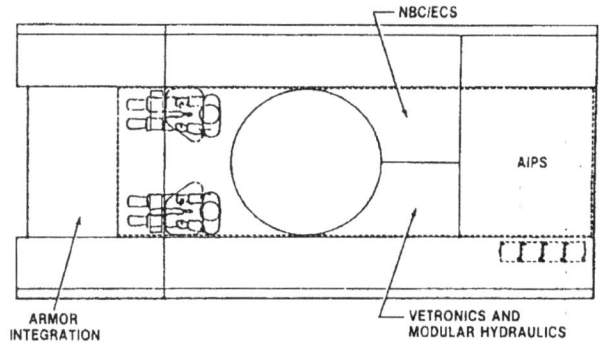

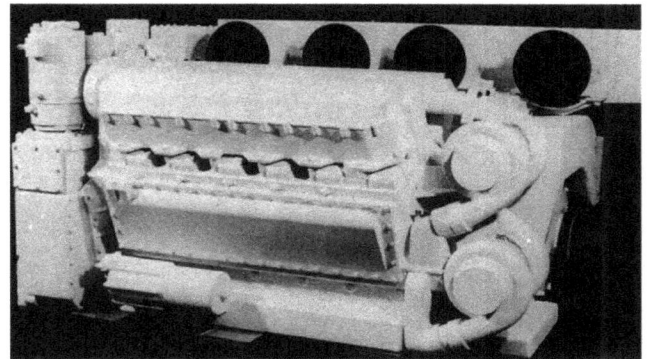

The mock-up of the General Electric—Textron Lycoming LV/100 AIPS candidate appears below at the left. The Cummins XAP-1000 also is shown in mock-up form above and below at the right.

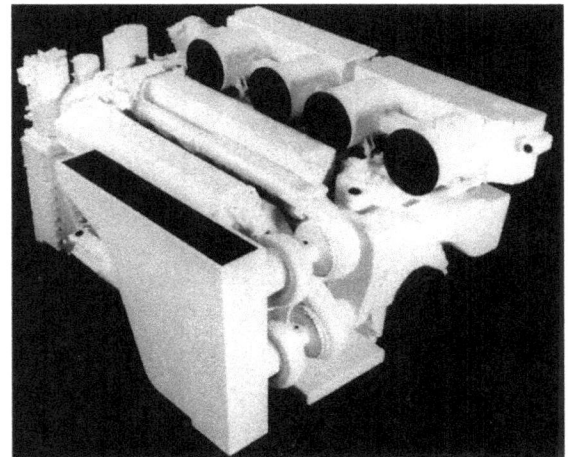

277

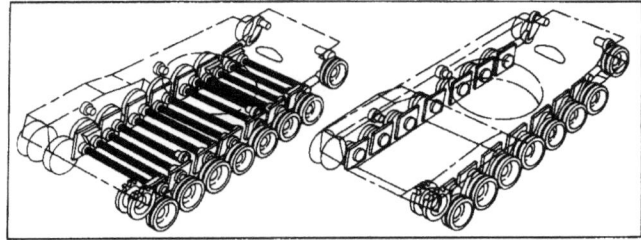

TORSION BAR SUSPENSION SYSTEM   IN-ARM SUSPENSION SYSTEM

The Cadillac Gage in-arm hydropneumatic suspension is installed on the XM1 prototype tank at the left. The installation of this suspension is compared above with the torsion bar system. Note that the in-arm suspension does not require any space inside the tank hull.

An in-arm hydropneumatic suspension developed by the Cadillac Gage Division of Textron, Inc. was successfully tested earlier on one of the XM1 prototype tanks. Weighing about a ton less than a conventional torsion bar system, its external bolt-on mount did not require any space inside the hull. With a total wheel travel of 21.0 inches, it also was under consideration for application to the Block III tank.

The tank test bed program not only produced data for future modifications of the Abrams, but also for concept studies of completely new fighting vehicles. It also provided valuable material for use by the Armored Family of Vehicles Task Force (AFVTF) in their effort to propose common chassis for several classes of armored vehicles. Although the Task Force report indicated that it would be possible to provide common chassis for the heavy, medium,

The drawing below depicts a concept of a future main battle tank with a remote control gun externally mounted in a small low silhouette turret. With the crew separated from the weapon, a bore evacuator was not required.

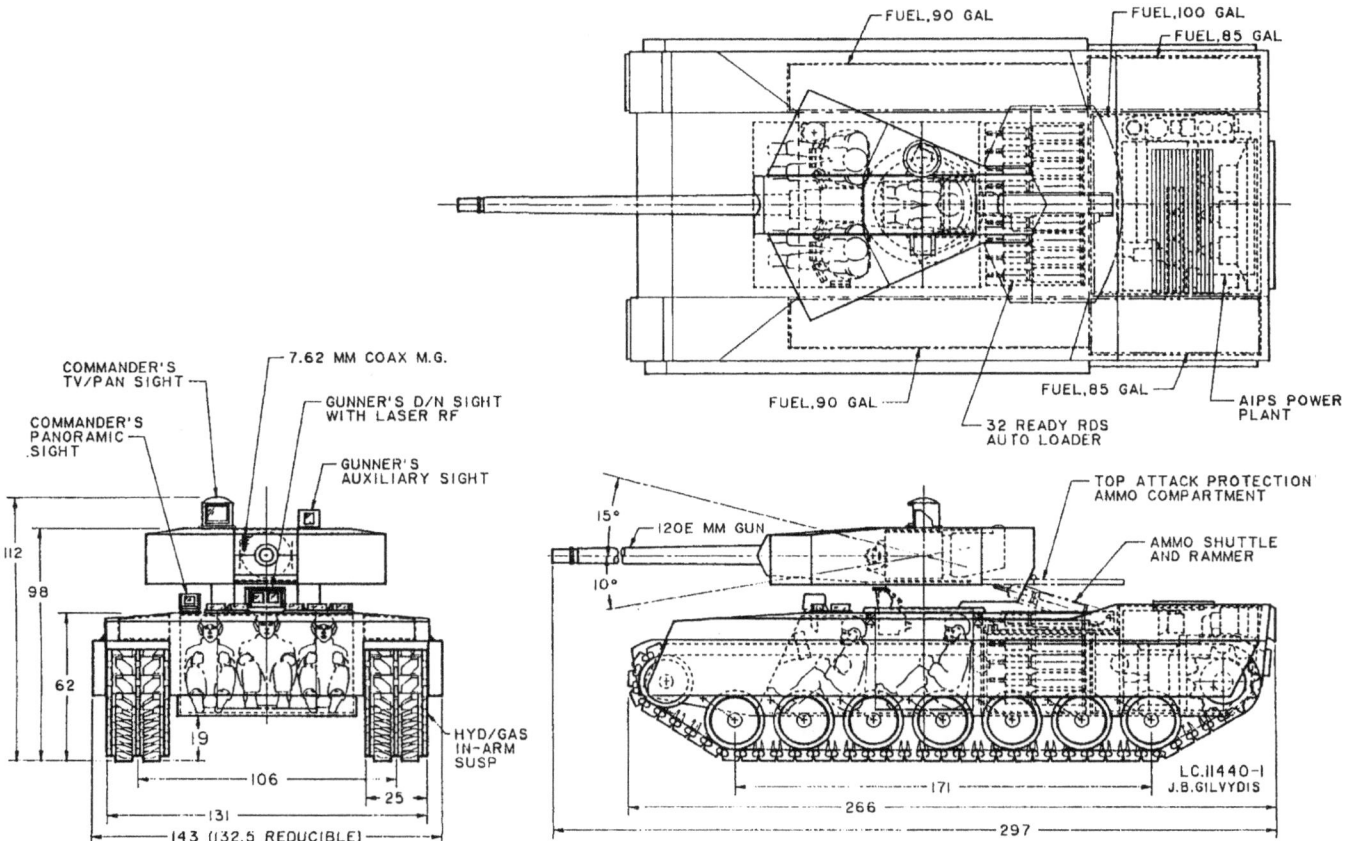

278

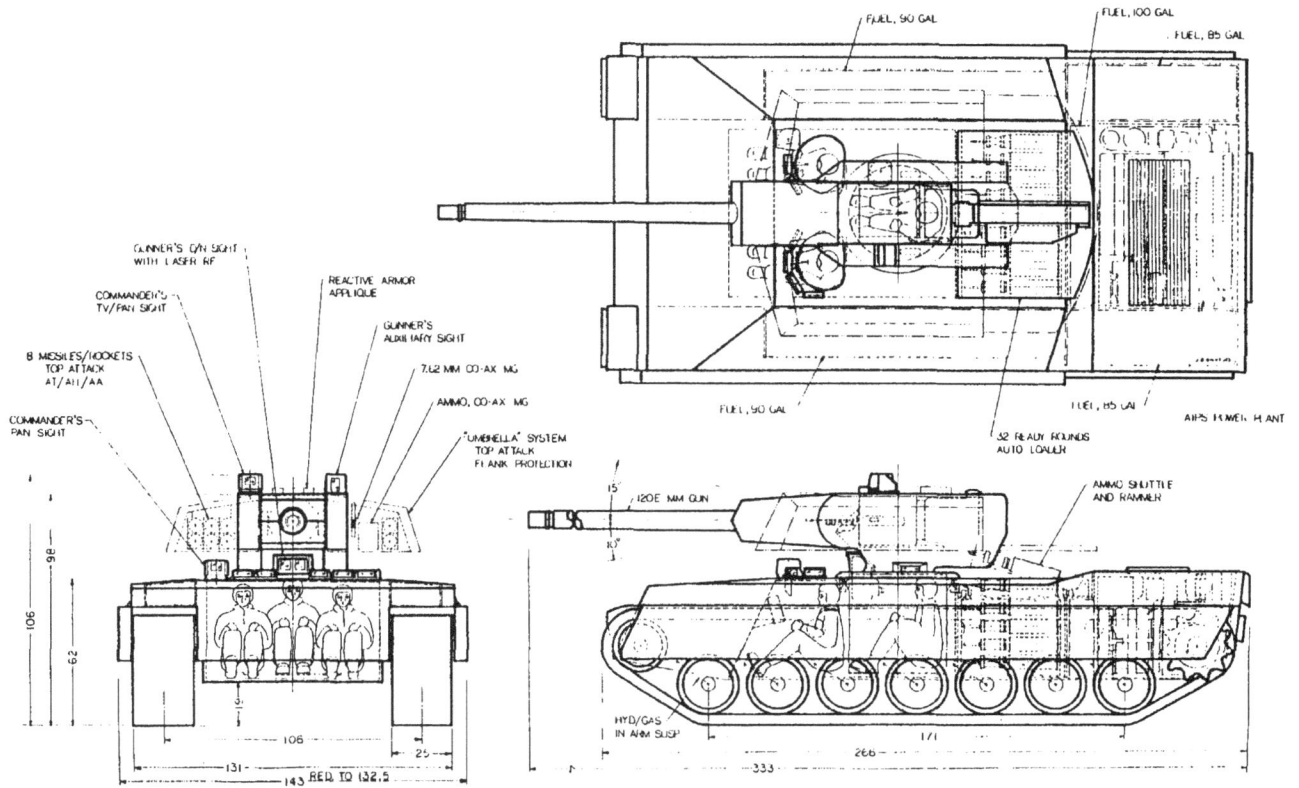

Above is a concept drawing for a future main battle tank with an external gun mount. An optional missile installation also is considered with improved top attack protection. The future main battle tank concept below features a manned low silhouette turret with an automatic loader in the bustle. Note the bore evacuator on the 120mm gun.

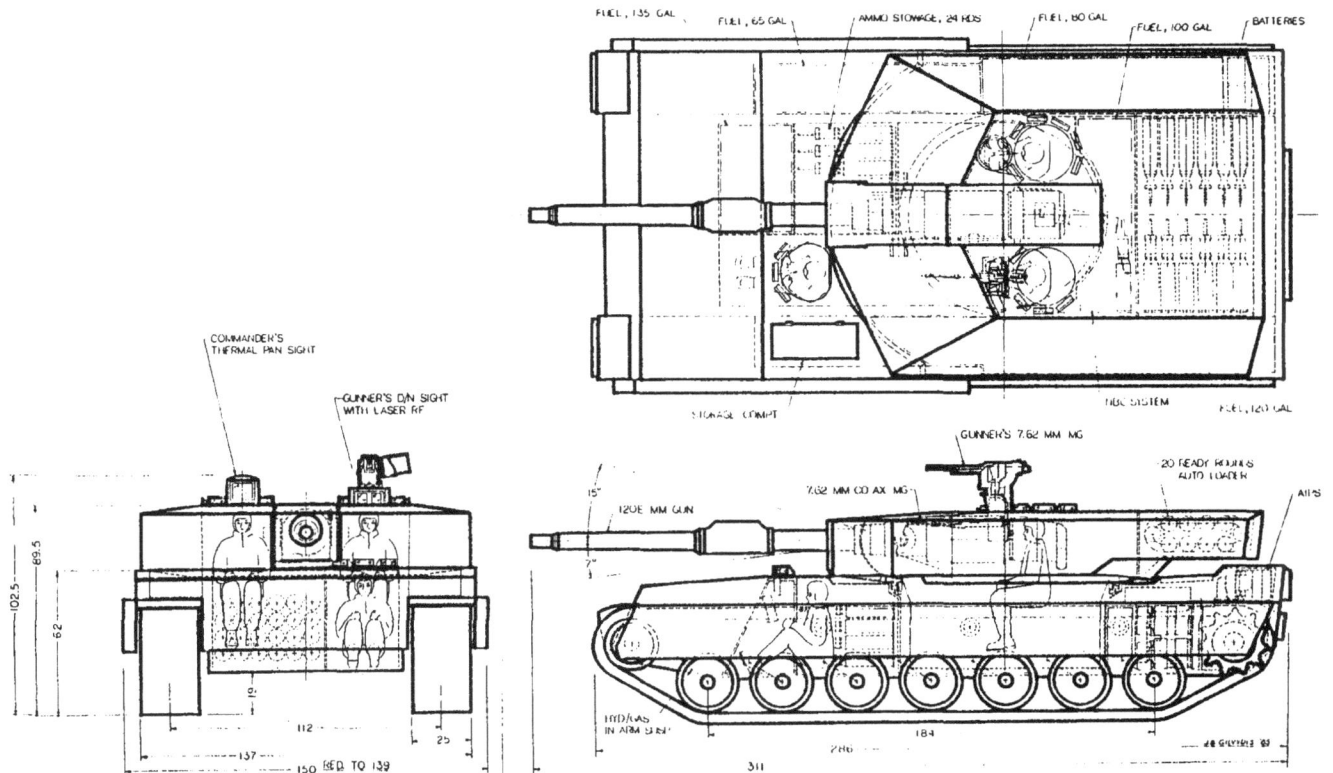

279

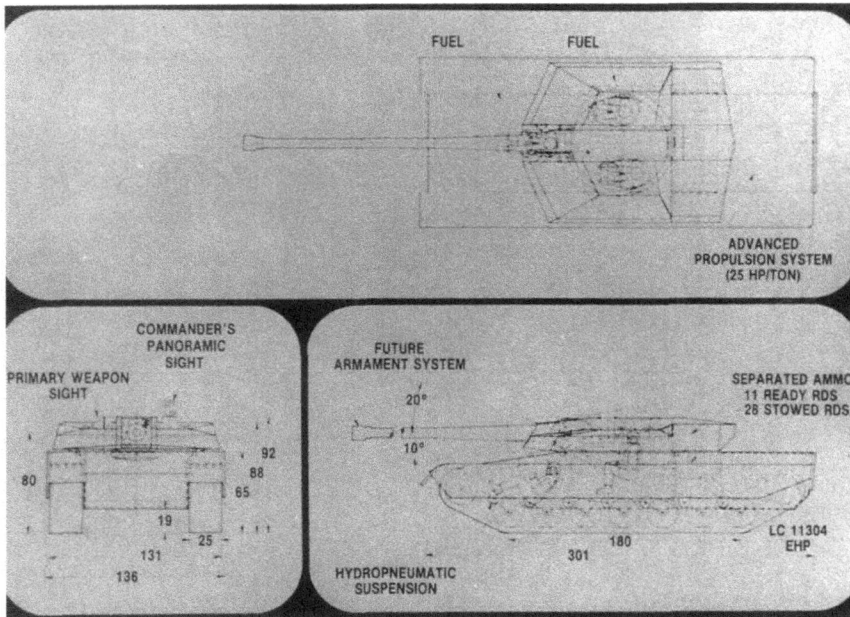

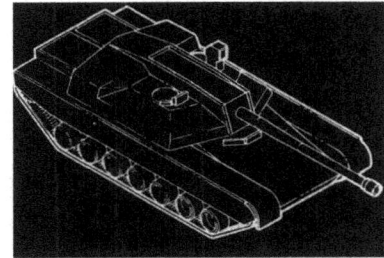

Above is another future main battle tank concept featuring a manned low silhouette turret. The armament proposed ranged in caliber from 120mm to 145mm and it was fitted with an automatic loader.

and light armored vehicles, modernization on this scale far exceeded the funds available. The program was then reduced to six vehicle types referred to as the Heavy Forces Modernization Plan (HFMP) with four vehicles on a common heavy protection chassis and two on a medium protection chassis. Under this plan, a common heavy

The proposed design below is armed with a 145mm gun in an external mount. The three man crew is seated in the hull behind the front mounted engine and power train. The 145mm ammunition stowage is in the rear hull.

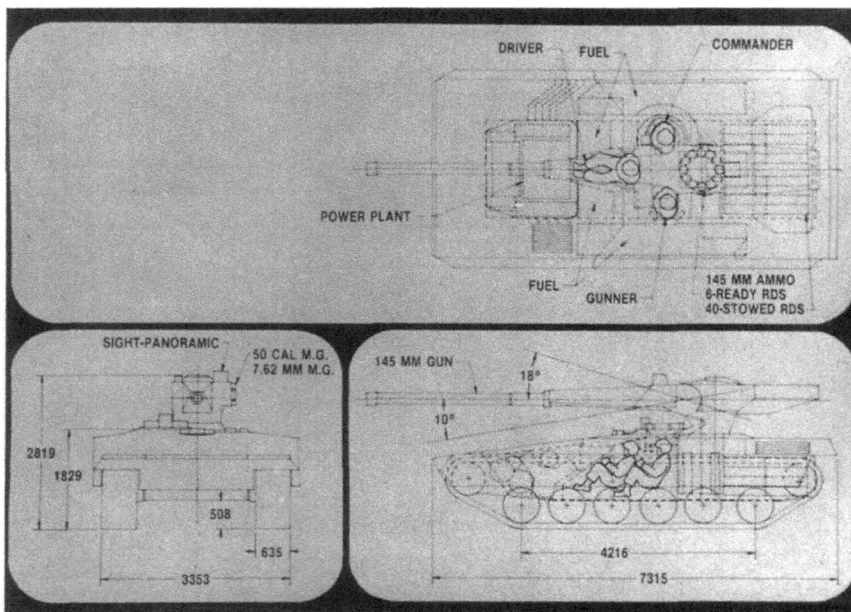

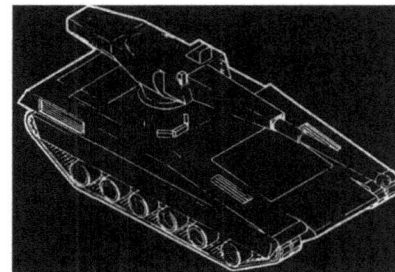

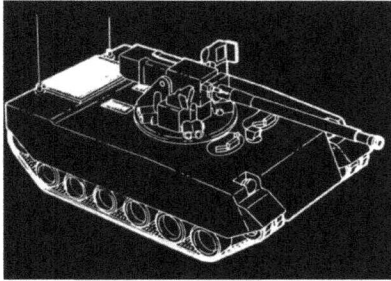

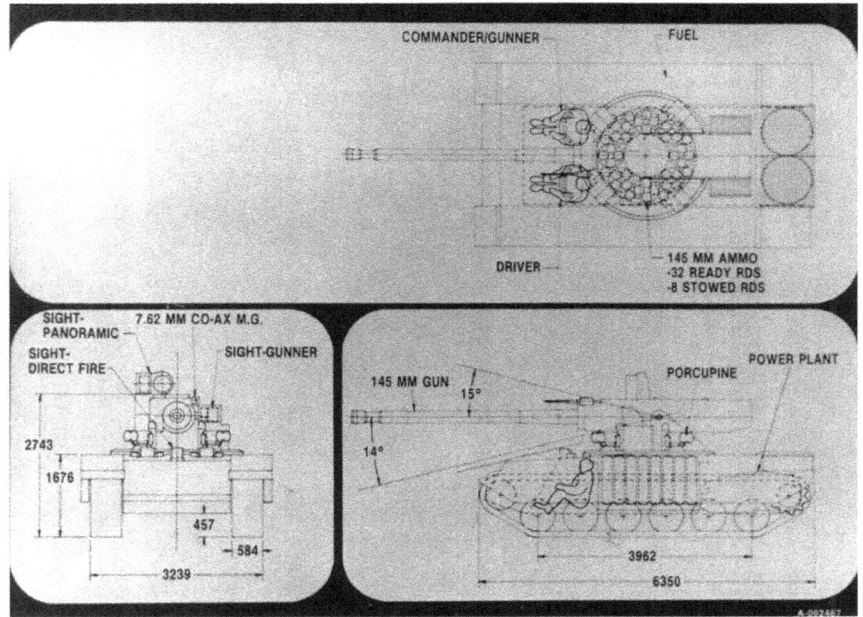

The future tank concept presented above also is armed with a 145mm gun in an external mount. In this case, the 120mm ammunition is stowed in a circular magazine directly under the gun mount. The crew is reduced to two and they are seated in the front hull. The engine and power train are in the rear hull behind the ammunition magazine. The four vehicles based on the common heavy chassis under the heavy forces modernization plan can be seen below at the right in these sketches from General Dynamics Land Systems Division.

protection chassis was proposed for the Block III tank, the future infantry fighting vehicle (FIFV), the advanced field artillery system (AFAS), and the combat mobility vehicle (CMV). The latter was a successor to the experimental counter obstacle vehicle discussed previously.

The Defense Science Board gave further emphasis to the use of test beds as a result of its 1987 Summer Study on Technology Base Management. It recommended that advanced technology transition demonstrators (ATTDs) be used to build and test experimental systems in a field environment. This would replace the performance and validation phase test programs then in use for new systems. This approach has been adopted under the Heavy Forces Modernization Plan. Under this plan, two competitively selected contractor teams will use ATTDs to demonstrate the feasibility of a common heavy protection chassis for the different vehicles required. Thus it is possible that the Abrams in a modified form could become the basic chassis for heavy vehicles in such a future armored vehicle family. This would follow a precedent set during World War II when tank chassis were adapted for use as self-propelled artillery, tank destroyers, and numerous tracked carriers. Regardless of what course the future takes, the Abrams will be a key element in the United States armored force for many years to come.

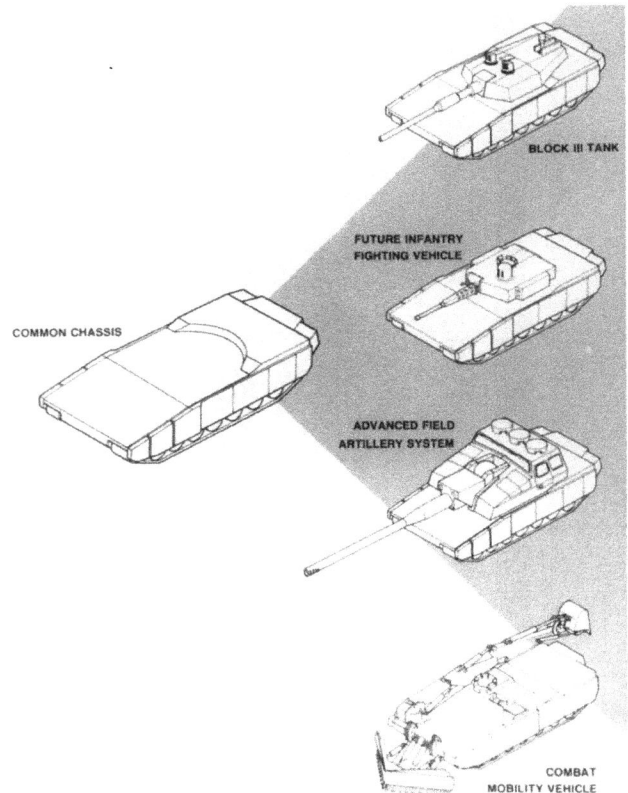

281

PART V

REFERENCE DATA

The M1A1 above is from A Company, 4th Battalion, 66th Armor of the 3rd Infantry Division with the Seventh Army in Germany.

These M1A1s in Germany are fitted with the multiple integrated laser engagement system (MILES) training equipment and a blank adapter is installed on the tank commander's .50 caliber machine gun.

The M1A1 tanks in these photographs are part of the Seventh U. S. Army in Germany.

This lineup of new M1 tanks is remarkably free of the extra stowage usually added by the troops.

The tank commander and the loader can be seen manning the .50 caliber and the 7.62mm machine guns on the M1 above. Below, this lineup of M1s is engaged in night firing at Fort Bragg, North Carolina.

The M1 above with the MILES training equipment was photographed by Greg Stewart at the National Training Center in March 1987. The ninth pilot XM1 tank (PV9) is shown in the view below.

These photographs of MBT70 pilot number two, registration number 09A002 67, show the tank with the fire control equipment installed. Note that the bore evacuator has been replaced by a sleeve since it was no longer required after installation of the closed breech scavenging system.

The T95E1 chassis was converted to the test rig shown here. Fitted with a mock-up turret, it was used to evaluate the hydropneumatic suspension. The vehicle is in the fully raised position above and lowered to the minimum height below.

VEHICLE DATA SHEETS

A wide range of experimental vehicles and design concepts are included in this book. Some of the former and most of the latter were never fully developed and detailed specifications are not available. Thus the data sheets in this section are limited to those vehicles for which at least pilots were built and tested. Even then, accurate information is often difficult to obtain. Since many of the vehicles were experimental test beds, changes were constantly being made and the descriptive data varied widely depending upon the configuration of the vehicle during the tests.

When possible, dimensional data were obtained from the original vehicle drawings. If these were not available, test reports from Aberdeen Proving Ground, Fort Knox or other test agencies usually provided the required information. Frequently, several sources were used and often they did not agree. In that case, the reports and photographs were reviewed to determine if the vehicle had been modified from the original design. Some dimensions are for reference only and would obviously vary from time to time. Thus the ground clearance would change with the load on the vehicle. The same would apply to the fire height, which is the distance from the ground to the centerline of the main weapon bore at zero elevation.

Although most of the items in the data sheets are self-explanatory, a few may need clarification. For example, the ground contact length at zero penetration is equal to the distance between the centers of the front and rear road wheels. This value is used to calculate the ground contact area and then the ground pressure using the combat weight of the vehicle. The tread is the distance between the centerlines of the two tracks. When available, the maximum values are listed for gross and net engine horsepower and torque. The gross horsepower and torque are the values obtained with only those accessories essential to engine operation. The net horsepower and torque reflect actual operation in the vehicle with all of the normal accessories such as generators and air cleaners. If available, weights

are provided in the tables for each vehicle unstowed and combat loaded. The latter included the crew as well as a full load of fuel and ammunition. The combat weight was used to calculate the power to weight ratio.

For experimental vehicles, the actual weight is quoted if it was obtainable. For some of these, only an approximate weight was available. Production vehicles often had the weights specified to the nearest 1000 pounds. Stowage arrangements frequently were modified during the period of service. In that case, the stowage shown usually is for the period of greatest service.

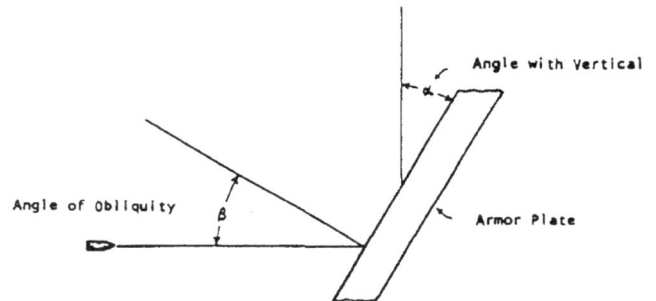

For the earlier vehicles, the armor protection is specified by type, thickness, and the angle with the vertical. This angle is measured between a vertical plane and the armor plate surface as indicated by the angle alpha in the sketch. Also, in this two dimensional drawing, the angle of obliquity is shown as the angle beta. This is defined as the angle between a line perpendicular to the armor plate and the path of a projectile impacting the armor. Because of security restrictions, particularly on composite special armor, detailed data on the armor protection of the later vehicles were not available.

90mm GUN TANK T95

GENERAL DATA

Crew:	4	men
Length: Gun forward, pilots 1 and 2 *	400.9	inches
pilots 3 and 4 *	403.9	inches
Length: Gun in travel position, pilots 1 and 2	388.9	inches
pilots 3 and 4	391.9	inches
Length: Without gun *	275.5	inches
Gun Overhang: Gun forward, pilots 1 and 2	125.4	inches
pilots 3 and 4	128.4	inches
Width: Over tracks	124.0	inches
Height: Over cupola periscope guard	112.0	inches
Tread:	103.0	inches
Ground Clearance:	17.0	inches
Fire Height:	approx. 69	inches
Turret Ring Diameter: (inside)	85.0	inches
Weight, Combat Loaded:	84,300	pounds
Weight, Unstowed:	77,300	pounds
Power to Weight Ratio: Net	10.9	hp/ton
Gross	13.3	hp/ton
Ground Pressure: Zero penetration, T114 track	12.1	psi
T127 track	10.6	psi

ARMOR

Type: Turret, cast homogeneous steel; Hull, rolled and cast
homogeneous steel; Welded assembly

Hull Thickness:		Actual	Angle w/Vertical
Front,	Upper	3.75 inches(95mm)	65 degrees
	Lower	5.0 to 3.0 inches(127-76mm)	45 degrees
Sides,	Front	4.0 to 2.0 inches(102-51mm)	0 degrees
	Rear	1.25 inches(32mm)	0 degrees
Rear,	Upper	0.75 inches(19mm)	20 degrees
	Lower	1.0 inches(25mm)	0 degrees
Top,	Front	2.0 inches(51mm)	80 degrees
	Rear	1.0 inches(25mm)	90 degrees
Floor,	Front	0.75 inches(19mm)	90 degrees
	Rear	0.5 inches(13mm)	90 degrees

Turret Thickness:			
Gun Shield		equals 15 inches(381mm)	0 degrees
Front		equals 7 inches(178mm)	60 degrees
Sides		3.0 inches(76mm)	45 degrees
Rear		2.0 inches(51mm)	0 degrees
Top,	Front	2.0 inches(51mm)	80 degrees
	Rear	1.5 inches(38mm)	90 degrees

ARMAMENT

Primary: 90mm Gun T208 or T208E13 in Mount T191 (rigid) in turret

Traverse: Electric-hydraulic and manual	360 degrees
Traverse Rate: (max)	10 seconds/360 degrees
Elevation: Electric-hydraulic and manual	+20 to −10 degrees
Elevation Rate: (max)	4 degrees/second
Firing Rate: (max)	7 rounds/minute
Loading System:	Manual
Stabilizer System:	Azimuth and elevation

Secondary:
(1) .50 caliber MG HB M2 in cupola mount on turret
(1) .30 caliber MG M37 coaxial w/90mm gun in turret
Provision for (1) .45 caliber SMG M3A1
Provision for (1) .30 caliber Carbine M2

AMMUNITION

50 rounds 90mm	8 hand grenades
1700 rounds .50 caliber	
180 rounds .45 caliber	
4750 rounds .30 caliber	
180 rounds .30 caliber (carbine)	

FIRE CONTROL AND VISION EQUIPMENT

Primary Weapon:	Direct	Indirect
	Range Finder T53 (Optar)	Azimuth Indicator M28A1
	Periscope T50	Elevation Quadrant M13
	Ballistic Drive T50	Gunner's Quadrant M1A1
	Ballistic Computer T37	
	Telescope T183	
	Telescope T171E1	

Vision Devices:	Direct	Indirect
Driver	Hatch	Periscope T48 (3) and Periscope T161 infrared (1)
Commander	Vision blocks (5) in cupola, hatch	Periscope M28 (AA MG sight) and Telescope T183
Gunner	None	Periscope T50 and Telescope T171E1
Loader	Hatch	None

Total Periscopes: T48 (3), T161 infrared (1), T50 (1), M28 (1)
Total Vision Blocks: (5) in Aircraft Armaments Model 108 cupola

ENGINE

Make and Model: Continental AOI-1195-5
Type: 8 cylinder, 4 cycle, horizontally opposed, fuel injection
Cooling System: Air   Ignition: Magneto

Displacement:	1194.5 cubic inches
Bore and Stroke:	5.75 x 5.75 inches
Compression Ratio:	6.5:1
Net Horsepower (max):	460 hp at 2800 rpm
Gross Horsepower (max):	560 hp at 2800 rpm
Net Torque (max):	1045 ft-lb at 1900 rpm
Gross Torque (max):	1170 ft-lb at 2100 rpm
Weight:	approx. 2200 pounds, dry
Fuel: 80 octane gasoline	205 gallons
Engine Oil:	40 quarts

POWER TRAIN

Transmission: X-Drive XTG-410-1, 4 ranges forward, 1 reverse
    Allison 4 element hydraulic torque convertor
    Stall multiplication: 3.05:1

Overall power train ratios:	Low 5.23:1	2nd 1.85:1
	1st 3.70:1	3rd 0.924:1
		Reverse 4.18:1

Steering Control: Mechanical, T-bar
    Clutch-brake steering in ranges low, 1st, and reverse
    Geared steering in ranges 2nd and 3rd
Brakes: Multiple disc
Final Drive: Planetary gear   Gear Ratio: 5.40:1
Drive Sprocket: At rear of vehicle with 10 teeth
    Pitch Diameter: 22.187 inches (T114 track)

RUNNING GEAR

Suspension: Torsion bar
    10 individually sprung dual road wheels (5/track)
    Tire Size: 32 x 4 inches
    Flat track suspension w/o support rollers
    Dual compensating idler at front of each track
    Idler Tire Size: 21 x 4 inches
    Shock absorbers fitted on first and last road wheels on each side
Tracks: Center guide T114 and T127
    Type: (T114) Double pin, 21 inch width, rubber chevron
        (T127) Single pin, 24 inch width, removable rubber pad
    Pitch: 6.94 inches (T114)
        6.5 inches (T127)
    Shoes per Vehicle: 160 (80/track, T114)
    Ground Contact Length: 165.5 inches

ELECTRICAL SYSTEM

Nominal Voltage: 24 volts DC
Main Generator: (1) 30 volts, 300 amperes, gear driven by main engine
Auxiliary Generator: (1) 28 volts, 300 amperes, driven by auxiliary engine
Battery: (2) 12 volts in series

COMMUNICATIONS

Radio: AN/GRC-3,4,7,8 in turret bustle
    AN/VRC-24 (air to ground) also may be fitted
Interphone: AN/UIC-1 (4 stations) plus external box AN/VIA-4

FIRE PROTECTION

(3) 10 pound carbon dioxide, fixed
(1) 5 pound carbon dioxide, portable

PERFORMANCE

Maximum Speed: Sustained, level road	35 miles/hour
Maximum Tractive Effort: TE at stall	65,000 pounds
Per Cent of Vehicle Weight: TE/W	77 per cent
Maximum Grade:	60 per cent
Maximum Trench:	8.5 feet
Maximum Vertical Wall:	36 inches
Maximum Fording Depth:	48 inches
Minimum Turning Circle: (diameter)	pivot
Cruising Range: Roads	145 miles

* Over pintle bracket w/o pintle

90mm GUN TANK T95E1

GENERAL DATA

Crew:	4	men
Length: Gun forward *	405.9	inches
Length: Gun in travel position	393.9	inches
Length: Without gun *	275.5	inches
Gun Overhang: Gun forward	130.4	inches
Width: Over tracks	124.0	inches
Height: Over cupola periscope guard	112.0	inches
Tread:	103.0	inches
Ground Clearance:	17.0	inches
Fire Height:	approx. 69	inches
Turret Ring Diameter: (inside)	85.0	inches
Weight, Combat Loaded:	84,800	pounds
Weight, Unstowed:	77,800	pounds
Power to Weight Ratio: Net	10.8	hp/ton
Gross	13.2	hp/ton
Ground Pressure: Zero penetration, T114 track	12.2	psi
T127 track	10.7	psi

ARMOR

Type: Turret, cast homogeneous steel; Hull, rolled and cast homogeneous steel; Welded assembly

Hull Thickness:	Actual	Angle w/Vertical
Front, Upper	3.75 inches(95mm)	65 degrees
Lower	5.0 to 3.0 inches(127-76mm)	45 degrees
Sides, Front	4.0 to 2.0 inches(102-51mm)	0 degrees
Rear	1.25 inches(32mm)	0 degrees
Rear, Upper	0.75 inches(19mm)	20 degrees
Lower	1.0 inches(25mm)	0 degrees
Top, Front	2.0 inches(51mm)	80 degrees
Rear	1.0 inches(25mm)	90 degrees
Floor, Front	0.75 inches(19mm)	90 degrees
Rear	0.5 inches(13mm)	90 degrees
Turret Thickness:		
Gun Shield	equals 15 inches(381mm)	0 degrees
Front	equals 7 inches(178mm)	60 degrees
Sides	3.0 inches(76mm)	45 degrees
Rear	2.0 inches(51mm)	0 degrees
Top, Front	2.0 inches(51mm)	80 degrees
Rear	1.5 inches(38mm)	90 degrees

ARMAMENT

Primary: 90 mm Gun T208E9 in Mount T192 in turret

Traverse: Hydraulic and manual	360 degrees
Traverse Rate: (max)	10 seconds/360 degrees
Elevation: Hydraulic and manual	+20 to −10 degrees
Elevation Rate: (max)	4 degrees/second
Firing Rate: (max)	7 rounds/minute
Loading System:	Manual
Stabilizer System:	None

Secondary:
(1) .50 caliber MG HB M2 in cupola mount on turret
(1) .30 caliber MG M37 coaxial w/90mm gun in turret
Provision for (1) .45 caliber SMG M3A1
Provision for (1) .30 caliber Carbine M2

AMMUNITION

50 rounds 90mm	8 hand grenades
1700 rounds .50 caliber	
180 rounds .45 caliber	
4750 rounds .30 caliber	
180 rounds .30 caliber (carbine)	

FIRE CONTROL AND VISION EQUIPMENT

Primary Weapon:

Primary Weapon:	Direct	Indirect
	Periscope T50E2	Azimuth Indicator M28A1
	Telescope T183E1	Elevation Quadrant M13
	Telescope T171E1	Gunner's Quadrant M1A1
Vision Devices:	Direct	Indirect
Driver	Hatch	Periscope T48 (3) and Periscope T161 infrared (1)
Commander	Vision blocks (5) in cupola, hatch	Periscope M28 (AA MG sight) and Telescope T183E1
Gunner	None	Periscope T50E2 and Telescope T171E1
Loader	Hatch	None

Total Periscopes: T48 (3), T161 infrared (1), T50E2 (1), M28 (1)
Total Vision Blocks: (5) in Aircraft Armaments Model 108 cupola

ENGINE

Make and Model: Continental AOI-1195-5	
Type: 8 cylinder, 4 cycle, horizontally opposed, fuel injection	
Cooling System: Air Ignition: Magneto	
Displacement:	1194.5 cubic inches
Bore and Stroke:	5.75 x 5.75 inches
Compression Ratio:	6.5:1
Net Horsepower (max):	460 hp at 2800 rpm
Gross Horsepower (max):	560 hp at 2800 rpm
Net Torque (max):	1045 ft-lb at 1900 rpm
Gross Torque (max):	1170 ft-lb at 2100 rpm
Weight:	approx. 2200 pounds, dry
Fuel: 80 octane gasoline	205 gallons
Engine Oil:	40 quarts

POWER TRAIN

Transmission: X-Drive XTG-410-1, 4 ranges forward, 1 reverse
   Allison 4 element hydraulic torque convertor
   Stall multiplication: 3.05:1

Overall power train ratios:	Low 5.23:1	2nd 1.85:1
	1st 3.70:1	3rd 0.924:1
		Reverse 4.18:1

Steering Control: Mechanical, T-bar
   Clutch-brake steering in ranges low, 1st, and reverse
   Geared steering in ranges 2nd and 3rd
Brakes: Multiple disc
Final Drive: Planetary gear   Gear Ratio: 5.40:1
Drive Sprocket: At rear of vehicle with 10 teeth
   Pitch Diameter: 22.187 inches (T114 track)

RUNNING GEAR

Suspension: Torsion bar
   10 individually sprung dual road wheels (5/track)
   Tire Size: 32 x 4 inches
   Flat track suspension w/o support rollers
   Dual compensating idler at front of each track
   Idler Tire Size: 21 x 4 inches
   Shock absorbers fitted on first and last road wheels on each side
Tracks: Center guide T114 and T127
   Type: (T114) Double pin, 21 inch width, rubber chevron
      (T127) Single pin, 24 inch width, removable rubber pad
   Pitch: 6.94 inches (T114)
      6.5 inches (T127)
   Shoes per Vehicle: 160 (80/track, T114)
   Ground Contact Length: 165.5 inches

ELECTRICAL SYSTEM

Nominal Voltage: 24 volts DC
Main Generator: (1) 30 volts, 300 amperes, gear driven by main engine
Auxiliary Generator: (1) 28 volts, 300 amperes, driven by auxiliary engine
Battery: (2) 12 volts in series

COMMUNICATIONS

Radio: AN/GRC-3,4,7,8 in turret bustle
   AN/VRC-24 (air to ground) also may be fitted
Interphone: AN/UIC-1 (4 stations) plus external box AN/VIA-4

FIRE PROTECTION

(3) 10 pound carbon dioxide, fixed
(1) 5 pound carbon dioxide, portable

PERFORMANCE

Maximum Speed: Sustained, level road	35 miles/hour
Maximum Tractive Effort: TE at stall	65,000 pounds
Per Cent of Vehicle Weight: TE/W	77 per cent
Maximum Grade:	60 per cent
Maximum Trench:	8.5 feet
Maximum Vertical Wall:	36 inches
Maximum Fording Depth:	48 inches
Minimum Turning Circle: (diameter)	pivot
Cruising Range: Roads	145 miles

* Over pintle bracket w/o pintle

90mm GUN TANK T95E2

GENERAL DATA

Crew:	4	men
Length: Gun forward *	330.1	inches
Length: Gun in travel position	313.6	inches
Length: Without gun *	275.5	inches
Gun Overhang: Gun forward	54.6	inches
Width: Over tracks	124.0	inches
Height: Over cupola periscope guard	114.0	inches
Tread:	103.0	inches
Ground Clearance:	17.0	inches
Fire Height:	approx. 71	inches
Turret Ring Diameter: (inside)	85.0	inches
Weight, Combat Loaded:	81,400	pounds
Weight, Unstowed:	75,130	pounds
Power to Weight Ratio: Net	11.3	hp/ton
Gross	13.8	hp/ton
Ground Pressure: Zero penetration, T114 track	11.7	psi
T127 track	10.2	psi

ARMOR

Type: Turret, cast homogeneous steel; Hull, rolled and cast homogeneous steel; Welded assembly

Hull Thickness:

		Actual	Angle w/Vertical
Front,	Upper	3.75 inches(95mm)	65 degrees
	Lower	5.0 to 3.0 inches(127-76mm)	45 degrees
Sides,	Front	4.0 to 2.0 inches(102-51mm)	0 degrees
	Rear	1.25 inches(32mm)	0 degrees
Rear,	Upper	0.75 inches(19mm)	20 degrees
	Lower	1.0 inches(25mm)	0 degrees
Top,	Front	2.0 inches(51mm)	80 degrees
	Rear	1.0 inches(25mm)	90 degrees
Floor,	Front	0.75 inches(19mm)	90 degrees
	Rear	0.5 inches(13mm)	90 degrees

Turret Thickness:

Gun Shield	4.5 inches(114mm)	30 degrees
Front	equals 7.0 inches(178mm)	0 degrees
Sides	equals 3.0 inches(76mm)	0 degrees
Rear	equals 2.0 inches(51mm)	0 degrees
Top	1.0 inches(25mm)	90 degrees

ARMAMENT

Primary: 90mm Gun M41 in Mount M87 in turret

Traverse: Hydraulic and manual	360 degrees
Traverse Rate: (max)	15 seconds/360 degrees
Elevation: Hydraulic and manual	+19 to −degrees
Elevation Rate: (max)	4 degrees/second
Firing Rate: (max)	8 rounds/minute
Loading System:	Manual
Stabilizer System:	None

Secondary:

(1) .50 caliber MG HB M2 in cupola mount on turret
(1) .30 caliber MG M37 coaxial w/90mm gun in turret
Provision for (1) .45 caliber SMG M3A1
Provision for (1) .30 caliber Carbine M2

AMMUNITION

64 rounds 90mm	8 hand grenades
1265 rounds .50 caliber	
180 rounds .45 caliber	
5950 rounds .30 caliber	
180 rounds .30 caliber (carbine)	

FIRE CONTROL AND VISION EQUIPMENT

Primary Weapon	Direct	Indirect
	Range Finder M13A1	Azimuth Indicator M28A1
	Periscope M20A3	Elevation Quadrant M13
	Ballistic Drive M5A1	Gunner's Quadrant M1A1
	Ballistic Computer M13A1	
	Telescope M97C	

Vision Devices:	Direct	Indirect
Driver	Hatch	Periscope T48 (3) and
		Periscope T161 (1) infrared
Commander	Vision blocks (5)	Periscope M28 (AA MG sight)
	in cupola, hatch	
Gunner	None	Periscope M20A3 and
		Telescope M97C
Loader	Hatch	None

Total Periscopes: M27 (3), M24 infrared (1), M20A3 (1), M28 (1) Total Vision Blocks: (5) in M1 cupola on turret top

ENGINE

Make and Model: Continental AOI-1195-5	
Type: 8 cylinder, 4 cycle, horizontally opposed, fuel injection	
Cooling System: Air Ignition: Magneto	
Displacement:	1194.5 cubic inches
Bore and Stroke:	5.75 x 5.75 inches
Compression Ratio:	6.5:1
Net Horsepower (max):	460 hp at 2800 rpm
Gross Horsepower (max):	560 hp at 2800 rpm
Net Torque (max):	1045 ft-lb at 1900 rpm
Gross Torque (max):	1170 ft-lb at 2100 rpm
Weight:	approx. 2200 pounds, dry
Fuel: 80 octane gasoline	205 gallons
Engine Oil:	40 quarts

POWER TRAIN

Transmission: X-Drive XTG-410-1, 4 ranges forward, 1 reverse
   Allison 4 element hydraulic torque convertor
   Stall multiplication: 3.05:1

Overall power train ratios:	Low	5.23:1	2nd	1.85:1
	1st	3.70:1	3rd	0.924:1
			Reverse	4.18:1

Steering Control: Mechanical, T-bar
   Clutch-brake steering in ranges low, 1st, and reverse
   Geared steering in ranges 2nd and 3rd
Brakes: multiple disc
Final Drive: Planetary gear   Gear Ratio: 5.40:1
Drive Sprocket: At rear of vehicle with 10 teeth
   Pitch Diameter: 22.187 inches (T114 track)

RUNNING GEAR

Suspension: Torsion bar
   10 individually sprung dual road wheels (5/track)
   Tire Size: 32 x 4 inches
   Flat track suspension w/o support rollers
   Dual compensating idler at front of each track
   Idler Tire Size: 21 x 4 inches
   Shock absorbers fitted on first and last road wheels on each side
Tracks: Center guide T114 and T127
   Type: (T114) Double pin, 21 inch width, rubber chevron
      (T127) Single pin, 24 inch width, removable rubber pad
   Pitch: 6.94 inches (T114)
      6.5 inches (T127)
   Shoes per Vehicle: 160 (80/track, T114)
   Ground Contact Length: 165.5 inches

ELECTRICAL SYSTEM

Nominal Voltage: 24 volts DC
Main Generator: (1) 30 volts, 300 amperes, gear driven by main engine
Auxiliary Generator: (1) 28 volts, 300 amperes, driven by auxiliary engine
Battery: (2) 12 volts in series

COMMUNICATIONS

Radio: AN/GRC-3,4,7,8 in turret bustle
   AN/VRC-24 (air to ground) also may be fitted
Interphone: AN/UIC-1 (4 stations) plus external box AN/VIA-4

FIRE PROTECTION

(3) 10 pound carbon dioxide, fixed
(1) 5 pound carbon dioxide, portable

PERFORMANCE

Maximum Speed: Sustained, level road	35 miles/hour
Maximum Tractive Effort: TE at stall	65,000 pounds
Per Cent of Vehicle Weight: TE/W	80 per cent
Maximum Grade:	60 per cent
Maximum Trench:	8.5 feet
Maximum Vertical Wall:	36 inches
Maximum Fording Depth:	48 inches
Minimum Turning Circle: (diameter)	pivot
Cruising Range: Roads	145 miles

* Over pintle bracket w/o pintle

105mm GUN TANK T95E3

GENERAL DATA

Crew:	4	men
Length: Gun forward *	448.5	inches
Length: Gun in travel position	418.8	inches
Length: Without gun *	275.5	inches
Gun Overhang: Gun forward	173.0	inches
Width: Over tracks	124.0	inches
Height: Over cupola periscope guard	115.0	inches
Tread:	103.0	inches
Ground Clearance:	17.0	inches
Fire Height:	approx. 71	inches
Turret Ring Diameter: (inside)	85.0	inches
Weight, Combat Loaded:	90,250	pounds
Weight, Unstowed:	83,250	pounds
Power to Weight Ratio: Net	10.2	hp/ton
Gross	12.4	hp/ton
Ground Pressure: Zero penetration, T114 track	13.0	psi
T127 track	11.4	psi

ARMOR

Type: Turret, cast homogeneous steel; Hull, rolled and cast homogeneous steel; Welded assembly

Hull Thickness:

		Actual	Angle w/Vertical
Front,	Upper	3.75 inches(95mm)	65 degrees
	Lower	5.0 to 3.0 inches(127-76mm)	45 degrees
Sides,	Front	4.0 to 2.0 inches(102-51mm)	0 degrees
	Rear	1.25 inches(32mm)	0 degrees
Rear,	Upper	0.75 inches(19mm)	20 degrees
	Lower	1.0 inches(25mm)	0 degrees
Top,	Front	2.0 inches(51mm)	80 degrees
	Rear	1.0 inches(25mm)	90 degrees
Floor,	Front	0.75 inches(19mm)	90 degrees
	Rear	0.5 inches(13mm)	90 degrees

Turret Thickness:

	Actual	Angle w/Vertical
Gun Shield	equals 4.0 inches(102mm)	60 degrees
Front	4.0 inches(102mm)	60 degrees
Sides	2.0 to 3.5 inches(51-89mm)	10 degrees
Rear	2.5 inches(64mm)	0 degrees
Top	1.0 inches(25mm)	90 degrees

ARMAMENT

Primary: 105mm Gun T140E3 in Mount T174 in turret

Traverse: Hydraulic and manual	360 degrees
Traverse Rate: (max)	17 seconds/360 degrees
Elevation: Hydraulic and manual	+20 to −10 degrees
Elevation Rate: (max)	4 degrees/second
Firing Rate: (max)	6 rounds/minute
Loading System:	Manual
Stabilizer System:	None

Secondary:
(1) .50 caliber MG HB M2 in cupola mount on turret
(1) .30 caliber MG M1919A4E1 coaxial w/105mm gun in turret
Provision for (1) .45 caliber SMG M3A1
Provision for (1) .30 caliber Carbine M2

AMMUNITION

39 rounds 105mm	8 hand grenades
1850 rounds .50 caliber	
180 rounds .45 caliber	
4750 rounds .30 caliber	
180 rounds .30 caliber (carbine)	

FIRE CONTROL AND VISION EQUIPMENT

Primary Weapon:	Direct	Indirect
	Range Finder T46E3	Azimuth Indicator M28
	Periscope M16E1	Elevation Quadrant M13 (T21)
	Ballistic Drive T37	Gunner's Quadrant M1A1
	Ballistic Computer T32	
	Telescope T156E2	

Vision Devices:	Direct	Indirect
Driver	Hatch	Periscope T48 (3) and
		Periscope T161 infrared (1)
Commander	Vision blocks (5)	Periscope T42 (AA MG sight)
	in cupola, hatch	
Gunner	None	Periscope M16E1
Loader	Hatch	None

Total Periscopes: T48 (3), T161 infrared (1), M16E1 (1), T42 (1)
Total Vision Blocks: (5) in Aircraft Armaments Model 108 cupola

ENGINE

Make and Model: Continental AOI-1195-5		
Type: 8 cylinder, 4 cycle, horizontally opposed, fuel injection		
Cooling System: Air	Ignition: Magneto	
Displacement:		1194.5 cubic inches
Bore and Stroke:		5.75 x 5.75 inches
Compression Ratio:		6.5:1
Net Horsepower (max):		460 hp at 2800 rpm
Gross Horsepower (max):		560 hp at 2800 rpm
Net Torque (max):		1045 ft-lb at 1900 rpm
Gross Torque (max):		1170 ft-lb at 2100 rpm
Weight:	approx.	2200 pounds, dry
Fuel: 80 octane gasoline		205 gallons
Engine Oil:		40 quarts

POWER TRAIN

Transmission: X-Drive XTG-410-1, 4 ranges forward, 1 reverse
Allison 4 element hydraulic torque convertor
Stall multiplication: 3.05:1

Overall power train ratios:	Low	5.23:1	2nd	1.85:1
	1st	3.70:1	3rd	0.924:1
			Reverse	4.18:1

Steering Control: Mechanical, T-bar
Clutch-brake steering in ranges low, 1st, and reverse
Geared steering in ranges 2nd and 3rd
Brakes: Multiple disc
Final Drive: Planetary gear   Gear Ratio: 5.40:1
Drive Sprocket: At rear of vehicle with 10 teeth
Pitch Diameter: 22.187 inches (T114 track)

RUNNING GEAR

Suspension: Torsion bar
10 individually sprung dual road wheels (5/track)
Tire Size: 32 x 4 inches
Flat track suspension w/o support rollers
Dual compensating idler at front of each track
Idler Tire Size: 21 x 4 inches
Shock absorbers fitted on first and last road wheels on each side
Tracks: Center guide T114 and T127
Type: (T114) Double pin, 21 inch width, rubber chevron
(T127) Single pin, 24 inch width, removable rubber pad
Pitch: 6.94 inches (T114)
6.5 inches (T127)
Shoes per Vehicle: 160 (80/track, T114)
Ground Contact Length: 165.5 inches

ELECTRICAL SYSTEM

Nominal Voltage: 24 volts DC
Main Generator: (1) 30 volts, 300 amperes, gear driven by main engine
Auxiliary Generator: (1) 28 volts, 300 amperes, driven by auxiliary engine
Battery: (2) 12 volts in series

COMMUNICATIONS

Radio: AN/GRC-3,4,7,8 in turret bustle
AN/VRC-24 (air to ground) also may be fitted
Interphone: AN/UIC-1 (4 stations) plus external box AN/VIA-4

FIRE PROTECTION

(3) 10 pound carbon dioxide, fixed
(1) 5 pound carbon dioxide, portable

PERFORMANCE

Maximum Speed: Sustained, level road	35 miles/hour
Maximum Tractive Effort: TE at stall	65,000 pounds
Per Cent of Vehicle Weight: TE/W	72 per cent
Maximum Grade:	60 per cent
Maximum Trench:	8.5 feet
Maximum Vertical Wall:	36 inches
Maximum Fording Depth:	48 inches
Minimum Turning Circle: (diameter)	pivot
Cruising Range: Roads	145 miles

* Over pintle bracket w/o pintle

120mm GUN TANK T95E6

GENERAL DATA

Crew:	4	men
Length: Gun forward *	426.1	inches
Length: Gun in travel position	406.6	inches
Length: Without gun *	275.5	inches
Gun Overhang: Gun forward	150.6	inches
Width: Over tracks	124.0	inches
Height: Over cupola periscope guard	115.0	inches
Tread:	103.0	inches
Ground Clearance:	17.0	inches
Fire Height:	approx. 71	inches
Turret Ring Diameter: (inside)	85.0	inches
Weight, Combat Loaded:	90,200	pounds
Weight, Unstowed:	85,200	pounds
Power to Weight Ratio: Net	10.2	hp/ton
Gross	12.4	hp/ton
Ground Pressure: Zero penetration, T114 track	13.0	psi
T127 track	11.4	psi

ARMOR

Type: Turret, cast homogeneous steel; Hull, rolled and cast homogeneous steel; Welded assembly

Hull Thickness:		Actual	Angle w/Vertical
Front,	Upper	3.75 inches(95mm)	65 degrees
	Lower	5.0 to 3.0 inches(127-76mm)	45 degrees
Sides,	Front	4.0 to 2.0 inches(102-51mm)	0 degrees
	Rear	1.25 inches(32mm)	0 degrees
Rear,	Upper	0.75 inches(19mm)	20 degrees
	Lower	1.0 inches(25mm)	0 degrees
Top,	Front	2.0 inches(51mm)	80 degrees
	Rear	1.0 inches(25mm)	90 degrees
Floor,	Front	0.75 inches(19mm)	90 degrees
	Rear	0.5 inches(13mm)	90 degrees

Turret Thickness:			
Gun Shield		equals 15 inches(381mm)	0 degrees
Front		equals 7 inches(178mm)	60 degrees
Sides		3.0 inches(76mm)	45 degrees
Rear		2.0 inches(51mm)	0 degrees
Top,	Front	2.0 inches(51mm)	80 degrees
	Rear	1.5 inches(38mm)	90 degrees

ARMAMENT

Primary: 120mm Gun T123E6 in combination mount in turret

Traverse: Hydraulic and manual	360 degrees
Traverse Rate: (max)	15 seconds/360 degrees
Elevation:Hydraulic and manual	+20 to −9 degrees
Elevation Rate: (max)	4 degrees/second
Firing Rate: (max) one loader w/assist	4 rounds/minute
Loading System:	Manual
Stabilizer System:	None

Secondary:
(1) .50 caliber MG T175E2 in cupola mount on turret
(1) .30 caliber MG M37 coaxial w/120mm gun in turret
Provision for (1) .45 caliber SMG M3A1
Provision for (1) .30 caliber Carbine M2

AMMUNITION

36 rounds 120mm	8 hand grenades
1500 rounds .50 caliber	
180 rounds .45 caliber	
4500 rounds .30 caliber	
30 rounds .30 caliber (carbine)	

FIRE CONTROL AND VISION EQUIPMENT

Primary Weapon:	Direct	Indirect
	Range Finder T53 (Optar)	Azimuth Indicator M28A1
	Periscope T50	Elevation Quadrant M13
	Ballistic Drive T51	Gunner's Quadrant M1A1
	Ballistic Computer T37E1	
	Telescope T183	
	Telescope T171E2	

Vision Devices:	Direct	Indirect
Driver	Hatch	Periscope T48 (3) and Periscope T161 infrared (1)
Commander	Vision blocks (8) in cupola, hatch	Periscope M28C (AA MG sight) and Telescope T183
Gunner	None	Periscope T50 and Telescope T171E2
Loader	Hatch	None

Total Periscopes: T48 (3), T161 infrared (1), T50 (1), M28C (1)
Total Vision Blocks: (8) in cupola T9 on turret top

ENGINE

Make and Model: Continental AOI-1195-5
Type: 8 cylinder, 4 cycle, horizontally opposed, fuel injection
Cooling System: Air   Ignition: Magneto

Displacement:	1194.5 cubic inches
Bore and Stroke:	5.75 x 5.75 inches
Compression Ratio:	6.5:1
Net Horsepower (max):	460 hp at 2800 rpm
Gross Horsepower (max):	560 hp at 2800 rpm
Net Torque (max):	1045 ft-lb at 1900 rpm
Gross Torque (max):	1170 ft-lb at 2100 rpm
Weight:	approx. 2200 pounds, dry
Fuel: 80 octane gasoline	205 gallons
Engine Oil:	40 quarts

POWER TRAIN

Transmission: X-Drive XTG-410-1, 4 ranges forward, 1 reverse
Allison 4 element hydraulic torque convertor
Stall multiplication: 3.05:1

Overall power train ratios:	Low	5.23:1	2nd	1.85:1
	1st	3.70:1	3rd	0.924:1
			Reverse	4.18:1

Steering Control: Mechanical, T-bar
Clutch-brake steering in ranges low, 1st, and reverse
Geared steering in ranges 2nd and 3rd
Brakes: Multiple disc
Final Drive: Planetary gear   Gear Ratio: 5.40:1
Drive Sprocket: At rear of vehicle with 10 teeth
Pitch Diameter: 22.187 inches (T114 track)

RUNNING GEAR

Suspension: Torsion bar
10 individually sprung dual road wheels (5/track)
Tire Size: 32 x 4 inches
Flat track suspension w/o support rollers
Dual compensating idler at front of each track
Idler Tire Size: 21 x 4 inches
Shock absorbers fitted on first and last road wheels on each side
Tracks: Center guide T114 and T127
Type: (T114) Double pin, 21 inch width, rubber chevron
(T127) Single pin, 24 inch width, removable rubber pad
Pitch: 6.94 inches (T114)
6.5 inches (T127)
Shoes per Vehicle: 160 (80/track, T114)
Ground Contact Length: 165.5 inches

ELECTRICAL SYSTEM

Nominal Voltage: 24 volts DC
Main Generator: (1) 30 volts, 300 amperes, gear driven by main engine
Auxiliary Generator: (1) 28 volts, 300 amperes, driven by auxiliary engine
Battery: (2) 12 volts in series

COMMUNICATIONS

Radio: AN/GRC-3,4,7,8 in turret bustle
AN/VRC-24 (air to ground) also may be fitted
Interphone: AN/UIC-1 (4 stations) plus external box AN/VIA-4

FIRE PROTECTION

(3) 10 pound carbon dioxide, fixed
(1) 5 pound carbon dioxide, portable

PERFORMANCE

Maximum Speed: Sustained, level road	35 miles/hour
Maximum Tractive Effort: TE at stall	65,000 pounds
Per Cent of Vehicle Weight: TE/W	72 per cent
Maximum Grade:	60 per cent
Maximum Trench:	8.5 feet
Maximum Vertical Wall:	36 inches
Maximum Fording Depth:	48 inches
Minimum Turning Circle: (diameter)	pivot
Cruising Range: Roads	145 miles

* Over pintle bracket w/o pintle

GENERAL DATA

Crew:	3	men
Length: Gun forward	366.0	inches
Length: Gun in travel position *	291	inches
Length: Without gun *	283	inches
Gun Overhang: Gun forward *	82	inches
Width: Over tracks	138.0	inches
Height: Over commander's panoramic sight	116.7	inches
Tread:	113.2	inches
Ground Clearance: Variable 6 to 25 inches **	normal 21.0	inches
Fire Height: At 21.0 inches ground clearance	approx. 72	inches
Turret Ring Diameter: (inside ring gear)	101.0	inches
Weight, Combat Loaded:	approx. 114,000	pounds
Weight, Unstowed:	approx. 107,000	pounds
Power to Weight Ratio: Net	estimated 18.6	hp/ton
Gross	25.9	hp/ton
Ground Pressure: Zero penetration	12.9	psi

ARMOR

Turret: Cast homogeneous armor steel inner shell overlaid with spaced high hardness homogeneous rolled armor steel, welded assembly Hull: Welded assembly of rolled homogeneous armor steel plate with some armor steel castings and aluminum armor

ARMAMENT

Primary: 152mm Gun-Launcher XM150E5 in turret combination mount

Traverse: Electrohydraulic and manual	360 degrees
Traverse Rate: (max)	10/seconds 360 degrees
Elevation: Electrohydraulic and manual	+20 to −10 degrees
Elevation Rate: (max)	25 degrees/second
Firing Rate: Design rate, Rheinmetall loader	10 rounds/minute #
Actual rate, GM loader	6 rounds/minute #
Loading System:	Automatic
Stabilizer System:	Line of sight, Azimuth and elevation

Secondary:
(1) 20mm Rheinmetall RH202 cannon in remote control turret mount
(1) 7.62mm MG M73 coaxial w/152mm gun-launcher in turret
Provision for (1) .45 caliber SMG M3A1
Smoke grenade launchers on turret

AMMUNITION

46 rounds 152mm, w/Rheinmetall loader	8 hand grenades
48 rounds 152mm, w/GM loader	
750 rounds 20mm	
180 rounds .45 caliber	
6000 rounds 7.62mm	

FIRE CONTROL AND VISION EQUIPMENT

Primary Weapon:	Direct	Indirect
	Gunner's Primary Sight Group	Azimuth Indicator
	Commander's Panoramic Sight	Elevation Quadrant
	Commander's Night Sight	Gunner's Quadrant M1A1
	Gunner's Auxiliary Sight	
	Ballistic Computer	

Vision Devices:	Direct	Indirect
Driver	Hatch	Periscope vision blocks (3)
		Periscope night sight (1)
		Television night sight (1)
Commander	Hatch	Panoramic sight, direct night sight, TV night sight, periscope vision blocks (6)
Gunner	Hatch	Gunner's primary sight periscope, gunner's auxiliary sight, periscope vision block (1)

Total Periscopes: Vision block type (10)
Searchlight: (1) White light or infrared

ENGINE

Make and Model: Continental AVCR-1100-3		
Type: 12 cylinder, 4 cycle, 120 degree vee, variable compression ratio, supercharged		
Cooling System: Air Ignition: Compression		
Displacement:		1361.4 cubic inches
Bore and Stroke:		5.375 x 5.00 inches
Compression Ratio:		10:1 to 22:1
Net Horsepower (max):	estimated	1060 hp at 2800 rpm
Gross Horsepower (max):		1475 hp at 2800 rpm
Net Torque (max):	estimated	2750 ft-lb at 2200 rpm
Gross Torque (max):		2850 ft-lb at 2200 rpm
Weight:		approx. 4255 pounds.dry
Fuel: Diesel, DF-1 or DF-2		400 gallons
Engine Oil:		92 quarts

POWER TRAIN

Transmission: Renk HSWL 354, 4 ranges forward, 4 reverse
    Automatic and manual shift
    Two stage hydrodynamic torque converter w/lockup clutch
Steering Control: Driver or commander T-type control handles
Brakes: Multiple disc
Final Drive: Planetary gear   Gear Ratio: 4.5:1
Drive Sprocket: At rear of vehicle with 11 teeth
    Pitch Diameter: 25.63 inches

RUNNING GEAR

Suspension: Hydropneumatic (double piston), variable height **
    12 individually sprung dual road wheels (6/track)
    Tire Size: 26 x 6 inches
    6 dual track support rollers (3/track)
    Dual compensating idler at front of each track
    Idler Tire Size: 26 x 6 inches
Tracks: Center guide, Diehl 170
    Type: Double pin, 25 inch width, replaceable rubber pads
    Pitch: 7.22 inches
    Shoes per Vehicle: 156 (78/track)
    Ground Contact Length: 177.0 inches

ELECTRICAL SYSTEM

Nominal Voltage: 24 volts DC
Alternator, oil cooled, 28 volts, 700 amperes gear driven by main engine
Battery: (8) 12 volts, 4 sets of 2 in series connected in parallel

COMMUNICATIONS

Radio: AN/VRC-12 or AN/VRC-46 in turret
    AN/VRC-24 (air to ground) also may be fitted
Interphone: 3 stations plus external box

NUCLEAR, BIOLOGICAL, CHEMICAL PROTECTION

Overpressure environmental control system
Chemical and radiation detection equipment
Radiation shielding

FIRE PROTECTION

Automatic Halon fire detection and extinguisher system
(1) Portable Halon fire extinguisher in turret

PERFORMANCE

Maximum Speed: Sustained, level road	40 miles/hour
Maximum Grade:	60 per cent
Maximum Trench:	9.2 feet
Maximum Vertical Wall:	43 inches
Maximum Fording Depth: w/o kit	approx. 88 inches
w/kit	approx. 16 feet
Minimum Turning Circle: (diameter)	pivot
Cruising Range: Roads	400 miles

* With vehicle at normal 21 inch ground clearance
** National Water Lift suspension
\# With conventional ammunition, not missiles

152mm GUN-LAUNCHER TANK XM803

GENERAL DATA

Crew:	3	men
Length: Gun forward, over external phone box	369.5	inches
Length: Gun in travel position *	269	inches
Length: Without gun *	288	inches
Gun Overhang: Gun forward *	81	inches
Width: Over armor skirts	145.5	inches
Height: Over .50 caliber MG *	127.7	inches
Tread:	118.5	inches
Ground Clearance: Variable 6 to 25 inches **	normal 21.0	inches
Fire Height: At 21.0 inches ground clearance	approx. 74	inches
Turret Ring Diameter: (inside ring gear)	101.0	inches
Weight, Combat Loaded:	approx. 114,000	pounds
Weight, Unstowed:	approx. 107,000	pounds
Power to Weight Ratio: Net	estimated 15.8	hp/ton
Gross	21.9	hp/ton
Ground Pressure: Zero penetration	13.0	psi

ARMOR

Turret: Cast homogeneous armor steel inner shell overlaid with spaced high hardness homogeneous rolled armor steel, welded assembly Hull: Welded assembly of rolled homogeneous steel armor plate with some armor steel castings and aluminum armor

ARMAMENT

Primary: 152mm Gun-Launcher XM150E6 in turret combination mount

Traverse: Electrohydraulic and manual	360 degrees
Traverse Rate: (max)	10 seconds/360 degrees
Elevation: Electrohydraulic and manual	+20 to −10 degrees
Elevation Rate: (max)	25 degrees/second
Firing Rate: (max) Automatic loader	6 rounds/minute #
Loading System:	Automatic
Stabilizer System:	Line of sight, azimuth and elevation

Secondary:
(1) .50 caliber MG M85 on top of commander's day/night sight
(1) 7.62mm MG M73 coaxial w/152mm gun-launcher in turret
Provision for (1) .45 caliber SMG M3A1
(2) XM176 smoke grenade launchers (4 tubes each) on turret

AMMUNITION

50 rounds 152mm	8 hand grenades
900 rounds .50 caliber	
180 rounds .45 caliber	
6000 rounds 7.62mm	

FIRE CONTROL AND VISION EQUIPMENT

Primary

Weapon:	Direct	Indirect
	Gunner's Primary Sight Group	Azimuth Indicator
	Commander's Day/Night Sight	Elevation Quadrant
	Gunner's Auxiliary Sight	Gunner's Quadrant M1A1
	Ballistic Computer	

Vision Devices:	Direct	Indirect
Driver	Hatch	Periscope vision blocks (3)
		Periscope, infrared (1)
Commander	Hatch	Periscope vision blocks (7)
		Commander's day/night sight
Gunner	Hatch	Gunner's periscope in primary sight group and gunner's auxiliary sight

Total Periscopes: Vision block type (10)
Searchlight: AN/VSS-3, 1.0 KW, xenon white light or infrared

ENGINE

Make and Model: Continental AVCR-1100-3
Type: 12 cylinder, 4 cycle, 120 degree vee, variable compression ratio, supercharged

Cooling System: Air	Ignition: Compression	
Displacement:		1361.4 cubic inches
Bore and Stroke:		5.375 x 5.00 inches
Compression Ratio:		10:1 to 22:1
Net Horsepower (max):	estimated	1060 hp at 2800 rpm
Gross Horsepower (max):		1475 hp at 2800 rpm
Net Torque (max):	estimated	2750 ft-lb at 2200 rpm
Gross Torque (max):		2850 ft-lb at 2200 rpm
Weight:	approx.	4255 pounds, dry
Fuel: Diesel, DF-1 or DF-2		400 gallons
Engine Oil:		92 quarts

POWER TRAIN

Transmission: Renk HSWL 354, 4 ranges forward, 4 reverse
    Automatic and manual shift
    Two stage hydrodynamic torque converter w/lockup clutch
Steering Control: Driver or commander T-type control handles
Brakes: Multiple disc
Final Drive: Planetary gear   Gear Ratio: 4.5:1
Drive Sprocket: At rear of vehicle with 11 teeth
    Pitch Diameter: 25.63 inches

RUNNING GEAR

Suspension: Hydropneumatic (double piston), variable height **
    12 individually sprung dual road wheels (6/track)
    Tire Size: 26 x 6 inches
    6 dual track support rollers (3/track)
    Dual compensating idler at front of each track
    Idler Tire Size: 26 x 6 inches
Tracks: Center guide, Diehl 170
    Type: Double pin, 25 inch width, replaceable rubber pads
    Pitch: 7.22 inches
    Shoes per Vehicle: 156 (78/track)
    Ground Contact Length: 177.0 inches

ELECTRICAL SYSTEM

Nominal Voltage: 24 volts DC
Alternator, oil cooled, 28 volts, 700 amperes gear driven by main engine
Battery: (8) 12 volts, 4 sets of 2 in series connected in parallel

COMMUNICATIONS:

Radio: AN/VRC-12 or AN/VRC-46 in turret
    AN/VRC-24 (air to ground) also may be fitted
Interphone: 3 stations plus external box

NUCLEAR, BIOLOGICAL, CHEMICAL PROTECTION

    Overpressure environmental control system
    Chemical and radiation detection equipment
    Radiation shielding

FIRE PROTECTION

    Automatic Halon fire detection and extinguisher system
    (1) Portable Halon fire extinguisher in turret

PERFORMANCE

Maximum Speed: Sustained, level road	40 miles/hour
Maximum Grade:	60 per cent
Maximum Trench:	9.2 feet
Maximum Vertical Wall:	43 inches
Maximum Fording Depth: w/o kit	approx. 88 inches
w/kit	approx. 16 feet
Minimum Turning Circle: (diameter)	pivot
Cruising Range: Roads	400 miles

  * With vehicle at normal 21 inch ground clearance
 ** National Water Lift suspension
  # With conventional ammunition, not missiles

105mm GUN TANK XM1 (General Motors)

GENERAL DATA

Crew:	4	men
Length: Gun Forward	382.0	inches
Length: Gun to rear	351.0	inches
Length: Without gun	306.0	inches
Gun Overhang: Gun forward	76.0	inches
Width: Over armor skirts	143.5	inches
Height: Over .50 caliber MG	113.0	inches
Tread:	112.5	inches
Ground Clearance:	19.0	inches
Fire Height:	77.0	inches
Turret Ring Diameter: (inside ring gear)	85.7	inches
Weight, Combat Loaded:	approx. 117,000	pounds
Weight, Unstowed:	approx. 110,000	pounds
Power to Weight Ratio: Net	estimated 18.8	hp/ton
Gross	25.6	hp/ton
Ground Pressure: Zero penetration	13.0	psi

ARMOR

Turret: Welded assembly of rolled homogeneous steel armor with special armor arrays in the frontal area

Hull: Welded assembly of rolled homogeneous steel armor with special armor arrays in the frontal area and side skirts protecting the upper half of the suspension system

ARMAMENT

Primary: 105mm Gun M68 w/thermal shroud in turret combination mount

Traverse: Electrohydraulic and manual	360 degrees
Traverse Rate: (max)	9 seconds/360 degrees
Elevation: Electrohydraulic and manual	+20 to −10 degrees
Elevation Rate: (max)	25 degrees/second
Firing Rate: (max)	7 rounds/minute
Loading System:	Manual
Stabilizer System:	Line of sight, azimuth and elevation

Secondary:
  (1) .50 caliber MG M85 on commander's cupola, later (1) 40mm grenade launcher M85
  (1) Bushmaster coaxial w/105mm gun (proposed), .50 caliber MG M85 installed for validation phase later replaced by 7.62mm MG M240
  (1) 7.62mm MG M60D at loader's hatch
  Provision for (1) 5.56mm M16A1 Rifle
  (2) Smoke grenade launchers (4 tubes each) on turret

AMMUNITION (final version)
  55 rounds 105mm (compartmented)   8 hand grenades
  250 rounds 40mm
  8900 rounds 7.62mm (coaxial MG)
  1100 rounds 7.62mm (loader's MG)
  210 rounds 5.56mm
  24 smoke grenades

FIRE CONTROL AND VISION EQUIPMENT

Primary Weapon:	Direct	Indirect
	Gunner's Primary Sight	Azimuth Indicator
	Gunner's Auxiliary Sight	Elevation Quadrant
	Ballistic Computer	Gunner's Quadrant M1A1
Vision Devices:	Direct	Indirect
Driver	Hatch	Periscopes (3) and Periscope night vision (1)
Commander	Hatch	Periscope vision blocks (7) weapon sight (1)
Gunner	None	Gunner's Primary Sight Gunner's Auxiliary Sight Gunner's periscope (1)
Loader	Hatch	Loader's periscope (1)

ENGINE

Make and Model: Continental AVCR-1360-3
Type: 12 cylinder, 4 cycle, 120 degree vee, variable compression ratio, supercharged
Cooling System: Air   Ignition: Compression

Displacement:	1361.4 cubic inches
Bore and Stroke:	5.375 x 5.00 inches
Compression Ratio:	11.0:1 to 19.5:1
Net Horsepower (max):	estimated 1100 hp at 2600 rpm
Gross Horsepower (max):	1500 hp at 2600 rpm
Net Torque (max):	estimated 3400 ft-lb at 1900 rpm
Gross Torque (max):	3600 ft-lb at 1900 rpm
Weight:	4475 pounds, dry
Fuel: Diesel, DF-1 or DF-2	414 gallons
Engine Oil:	104 quarts

POWER TRAIN

Transmission: General Motors X-1100-1A 4 ranges forward, 2 reverse Hydrokinetic, fully automatic, 3 element torque convertor w/lockup clutch

Low (Automatic 1st, 2nd, 3rd, and 4th)	Neutral
Drive (Automatic 2nd, 3rd, and 4th)	Pivot steer
Reverse (Automatic 1st and 2nd)	

Drive Ratios: 1st 5.887.1   Reverse 1 8.305:1
                2nd 3.021:1   Reverse 2 2.354:1
                3rd 1.891:1
                4th 1.278:1

Steering Control: T-bar (Hydrostatic)
Brakes: Multiple disc
Final Drive: Planetary gear   Gear Ratio: 3.91:1
Drive Sprocket: At rear of vehicle with 11 teeth
  Pitch Diameter: 26.812 inches

RUNNING GEAR

Suspension: Hybrid with hydropneumatic units at road wheel stations 1, 2 and 6 and torsion bars on 3 and 5
  12 individually sprung dual road wheels (6/track)
  Tire Size: 31 x 6.5 inches
  4 dual track support rollers (2/track)
  Dual compensating idler at front of each track
  Idler Tire Size: 24 x 6.5 inches
Tracks: Center guide
  Type: Double pin, 24.0 inch width, replaceable rubber pads
  Pitch: 7.622 inches
  Shoes per Vehicle: 162 (81/track)
  Ground Contact Length: 188.0 inches

ELECTRICAL SYSTEM

Nominal Voltage: 24 volts DC
Alternator, oil cooled, 28 volts, 650 amperes, gear driven by main engine
Auxiliary Generator: None
Battery: (6) 12 volts, 3 sets of 2 in series connected in parallel

COMMUNICATIONS

Radio: AN/VRC-12 or AN/VRC-64
Interphone: AN/VIC-1, 4 stations plus external box

NUCLEAR, BIOLOGICAL, CHEMICAL PROTECTION

M13A1 gas, particulate filter unit w/four M25A1 masks
  (1) Chemical agent detector
  (1) AN/VDR Radiac nuclear agent detector
  (3) ABC M11 (1½ quart) decontamination apparatus

FIRE PROTECTION

Automatic Halon fire detection and extinguisher system
  (2) Portable Halon fire extinguishers

PERFORMANCE

Maximum Speed: Level road		48 miles/hour
Maximum Grade:		60 per cent
Maximum Trench:		8.6 feet
Maximum Vertical Wall:		42 inches
Maximum Fording Depth: w/o kit		48 inches
w/kit		7.5 feet
Minimum Turning Circle: (diameter)		pivot
Cruising Range: Roads		275 miles

105mm GUN TANK XM1 (Chrysler)

GENERAL DATA (original proposal data)

Crew:	4	men
Length: Gun forward	387.5	inches
Length: Gun to rear	348.5	inches
Length: Without gun	304	inches
Gun Overhang: Gun forward	83.5	inches
Width: Over armor skirts	140.0	inches
Height:	112.0	inches
Tread:	112.0	inches
Ground Clearance:	19.0	inches
Fire Height:	73.0	inches
Turret Ring Diameter: (inside)	85.0	inches
Weight, Combat Loaded:	approx. 116,000	pounds
Weight, Unstowed:	approx. 109,000	pounds
Power to Weight Ratio: Net	21.2	hp/ton
Gross	25.9	hp/ton
Ground Pressure: Zero penetration	12.7	psi

ARMOR

Turret: Welded assembly of rolled and cast homogeneous steel armor with special armor arrays in the frontal area

Hull: Welded assembly of rolled homogeneous steel armor with special armor arrays in the frontal area and skirts protecting the upper sides and part of the suspension system

ARMAMENT

Primary: 105mm Gun M68 w/thermal shroud in turret combination mount

Traverse: Electrohydraulic and manual	360 degrees
Traverse Rate: (max)	9 sec/360 degrees
Elevation: Electrohydraulic and manual	+20 to −10 degrees
Elevation Rate: (max)	25 degrees/second
Firing Rate: (max)	7 rounds/minute
Loading System:	Manual
Stabilizer System:	Turret, azimuth
	Line of sight, elevation

Secondary:

(1) .50 caliber MG M85 on commander's cupola

(1) Bushmaster coaxial w/105mm gun (proposed), .50 caliber MG M85 installed for validation phase later replaced by 7.62mm MG M240

(1) 7.62mm M60D MG or 40mm grenade launcher XM175 at loader's hatch

Provision for (1) 5.56mm M16A1 Rifle

AMMUNITION (original proposal)

40 rounds 105mm (compartmented)	8 hand grenades
500 rounds Bushmaster	
1092 rounds .50 caliber	
1600 rounds 7.62mm	
210 rounds 5.56mm	
24 smoke grenades	

FIRE CONTROL AND VISION EQUIPMENT

Primary Weapon:	Direct	Indirect
	Gunner's Primary Sight	Azimuth Indicator
	Gunner's Auxiliary Sight	Elevation Quadrant
	Ballistic Computer	Gunner's Quadrant M1A1
Vision Devices:	Direct	Indirect
Driver	Hatch	Periscopes (3) and Periscope, night vision (1)
Commander	Hatch	Periscope vision blocks (6) and weapon sight (1)
Gunner	None	Gunner's Primary Sight Gunner's Auxiliary Sight Periscope (1)
Loader	Hatch	Periscope (1)

ENGINE

Make and Model: Avco Lycoming AGT-1500	
Type: Free shaft power gas turbine with a two spool gasifier and recuperator	
Cooling System: Air	
Net Horsepower (max):	1232 hp at 3000 rpm
Gross Horsepower (max):	1500 hp at 3000 rpm
Net Torque (max):	3800 ft-lbs at 1000 rpm
Gross Torque (max):	3934 ft-lbs at 1000 rpm
Weight:	2528 pounds, dry
Fuel: Diesel, DF-1 or DF-2	500 gallons
Engine Oil:	28 quarts

POWER TRAIN

Transmission: X1100-3B, 4 ranges forward, 2 reverse
Hydrokinetic, fully automatic, 3 element torque convertor w/lockup clutch

Low (Automatic 1st, 2nd, 3rd, and 4th)	Neutral
Drive (Automatic 2nd, 3rd, and 4th)	Pivot steer
Reverse (Automatic 1st and 2nd)	

Drive Ratios:	1st 5.877:1	Reverse 1 8.305:1
	2nd 3.021:1	Reverse 2 2.354:1
	3rd 1.891:1	
	4th 1.278:1	

Steering Control: T-bar (Hydrostatic)
Brakes: Multiple disc
Final Drive: Planetary gear   Gear Ratio: 4.30:1
Drive Sprocket: At rear of vehicle with 11 teeth
Pitch Diameter: 26.8 inches

RUNNING GEAR

Suspension: Tube-over-bar proposed, later torsion bar (high strength on stations 1, 2, and 7)
14 individually sprung dual road wheels (7/track)
Tire Size: 25 x 6.5 proposed, later 25 x 5.18 inches
6 dual track support rollers (3/track)
Dual compensating idler at front of each track
Idler Tire Size: 25 x 6.5 proposed, later 25 x 5.18 inches
Rotary shock absorbers on first two and last road wheels

Tracks: Center guide, T156
Type: Double pin, 25 inch width, integral rubber pad
Pitch: 7.625 inches
Shoes per Vehicle: 156 (78/track)
Ground Contact Length: 180.1 inches

ELECTRICAL SYSTEM

Nominal Voltage: 24 volts DC
Alternator, oil cooled w/solid state regulator, 28 volts, 650 amperes, gear driven by main engine
Auxiliary Generator: None
Battery: (4) 12 volts, 2 sets of 2 in series connected in parallel

COMMUNICATIONS

Radio: AN/VRC-12
Interphone: AN/VIC-1, 4 stations

NUCLEAR, BIOLOGICAL, CHEMICAL PROTECTION

M13A1 gas, particulate filter unit w/four M25A1 masks
(1) M15A2 chemical agent detector
(1) AN/VDR Radiac nuclear agent detector
(1) ABC M11 (1½ quart) decontamination apparatus

FIRE PROTECTION

Automatic Halon fire detection and extinguisher system
(2) Portable Halon fire extinguishers

PERFORMANCE

Maximum Speed: Level road		47 miles/hour
Maximum Grade:		60 per cent
Maximum Trench:		9.0 feet
Maximum Vertical Wall:		49 inches
Maximum Fording Depth: w/o kit		48 inches
	w/kit	7.5 feet
Minimum Turning Circle: (diameter)		pivot
Cruising Range: Roads		275 miles

105mm GUN TANK M1

GENERAL DATA

Crew:	4	men
Length: Gun forward	384.5	inches
Length: Gun to rear	353.2	inches
Length: Without gun	311.7	inches
Gun Overhang: Gun forward	72.8	inches
Width: Over armor skirts	144.0	inches
Height: Over .50 caliber MG	113.6	inches
Tread:	112.0	inches
Ground Clearance:	19.0	inches
Fire Height:	74.5	inches
Turret Ring Diameter: (inside)	85.0	inches
Weight, Combat Loaded:	approx. 120,000	pounds
Weight, Unstowed:	approx. 113,000	pounds
Power to Weight Ratio: Net	20.5	hp/ton
Gross	25.0	hp/ton
Ground Pressure: Zero penetration	13.3	psi

ARMOR

Turret: Welded assembly of rolled homogeneous steel armor with special armor arrays in the frontal area

Hull: Welded assembly of rolled homogeneous steel armor with special armor arrays in the frontal area and special armor skirts protecting the upper sides and part of the suspension system

ARMAMENT

Primary: 105mm Gun M68A1 in turret combination mount

Traverse: Electrohydraulic and manual	360 degrees
Traverse Rate: (max)	9 seconds/360 degrees
Elevation: Electrohydraulic and manual	+20 to −10 degrees
Elevation Rate: (max)	25 degrees/second
Firing Rate: (max)	7 rounds/minute
Loading System:	Manual
Stabilizer System:	Turret, azimuth
	Line of sight, elevation

Secondary:

(1) .50 caliber MG HB M2 on commander's cupola
(1) 7.62mm MG M240 coaxial w/105mm gun in turret
(1) 7.62mm MG M240 at loader's hatch
Provision for (1) 5.56mm M16A1 Rifle
(2) Smoke grenade launchers on turret

AMMUNITION

55 rounds 105mm
900 rounds .50 caliber
1400 rounds 7.62mm (loader's MG)
10,000 rounds 7.62mm (coaxial MG)
210 rounds 5.56mm
24 smoke grenades

8 M67 hand grenades

FIRE CONTROL AND VISION EQUIPMENT

Primary Weapon:	Direct	Indirect
	Gunner's Primary Sight	Azimuth Indicator
	Gunner's Auxiliary Sight	Elevation Quadrant
	Ballistic Computer	Gunner's Quadrant M1A1
Vision Devices:	Direct	Indirect
Driver	Hatch	Periscopes (3) and
		Periscope, night vision (1)
Commander	Hatch	Periscope vision blocks (6)
		and weapon sight (1)
Gunner	None	Gunner's Primary Sight and
		Gunner's Auxiliary Sight
Loader	Hatch	Periscope (1)

ENGINE

Make and Model: Avco Lycoming AGT-1500
Type: Free shaft power gas turbine with a two spool gasifier and recuperator
Cooling System: Air

Net Horsepower (max):	1232 hp at 3000 rpm
Gross Horsepower (max):	1500 hp at 3000 rpm
Net Torque (max):	3800 ft-lbs at 1000 rpm
Gross Torque (max):	3934 ft-lbs at 1000 rpm
Weight:	2528 pounds, dry
Fuel: Diesel, DF-1, DF-2, or DF-A	505 gallons
Engine Oil:	28 quarts

POWER TRAIN

Transmission: X1100-3B, 4 ranges forward, 2 reverse
  Hydrokinetic, fully automatic, 3 element torque convertor w/lockup clutch

Low (Automatic 1st, 2nd, 3rd, and 4th)	Neutral
Drive (Automatic 2nd, 3rd, and 4th)	Pivot steer
Reverse (Automatic 1st and 2nd)	

Drive ratios:	1st 5.877:1	Reverse 1 8.305:1
	2nd 3.021:1	Reverse 2 2.354:1
	3rd 1.891:1	
	4th 1.278:1	

Steering Control: T-bar (Hydrostatic)
Brakes: Multiple disc
Final Drive: Planetary gear   Gear Ratio: 4.30:1
Drive Sprocket: At rear of vehicle with 11 teeth and retaining ring
  Pitch Diameter: 26.8 inches

RUNNING GEAR

Suspension: Torsion bar (high strength)
  14 individually sprung dual road wheels (7/track)
  Tire Size: 25 x 5.18 inches
  4 single track support rollers (2/track)
  Dual compensating idler at front of each track
  Idler Tire Size: 25 x 5.18 inches
  Rotary hydraulic shock absorbers on first 2 and last road wheels
Tracks: Center guide, T156
  Type: Double pin, 25 inch width, integral rubber pad
  Pitch: 7.625 inches
  Shoes per Vehicle: 156 (78/track)
  Ground Contact Length: 180.1 inches

ELECTRICAL SYSTEM

Nominal Voltage: 24 volts DC
Alternator, oil cooled w/solid state regulator, 28 volts, 650 amperes, gear driven by main engine
Auxiliary Generator: None
Battery: (6) 12 volts, 3 sets of 2 in series connected in parallel

COMMUNICATIONS

Radio: AN/VRC-12 or AN/VRC-64
Interphone: AN/VIC-1, 4 stations plus external box

NUCLEAR, BIOLOGICAL, CHEMICAL PROTECTION

M13A1 gas, particulate filter unit w/four M25A1 masks
(1) M15A2A chemical agent detector
(1) AN/VDR Radiac nuclear agent detector
(3) ABC M11 (1½ quart) decontamination apparatus

FIRE PROTECTION

Automatic Halon fire detection and extinguisher system
(2) Portable Halon fire extinguishers

PERFORMANCE

Maximum Speed: Level road (governed)	45 miles/hour
Maximum Tractive Effort: TE at stall	163,000 pounds
Per Cent of Vehicle Weight TE/W	136 per cent
Maximum Grade:	60 per cent
Maximum Trench:	9.0 feet
Maximum Vertical Wall:	49 inches
Maximum Fording Depth: w/o kit	48 inches
w/kit	7.5 feet
Minimum Turning Circle: (diameter)	pivot
Cruising Range: Roads	275 miles

105mm GUN TANK M1, IMPROVED PERFORMANCE

GENERAL DATA

Crew:	4	men
Length: Gun forward	384.5	inches
Length: Gun to rear	353.2	inches
Length: Without gun	311.7	inches
Gun Overhang: Gun forward	72.8	inches
Width: Over armor skirts	144.0	inches
Height: Over .50 caliber MG	113.6	inches
Tread:	112.0	inches
Ground Clearance:	19.0	inches
Fire Height:	74.5	inches
Turret Ring Diameter: (inside)	85.0	inches
Weight, Combat Loaded:	approx. 122,000	pounds
Weight, Unstowed:	approx. 115,000	pounds
Power to Weight Ratio: Net	20.2	hp/ton
Gross	24.6	hp/ton
Ground Pressure: Zero penetration	13.5	psi

ARMOR

Turret: Welded assembly of rolled homogeneous steel armor with special armor arrays in the frontal area

Hull: Welded assembly of rolled homogeneous steel armor with special armor arrays in the frontal area and special armor skirts protecting the upper sides and part of the suspension system

ARMAMENT

Primary: 105mm Gun M68A1 in turret combination mount

Traverse: Electrohydraulic and manual	360 degrees
Traverse Rate: (max)	9 seconds/360 degrees
Elevation: Electrohydraulic and manual	+20 to −10 degrees
Elevation Rate: (max)	25 degrees/second
Firing Rate: (max)	7 rounds/minute
Loading System:	Manual
Stabilizer System:	Turret, azimuth
	Line of sight, elevation

Secondary:

(1) .50 caliber MG HB M2 on commander's cupola
(1) 7.62mm MG M240 coaxial w/105mm gun in turret
(1) 7.62mm MG M240 at loader's hatch
Provision for (1) 5.56mm M16A1 Rifle
(2) Smoke grenade launchers on turret

AMMUNITION

55 rounds 105mm	8 M67 hand grenades
900 rounds .50 caliber	
1400 rounds 7.62mm (loader's MG)	
10,000 rounds 7.62mm (coaxial MG)	
210 rounds 5.56mm	
24 smoke grenades	

FIRE CONTROL AND VISION EQUIPMENT

Primary Weapon:	Direct	Indirect
	Gunner's Primary Sight	Azimuth Indicator
	Gunner's Auxiliary Sight	Elevation Quadrant
	Ballistic Computer	Gunner's Quadrant M1A1
Vision Devices:	Direct	Indirect
Driver	Hatch	Periscopes (3) and Periscope, night vision (1)
Commander	Hatch	Periscope vision blocks (6) and weapon sight (1)
Gunner	None	Gunner's Primary Sight and Gunner's Auxiliary Sight
Loader	Hatch	Periscope (1)

ENGINE

Make and Model: Avco Lycoming AGT-1500	
Type: Free shaft power gas turbine with a two spool gasifier and recuperator	
Cooling System: Air	
Net Horsepower: (max)	1232 hp at 3000 rpm
Gross Horsepower: (max)	1500 hp at 3000 rpm
Net Torque: (max)	3800 ft-lbs at 1000 rpm
Gross Torque: (max)	3934 ft-lbs at 1000 rpm
Weight:	2528 pounds, dry
Fuel: Diesel, DF-1, DF-2, or DF-A	505 gallons
Engine Oil:	28 quarts

POWER TRAIN

Transmission: X1100-3B, 4 ranges forward, 2 reverse
Hydrokinetic, fully automatic, 3 element torque convertor w/lockup clutch

Low (Automatic 1st, 2nd, 3rd, and 4th)	Neutral
Drive (Automatic 2nd, 3rd, and 4th)	Pivot steer
Reverse (Automatic 1st and 2nd)	

Drive Ratios:	1st	5.877:1	Reverse 1	8.305:1
	2nd	3.021:1	Reverse 2	2.354:1
	3rd	1.891:1		
	4th	1.278:1		

Steering Control: T-bar (Hydrostatic)
Brakes: Multiple disc
Final Drive: Planetary gear   Gear Ratio: 4.67:1
Drive Sprocket: At rear of vehicle with 11 teeth and retaining ring
Pitch Diameter: 26.8 inches

RUNNING GEAR

Suspension: Torsion bar (high strength)
14 individually sprung dual road wheels (7/track)
Tire Size: 25 x 5.69 inches
4 single track support rollers (2/track)
Dual compensating idler at front of each track
Idler Tire Size: 25 x 5.69 inches
Rotary hydraulic shock absorbers on first 2 and last road wheels
Tracks: Center guide, T156
Type: Double pin, 25 inch width, integral rubber pad
Pitch: 7.625 inches
Shoes per Vehicle: 156 (78/track)
Ground Contact Length: 180.1 inches

ELECTRICAL SYSTEM

Nominal Voltage: 24 volts DC
Alternator, oil cooled w/solid state regulator, 28 volts, 650 amperes, gear driven by main engine
Auxiliary Generator: None
Battery: (6) 12 volts, 3 sets of 2 in series connected in parallel

COMMUNICATIONS

Radio: AN/VRC-12 or AN/VRC-64
Interphone: AN/VIC-1, 4 stations plus external box

NUCLEAR, BIOLOGICAL, CHEMICAL PROTECTION

M13A1 gas, particulate filter unit w/four M25A1 masks
(1) M15A2A chemical agent detector
(1) AN/VDR Radiac nuclear agent detector
(3) ABC M11 (1½ quart) decontamination apparatus

FIRE PROTECTION

Automatic Halon fire detection and extinguisher system
(2) Portable Halon fire extinguishers

PERFORMANCE

Maximum Speed: Level road (governed)	41.5 miles/hour
Maximum Tractive Effort: TE at stall	177,000 pounds
Per Cent of Vehicle Weight: TE/W	145 per cent
Maximum Grade:	60 per cent
Maximum Trench:	9.0 feet
Maximum Vertical Wall:	49 inches
Maximum Fording Depth: w/o kit	48 inches
w/kit	7.5 feet
Minimum Turning Circle:	pivot
Cruising Range: Roads	275 miles

120mm GUN TANK M1E1

GENERAL DATA

Crew:	4	men
Length: Gun forward	386.9	inches
Length: Gun to rear	355.6	inches
Length: Without gun	311.7	inches
Gun Overhang: Gun forward	75.2	inches
Width: Over armor skirts	144.0	inches
Height: Over .50 caliber MG	113.7	inches
Tread:	112.0	inches
Ground Clearance:	19.0	inches
Fire Height:	74.5	inches
Turret Ring Diameter: (inside)	85.0	inches
Weight, Combat Loaded:	approx. 126,000	pounds
Weight, Unstowed:	approx. 119,000	pounds
Power to Weight Ratio: Net	19.6	hp/ton
Gross	23.8	hp/ton
Ground Pressure: Zero penetration	14.0	psi

ARMOR

Turret: Welded assembly of rolled homogeneous steel armor with special armor arrays in the frontal area

Hull: Welded assembly of rolled homogeneous steel armor with special armor arrays in the frontal area and special armor skirts protecting the upper sides and part of the suspension system

ARMAMENT

Primary: 120mm Gun XM256 in turret combination mount

Traverse: Electrohydraulic and manual	360 degrees
Traverse Rate: (max)	9 seconds/360 degrees
Elevation: Electrohydraulic and manual	+20 to −10 degrees
Elevation Rate: (max)	25 degrees/second
Firing Rate: (max)	6 rounds/minute
Loading System:	Manual
Stabilizer System:	Turret, azimuth
	Line of sight, elevation

Secondary:

(1) .50 caliber MG HB M2 on commander's cupola
(1) 7.62mm MG M240 coaxial w/120mm gun in turret
(1) 7.62mm MG M240 at loader's hatch
Provision for (1) 5.56mm M16A1 Rifle
(2) Smoke grenade launchers on turret

AMMUNITION

40 rounds 120mm	8 M67 hand grenades
900 rounds .50 caliber	
1400 rounds 7.62mm (loader's MG)	
10,000 rounds 7.62mm (coaxial MG)	
210 rounds 5.56mm	
24 smoke grenades	

FIRE CONTROL AND VISION EQUIPMENT

Primary Weapon:	Direct	Indirect
	Gunner's Primary Sight	Azimuth Indicator
	Gunner's Auxiliary Sight	Elevation Quadrant
	Ballistic Computer	Gunner's Quadrant M1A1
Vision Devices:	Direct	Indirect
Driver	Hatch	Periscopes (3) and
		Periscope, night vision (1)
Commander	Hatch	Periscope vision blocks (6)
		and weapon sight (1)
Gunner	None	Gunner's Primary Sight and
		Gunner's Auxiliary Sight
Loader	Hatch	Periscope (1)

ENGINE

Make and Model: Avco Lycoming AGT-1500
Type: Free shaft power gas turbine with a two spool gasifier and recuperator
Cooling System: Air

Net Horsepower (max):	1232 hp at 3000 rpm
Gross Horsepower (max):	1500 hp at 3000 rpm
Net Torque (max):	3800 ft-lbs at 1000 rpm
Gross Torque (max):	3934 ft-lbs at 1000 rpm
Weight:	2528 pounds, dry
Fuel: Diesel, DF-1, DF-2, or DF-A	505 gallons
Engine Oil:	25 quarts

POWER TRAIN

Transmission: X1100-3B, 4 ranges forward, 2 reverse
Hydrokinetic, fully automatic, 3 element torque convertor w/lockup clutch

Low (Automatic 1st, 2nd, 3rd, and 4th)	Neutral
Drive (Automatic 2nd, 3rd, and 4th)	Pivot steer
Reverse (Automatic 1st and 2nd)	

Drive Ratios:	1st 5.877:1	Reverse 1 8.305:1
	2nd 3.021:1	Reverse 2 2.354:1
	3rd 1.891:1	
	4th 1.278:1	

Steering Control: T-bar (Hydrostatic)
Brakes: Multiple disc
Final Drive: Planetary gear   Gear Ratio: 4.67:1
Drive Sprocket: At rear of vehicle with 11 teeth and retaining ring
Pitch Diameter: 26.8 inches

RUNNING GEAR

Suspension: Torsion bar (high strength)
14 individually sprung dual road wheels (7/track)
Tire Size: 25 x 5.69 inches
4 single track support rollers (2/track)
Dual compensating idler at front of each track
Idler Tire Size: 25 x 5.69 inches
Rotary hydraulic shock absorbers on first 2 and last road wheels
Tracks: Center guide, T156
Type: Double pin, 25 inch width, integral rubber pad
Pitch: 7.625 inches
Shoes per Vehicle: 156 (78/track)
Ground Contact Length: 180.1 inches

ELECTRICAL SYSTEM

Nominal Voltage: 24 volts DC
Alternator, oil cooled w/solid state regulator, 28 volts, 650 amperes, gear driven by main engine
Auxiliary Generator: None
Battery: (6) 12 volts, 3 sets of 2 in series connected in parallel

COMMUNICATIONS

Radio: AN/VRC-12 or AN/VRC-64
Interphone: AN/VIC-1, 4 stations plus external box

NUCLEAR, BIOLOGICAL, CHEMICAL PROTECTION

Primary: Overpressure system w/air supplied to face masks and air-cooled vests for the crew
Backup: M13A1 gas, particulate filter unit
(1) M43A1 chemical agent detector
(1) AN/VDR Radiac nuclear agent detector
(3) ABC M11 (1½ quart) decontamination apparatus

FIRE PROTECTION

Automatic Halon fire detection and extinguisher system
(2) Portable Halon fire extinguishers

PERFORMANCE

Maximum Speed: Level road (governed)	41.5 miles/hour
Maximum Tractive Effort: TE at stall	177,000 pounds
Per Cent of Vehicle Weight: TE/W	140 per cent
Maximum Grade:	60 per cent
Maximum Trench:	9.0 feet
Maximum Vertical Wall:	49 inches
Maximum Fording Depth: w/o kit	48 inches
w/kit	7.5 feet
Minimum Turning Circle: (diameter)	pivot
Cruising Range: Roads	275 miles

120mm GUN TANK M1A1

GENERAL DATA

Crew:	4	men
Length: Gun forward	386.9	inches
Length: Gun to rear	355.6	inches
Length: Without gun	311.7	inches
Gun Overhang: Gun forward	75.2	inches
Width: Over armor skirts	143.8	inches
Height: Over .50 caliber MG	113.6	inches
Tread:	112.0	inches
Ground Clearance:	19.0	inches
Fire Height:	74.5	inches
Turret Ring Diameter: (inside)	85.0	inches
Weight, Combat Loaded:	approx. 13,000	pounds
Weight, Unstowed:	approx. 123,000	pounds
Power to Weight Ratio: Net	19.0	hp/ton
Gross	23.1	hp/ton
Ground Pressure: Zero penetration	14.4	psi

ARMOR

Turret: Welded assembly of rolled homogeneous steel armor with special armor arrays in the frontal area

Hull: Welded assembly of rolled homogeneous steel armor with special armor arrays in the frontal area and special armor skirts protecting the upper sides and part of the suspension system

ARMAMENT

Primary: 120mm Gun M256 in turret combination mount

Traverse: Electrohydraulic and manual	360 degrees
Traverse Rate: (max)	9 seconds/360 degrees
Elevation: Electrohydraulic and manual	+20 to −10 degrees
Elevation Rate: (max)	25 degrees/second
Firing Rate: (max)	6 rounds/minute
Loading System:	Manual
Stabilizer System:	Turret, azimuth
	Line of sight, elevation

Secondary:
- (1) .50 caliber MG HB M2 on commander's cupola
- (1) 7.62mm MG M240 coaxial w/120mm gun in turret
- (1) 7.62mm MG M240 at loader's hatch
- Provision for (1) 5.56mm M16A1 Rifle
- (2) Smoke grenade launchers on turret

AMMUNITION

40 rounds 120mm	8 M67 hand grenades
900 rounds .50 caliber	
1400 rounds 7.62mm (loader's MG)	
10,000 rounds 7.62mm (coaxial MG)	
210 rounds 5.56mm	
24 smoke grenades	

FIRE CONTROL AND VISION EQUIPMENT

Primary Weapon:	Direct	Indirect
	Gunner's Primary Sight	Azimuth Indicator
	Gunner's Auxiliary Sight	Elevation Quadrant
	Ballistic Computer	Gunner's Quadrant M1A1
Vision Devices:	Direct	Indirect
Driver	Hatch	Periscopes (3) and
		Periscope, night vision (1)
Commander	Hatch	Periscope vision blocks (6)
		and weapon sight (1)
Gunner	None	Gunner's Primary Sight and
		Gunner's Auxiliary Sight
Loader	Hatch	Periscope (1)

ENGINE

Make and Model: Avco Lycoming AGT-1500
Type: Free shaft power gas turbine with a two spool gasifier and recuperator
Cooling System: Air

Net Horsepower (max):	1232 hp at 3000 rpm
Gross Horsepower (max):	1500 hp at 3000 rpm
Net Torque (max):	3800 ft-lbs at 1000 rpm
Gross Torque (max):	3934 ft-lbs at 1000 rpm
Weight:	2528 pounds, dry
Fuel: Diesel, DF-1, DF-2, or DF-A	505 gallons
Engine Oil:	25 quarts

POWER TRAIN

Transmission: X1100-3B, 4 ranges forward, 2 reverse
Hydrokinetic, fully automatic, 3 element torque convertor w/lockup clutch

Low (Automatic 1st, 2nd, 3rd, and 4th)	Neutral
Drive (Automatic 2nd, 3rd, and 4th)	Pivot steer
Reverse (Automatic 1st and 2nd)	

Drive Ratios:	1st 5.877:1	Reverse 1 8.305:1
	2nd 3.021:1	Reverse 2 2.354:1
	3rd 1.891:1	
	4th 1.278:1	

Steering Control: T-bar (Hydrostatic)
Brakes: Multiple disc
Final Drive: Planetary gear   Gear Ratio: 4.67:1
Drive Sprocket: At rear of vehicle with 11 teeth
Pitch Diameter: 26.8 inches

RUNNING GEAR

Suspension: Torsion bar (high strength)
14 individually sprung dual road wheels (7/track)
Tire Size: 25 x 5.69 inches
4 single track support rollers (2/track)
Dual compensating idler at front of each track
Idler Tire Size: 25 x 5.69 inches
Rotary hydraulic shock absorbers on first 2 and last road wheels
Tracks: Center guide, T156 and T158
Type: (T156) Double pin, 25 inch width, integral rubber pad
(T158) Double pin, 25 inch width, replaceable rubber pad
Pitch: 7.625 inches
Shoes per Vehicle: 156 (78/track)
Ground Contact Length: 180.1 inches

ELECTRICAL SYSTEM

Nominal Voltage: 24 volts DC
Alternator, oil cooled w/solid state regulator, 28 volts, 650 amperes, gear driven by main engine
Auxiliary Generator: None
Battery: (6) 12 volts, 3 sets of 2 in series connected in parallel

COMMUNICATIONS

Radio: AN/VRC-12 or AN/VRC-64
Interphone: AN/VIC-1, 4 stations plus external box

NUCLEAR, BIOLOGICAL, CHEMICAL PROTECTION

Primary: Overpressure system w/air supplied to face masks and air-cooled vests for the crew
Backup: M13A1 gas, particulate filter unit
- (1) M43A1 chemical agent detector
- (1) AN/VDR Radiac nuclear agent detector
- (3) ABC M11 (1½ quart) decontamination apparatus

FIRE PROTECTION

Automatic Halon fire detection and extinguisher system
(2) Portable Halon fire extinguishers

PERFORMANCE

Maximum Speed: Level Road (governed)	41.5 miles/hour
Maximum Tractive Effort: TE at stall	177,000 pounds
Per Cent of Vehicle Weight: TE/W	136 per cent
Maximum Grade:	60 per cent
Maximum Trench:	9.0 feet
Maximum Vertical Wall:	49 inches
Maximum Fording Depth: w/o kit	48 inches
w/kit	7.5 feet
Minimum Turning Circle: (diameter)	pivot
Cruising Range: Roads	289 miles

These data sheets cover the armament for both the early experimental tanks as well as the later production vehicles. Many of these weapons themselves were highly experimental and some were cancelled prior to the completion of their test program. Thus the data are incomplete in those cases. The penetration performance for the armor piercing ammunition has been omitted from the data sheets because many of the later weapons and their ammunition are still subject to security restrictions. Other details of some types of ammunition also have been left out for the same reasons.

The dimensions of the various weapons included in the data sheets are defined in the following sketches.

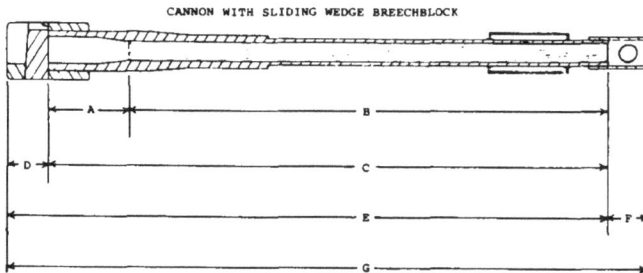

CANNON WITH SLIDING WEDGE BREECHBLOCK

A. Length of Chamber (to rifling)
B. Length of Rifling
C. Length of Bore
D. Depth of Breech Recess
E. Length, Muzzle to Rear Face of Breech
F. Additional Length, Blast Deflector, Etc.
G. Overall Length
H. Length, Breechblock and Firing Lock
I. Length of Tube
J. Length of Separable Chamber
K. Length of Tube and Chamber

The ammunition is listed according to the official U. S. Army nomenclature in use during its period of greatest service. Since this did change and sometimes needed clarification, a standard nomenclature is added in parentheses based on the following terms used separately and in combination.

APBC	Armor piercing with ballistic cap
APCR	Armor piercing, composite rigid
APDS	Armor piercing, discarding sabot
APFSDS	Armor piercing, fin stabilized, discarding sabot
HE	High explosive
HEAT	High explosive antitank, shaped charge
HESH	High explosive, squash head
APERS	Antipersonnel
TPBC	Target practice with ballistic cap
TPCR	Target practice, composite rigid
TPCSDS	Target practice, cone stabilized, discarding sabot
MP	Multipurpose
T	Tracer

The guns installed in the early experimental tanks were fitted with cylindrical blast deflectors often referred to as counterweights since they also balanced the gun in the mount. Later, the cylindrical type was replaced by the T-shape blast deflector as it came into use on the standard tanks in troop service. The blast deflectors were eliminated on the later weapons. Bore evacuators were included on all tank guns except for the later versions of the 152mm gun-launcher which were fitted with a closed breech scavenging system.

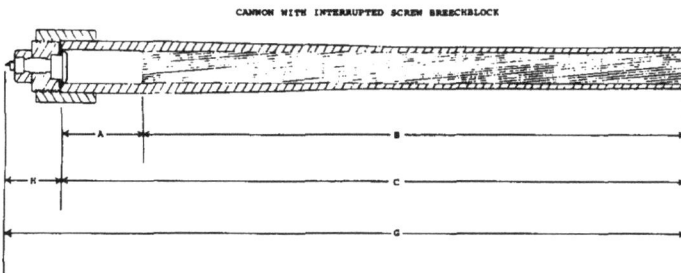

CANNON WITH INTERRUPTED SCREW BREECHBLOCK

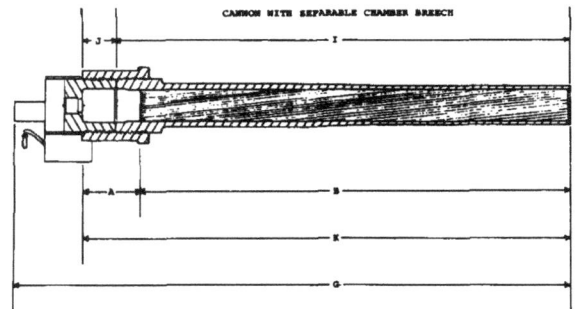

CANNON WITH SEPARABLE CHAMBER BREECH

90mm Gun M41

Carriage and Mount	90mm Gun Tanks T48, M48, M48A1, M48A2, and M48A3 in Mounts M87 or M87A1, 90mm Gun Tank T95E2 in Mount M87A1
Length of Chamber (to rifling)	24.4 inches
Length of Rifling	152.77 inches
Length of Chamber (to projectile base)	20.8 inches (boat-tailed projectiles)
Travel of Projectile in Bore	156.4 inches (boat-tailed projectiles)
Length of Bore	177.15 inches, 50.0 calibers
Depth of Breech Recess	9.00 inches
Length, Muzzle to Rear Face of Breech	186.15 inches, 52.5 calibers
Additional Length	7.2 inches, T-shape blast deflector
Overall Length	193.4 inches w/T-shape blast deflector
Diameter of Bore	3.543 inches (90mm)
Chamber Capacity	300 cubic inches
Weight, Tube	1580 pounds
Total Weight	2370 pounds, approx.
Type of Breechblock	Semiautomatic, vertical sliding wedge
Rifling	32 grooves, uniform right-hand twist, one turn in 25 calibers
Ammunition	Fixed
Primer	Percussion

Weight, Complete Round	AP-T M318(T33E7) Shot(APBC-T)	43.91 pounds(19.9 kg) **
	HEAT-T M431 Shell(HEAT-T)	32.25 pounds(14.6 kg)
	HE-T T91E3 Shell(HE-T)	36.25 pounds(16.5 kg) *
	HE-T M71A1 Shell(HE-T)	39.54 pounds(17.9 kg) #
	APERS-T XM580E1 (4100 flechettes)	41.25 pounds(18.7 kg)
	Canister M336 (1280 pellets)	42.50 pounds(19.3 kg) *
	Canister M377 (5600 flechettes)	39.30 pounds(17.8 kg) **
	TP-T M353(T225E1) Shot(TPBC-T)	43.91 pounds(19.9 kg) **
Weight, Projectile	AP-T M318(T33E7) Shot(APBC-T)	24.18 pounds(11.0 kg)
	HEAT-T M431 Shell(HEAT-T)	12.75 pounds(5.8 kg)
	HE-T T91E3 Shell(HE-T)	20.25 pounds(9.2 kg)
	HE-T M71A1 Shell(HE-T)	23.57 pounds(10.7 kg)
	APERS-T XM580E1 (4100 flechettes)	20 pounds(9 kg) approx.
	Canister M336 (1280 pellets)	23.24 pounds(10.5 kg)
	Canister M377 (5600 flechettes)	20.44 pounds(9.3 kg)
	TP-T M353(T225E1) Shot(TPBC-T)	24.18 pounds(11.0 kg)
Maximum Powder Pressure	47,000 psi	
Maximum Rate of Fire	8 rounds/minute	
Muzzle Velocity	AP-T M318(T33E7) Shot(APBC-T)	3000 ft/sec(914 m/sec)
	HEAT-T M431 Shell(HEAT-T)	4000 ft/sec(1219 m/sec)
	HE-T T91E3 Shell(HE-T)	2400 ft/sec(732 m/sec)
	HE-T M71A1 Shell(HE-T)	2400 ft/sec(732 m/sec)
	APERS-T XM580E1 (4100 flechettes)	3000 ft/sec(914 m/sec)
	Canister M336 (1280 pellets)	2870 ft/sec(875 m/sec)
	Canister M377 (5600 flechettes)	2950 ft/sec(899 m/sec)
	TP-T M353(T225E1) Shot(TPBC-T)	3000 ft/sec(914 m/sec)
Muzzle Energy of Projectile, KE=1/2MV² Rotational energy is neglected and values are based on long tons (2240 pounds)	AP-T M318(T33E7) Shot(APBC-T)	1509 ft-tons
	HEAT-T M431 Shell(HEAT-T)	1414 ft-tons
	HE-T T91E3 Shell(HE-T)	809 ft-tons
	HE-T M71A1 Shell(HE-T)	941 ft-tons
	APERS-T XM580E1 (4100 flechettes)	1250 ft-tons approx.
	Canister M336 (1280 pellets)	1327 ft-tons
	Canister M377 (5600 flechettes)	1230 ft-tons
	TP-T M353(T225E1) Shot(TPBC-T)	1509 ft-tons
Maximum Range (independent of mount)	AP-T M318(T33E7) Shot(APBC-T)	23,000 yards(21,031 m)
	HEAT-T M431 Shell(HEAT-T)	8900 yards(8138 m)
	HE-T T91E3 Shell(HE-T)	14,500 yards(13,259 m)
	HE-T M71A1 Shell(HE-T)	16,800 yards(15,362 m)
	APERS-T XM580E1 (4100 flechettes)	4800 yards(4389 m)
	Canister M336 (1280 pellets)	200 yards(183 m)
	Canister M377 (5600 flechettes)	440 yards(402 m)
	TP-T M353(T225E1) Shot(TPBC-T)	23,000 yards(21.031 m)

* Assembled with the M108(T24) brass cartridge case (weight 11.0 pounds)
** Assembled with the M108B1(T24B1) steel cartridge case (weight 10.3 pounds)
\# Assembled with the M19 brass cartridge case (weight 11.0 pounds)

The HEAT-T M431 and APERS-T XM580E1 rounds were assembled with the M114E1 and XM200 cartridge cases respectively. In addition to the ammunition assembled with the M108 or M108B1 cartridge cases, this weapon could use any of the rounds for the lower pressure M1, M2, and M3 series of 90mm guns fitted in the M19 or M19B1 cartridge cases.

90mm Guns T208, T208E9 and T208E13

Carriage and Mount	90mm Gun Tank T95 in rigid Mount T191 (T208 and T208E13 Guns) and 90mm Gun Tank T95E1 in recoiling Mount T192 (T208E9 Gun)
Length of Chamber	25 inches
Travel of Projectile in Bore	235 inches
Length of Bore	259.88 inches, 73.35 calibers
Depth of Breech Recess	9.00 inches
Length, Muzzle to Rear Face of Breech	268.88 inches, 75.9 calibers
Additional Length, Blast Deflector	7.5 inches
Overall Length	276.38 inches
Diameter of Bore	3.453 inches (90mm)
Chamber Capacity	498 cubic inches
Weight, Tube	2030 pounds(T208), 1965 pounds(T208E9)
Total Weight	3282 pounds(T208), 3324 pounds(T208E9)
Type of Breechblock	Semiautomatic, vertical sliding wedge, electrically operated on the T208 and T208E13
Rifling	None, smooth bore
Ammunition	Fixed
Primer	Percussion

Weight, Complete Round	APFSDS-T T320 Shot(APFSDS-T)	48.00 pounds(21.8 kg)
	HE T340E14H Shell(HE-T)	52.5 pounds(23.8 kg)
	WP T339E14 Shell(Smoke)	52.5 pounds(23.8 kg)
Weight, Projectile	APFSDS-T T320 Shot(APFSDS-T)	11.0 pounds(5.0 kg)
	HE T340E14H Shell(HE-T)	18.0 pounds(8.2 kg)
	WP T339E14 Shell(Smoke)	18.0 pounds(8.2 kg)
Maximum Powder Pressure	54,000 psi	
Maximum Rate of Fire	7 rounds/minute	
Muzzle Velocity	APFSDS-T T320 Shot(APFSDS-T)	5150 ft/sec(1570 m/sec)
	HE T340E14H Shell(HE-T)	3750 ft/sec(1143 m/sec)
	WP T339E14 Shell(Smoke)	3750 ft/sec(1143 m/sec)
Muzzle Energy of Projectile, $KE=1/2MV^2$ Rotational energy is neglected and values are based on long tons (2240 pounds)	APFSDS-T T320 Shot(APFSDS-T)	2022 ft-tons
	HE T340E14H Shell(HE-T)	1755 ft-tons
	WP T339E14 Shell(Smoke)	1755 ft-tons
Maximum Range (independent of mount)	Not available *	

*The ammunition development program was cancelled prior to completion.

When an external sleeve was added to the tube of the T208 gun to balance the weapon, it was redesignated as the 90mm gun T208E13. The T208 and the T208E13 were intended for installation in a rigid mount and the T208E9 was designed for use in a recoiling mount.

105mm Guns T140, T140E2, and T140E3

Carriage and Mount	105mm Gun Tanks T54(T140 Gun), T54E1(T140E2 Gun), and T54E2(T140E3 Gun) in Mounts T156, T157, and T174, 105mm Gun Tank T95E3(T140E3 Gun) in Mount T174
Length of Chamber (to rifling)	32.27 inches
Length of Rifling	236.54 inches
Length of Chamber (to projectile base)	28.81 inches
Travel of Projectile in Bore	240.00 inches
Length of Bore	268.81 inches, 65.0 calibers
Depth of Breech Recess	9.50 inches
Length, Muzzle to Rear Face of Breech	278.31 inches, 67.3 calibers
Additional Length, Muzzle Brake	14.25 inches
Overall Length	292.56 inches
Diameter of Bore	4.134 inches (105mm)
Chamber Capacity	615 cubic inches
Weight, Tube	3500 pounds approx.
Total Weight	4800 pounds approx.
Type of Breechblock	Semiautomatic, vertical sliding wedge
Rifling	36 grooves, uniform right-hand twist, one turn in 25 calibers
Ammunition	Fixed
Primer	Percussion-electric M67

Weight, Complete Round	AP-T T182E1 Shot(APBC-T)	72.8 pounds(33.1 kg)
	HVAPDS-T T279 Shot(APDS-T)	50.50 pounds(23.0 kg)
	HEAT-T T298E1 Shell(HEAT-T)	54.8 pounds(24.9 kg)
	TP-T T79E1 Shot(TP-T)	72.8 pounds(33.1 kg)
Weight, Projectile	AP-T T182E1 Shot(APBC-T)	35.04 pounds(15.9 kg)
	HVAPDS-T T279 Shot(APDS-T)	13.60 pounds(6.2 kg)
	HEAT-T T298E1 Shell(HEAT-T)	22.5 pounds(10.2 kg)
	TP-T T79E1 Shot(TP-T)	35.04 pounds(15.9 kg)
Maximum Powder Pressure	48,000 psi	
Maximum Rate of Fire	6 rounds/minute, manual loading (T140E3)	
Muzzle Velocity	AP-T T182E1 Shot(APBC-T)	3500 ft/sec(1067 m/sec)
	HVAPDS-T T279 Shot(APDS-T)	5100 ft/sec(1554 m/sec)
	HEAT-T T298E1 Shell(HEAT-T)	3700 ft/sec(1128 m/sec)
	TP-T T79E1 Shot(TP-T)	3500 ft/sec(1067 m/sec)
Muzzle Energy of Projectile, $KE=1/2MV^2$	AP-T T182E1 Shot(APBC-T)	2976 ft-tons
Rotational energy is neglected and	HVAPDS-T T279 Shot(APDS-T)	2452 ft-tons
values are based on long tons	HEAT-T T298E1 Shell(HEAT-T)	2135 ft-tons
(2240 pounds)	TP-T T79E1 Shot(TP-T)	2976 ft-tons
Maximum Range (independent of mount)	Not available *	

* The ammunition development was terminated for the T140 series of guns prior to completion and the test data were incomplete.

The T140 and the T140E2 guns were intended for use with an automatic loader and were mounted with the vertical sliding breechblock moving up to open and down to close. The T140E3 was mounted for manual loading with the breechblock moving down to open and up to close. The ammunition for the T140 series of 105mm guns was assembled with the T43 cartridge case.

105mm Gun T210E4

Carriage and Mount	105mm Gun Tank T95E4 in recoiling mount *	
Length of Chamber	33 inches	
Travel of Projectile in Bore	270 inches	
Length of bore	302.55 inches, 73.19 calibers	
Depth of Breech Recess	9.75 inches	
Length, Muzzle to Rear Face of Breech	312.30 inches, 75.54 calibers	
Diameter of Bore	4.134 inches (105mm)	
Chamber Capacity	800 cubic inches	
Weight, Tube	2844 pounds, estimated	
Total Weight	4456 pounds, estimated	
Type of Breechblock	Semiautomatic, vertical sliding wedge	
Rifling	None, smooth bore	
Ammunition	Fixed	
Primer	Percussion	
Weight, Complete Round	APFSDS T346 Shot(APFSDS-T)	64.00 pounds(29.0 kg)
	HE T344E2 Shell(HE-T)	65.20 pounds(29.6 kg)
	WP T343E2 Shell(Smoke)	65.20 pounds(29.6 kg)
Weight, Projectile	APFSDS T346 Shot(APFSDS-T)	12.50 pounds(5.7 kg)
	HE T344E2 Shell(HE-T)	27.20 pounds(12.3 kg)
	WP T343E2 Shell(Smoke)	27.20 pounds(12.3 kg)
Maximum Powder Pressure	56,000 psi	
Maximum Rate of Fire	6 rounds/minute	
Muzzle Velocity **	APFSDS T346 Shot(APFSDS-T)	5700 ft/sec(1737 m/sec)
	HE T344E2 Shell(HE-T)	3750 ft/sec(1143 m/sec)
	WP T343E2 Shell(Smoke)	3750 ft/sec(1143 m/sec)
Muzzle Energy of Projectile, $KE=1/2MV^2$	APFSDS T346 Shot(APFSDS-T)	2815 ft-tons
Rotational energy is neglected and	HE T344E2 Shell(HE-T)	2652 ft-tons
values are based on long tons	WP T343E2 Shell(Smoke)	2652 ft-tons
(2240 pounds)		
Maximum Range (independent of mount)	Not available **	

* The 105mm gun T210 was originally intended for installation in a rigid non-recoiling mount in the T95E4 tank. However, the T210E4 was modified to utilize a concentric recoil system.
** The values shown were design goals as the gun and ammunition development program was cancelled prior to completion.

105mm Guns T254, M68(T254E2), and M68A1 (M68E1)

Carriage and Mount	105mm Gun Tank T95E5 (T254 Gun), 105mm Gun Tanks M60, M60A1, and M60A3 in Mounts M116 and M140 (M68 and M68E1 Guns), 105mm Gun Tank M1 and IPM1 in a combination mount (M68A1 Gun)
Length of Chamber (to rifling)	24.9 inches
Length of Rifling	185.557 inches
Length of Chamber (to projectile base)	23.42 inches (APDS Shot)
Travel of Projectile in Bore	187.08 inches (APDS Shot)
Length of Bore	210.50 inches, 50.92 calibers
Depth of Breech Recess	8.00 inches
Length, Muzzle to Rear Face of Breech	218.50 inches, 52.85 calibers
Diameter of Bore	4.134 inches (105mm)
Chamber Capacity	403 cubic inches
Weight, Tube	1534 pounds (T254), 1660 pounds (M68)
Total Weight	2475 pounds (T254), 2492 pounds (M68)
Type of Breechblock	Semiautomatic, vertical sliding wedge
Rifling	28 grooves, uniform right-hand twist, one turn in 18 calibers
Ammunition	Fixed
Primer	Electric

Weight, Complete Round	APDS-T M392A2 Shot(APDS-T)	41.0 pounds(18.6 kg)
	APFSDS-T M735 Shot(APFSDS-T)	38 pounds(17 kg)
	HEP-T M393A1 Shell(HESH-T)	46.7 pounds(21.2 kg)
	HEAT-T M456(T384E4) Shell(HEAT-T)	48.0 pounds(21.8 kg)
	APERS-T XM494E3 (5000 flechettes)	55.0 pounds(25.0 kg)
	WP-T M416 Shell(Smoke)	45.5 pounds(20.7 kg)
	TP-T M393A1 Shell(TP-T)	46.7 pounds(21.2 kg)
	TP-T M490 Shell(TP-T)	48.0 pounds(21.8 kg)
Weight, Projectile	APDS-T M392A2 Shot(APDS-T)	12.75 pounds(5.8 kg)
	APFSDS-T M735 Shot(APFSDS-T)	12.78 pounds(5.8 kg)
	HEP-T M393A1 Shell(HESH-T)	24.8 pounds(11.3 kg)
	HEAT-T M456(T384E4) Shell(HEAT-T)	22.4 pounds(10.2 kg)
	APERS-T XM494E3 (5000 flechettes)	approx. 31 pounds(14 kg)
	WP-T M416 Shell(Smoke)	25.17 pounds(11.4 kg)
	TP-T M393A1 Shell(TP-T)	24.8 pounds(11.3 kg)
	TP-T M490 Shell(TP-T)	22.4 pounds(10.2 kg)
Maximum Powder Pressure	51,540 psi	
Maximum Rate of Fire	7 rounds/minute	
Muzzle Velocity	APDS-T M392A2 Shot(APDS-T)	4850 ft/sec(1478 m/sec)
	APFSDS-T M735 Shot(APFSDS-T)	4925 ft/sec(1501 m/sec)
	HEP-T M393A1 Shell(HESH-T)	2400 ft/sec(732 m/sec)
	HEAT-T M456(T384E4) Shell(HEAT-T)	3850 ft/sec(1173 m/sec)
	APERS-T XM494E3 (5000 flechettes)	2700 ft/sec(823 m/sec)
	WP-T M416 Shell(Smoke)	2400 ft/sec(732 m/sec)
	TP-T M393A1 Shell(TP-T)	2400 ft/sec(732 m/sec)
	TP-T M490 Shell(TP-T)	3850 ft/sec(1173 m/sec)
Muzzle Energy of Projectile, $KE=1/2MV^2$ Rotational energy is neglected and values are based on long tons (2240 pounds)	APDS-T M392A2 Shot(APDS-T)	2079 ft-tons
	APFSDS-T M735 Shot(APFSDS-T)	2149 ft-tons
	HEP-T M393A1 Shell(HESH-T)	990 ft-tons
	HEAT-T M456(T384E4) Shell(HEAT-T)	2302 ft-tons
	APERS-T XM494E3 (5000 flechettes)	1567 ft-tons
	WP-T M416 Shell(Smoke)	1005 ft-tons
	TP-T M393A1 Shell(TP-T)	990 ft-tons
	TP-T M490 Shell(TP-T)	2302 ft-tons
Maximum Range (independent of mount)	APDS-T M392A2 Shot(APDS-T)	40,162 yards(36,724 m)
	HEP-T M393A1 Shell(HESH-T)	
	HEAT-T M456(T384E4) Shell(HEAT-T)	8975 yards(8207 m)
	APERS-T XM494E3 (5000 flechettes)	4800 yards(4389 m)
	WP-T M416 Shell(Smoke)	10,400 yards(9510 m)
	TP-T M393A1 Shell(TP-T)	
	TP-T M490 Shell(TP-T)	8975 yards(8207 m)

The T254 gun was ballistically identical with the British 105mm X15E8 or L7 cannon, but it utilized a different tube and breech design. The T254E2 weapon retained the vertical sliding breechblock of the T254, but it featured a tube that was interchangeable with that on the British gun. On the original T254E2, the U.S. tube was fitted with a concentric bore evacuator. After the British tube was adopted with its eccentric bore evacuator, the T254E2 designation was retained and the weapon was standardized as the M68. The M68A1 (M68E1) differed only in minor details from the M68 and it could be fitted with a muzzle reference system as on the M1 tank. Ammunition for these weapons was assembled with the cartridge cases M115 (brass), M150 (brass), M150B1 (steel), M148A1 (brass), and M148A1B1 (steel).

120mm Guns M58(T123E1) and T123E6

Carriage and Mount	120mm Gun Tanks M103A1 and M103A2 in Mount M89A1(M58 Gun) and 120mm Gun Tank T95E6 in Mount XM109(T123E6 Gun)	
Length of Chamber (to rifling)	38.05 inches	
Length of Rifling	243.95 inches	
Length of Chamber (to projectile base)	33.7 inches	
Travel of Projectile in Bore	248.3 inches	
Length of Bore	282.00 inches, 60.0 calibers	
Depth of Breech Recess	9.50 inches(M58)	9.19 inches(T123E6)
Length, Muzzle to Rear Face of Breech	291.50 inches(M58)	291.19 inches(T123E6)
Additional Length, Blast Deflector	7.25 inches(M58)	none(T123E6)
Overall Length	298.75 inches(M58)	291.19 inches(T123E6)
Diameter of Bore	4.7 inches (120mm)	
Chamber Capacity	1021 cubic inches	
Weight, Tube	4600 pounds(M58)	2845 pounds(T123E6)
Total Weight	6280 pounds(M58)	4244 pounds(T123E6)
Type of Breechblock	Semiautomatic, vertical sliding wedge	
Rifling	42 grooves, uniform right-hand twist, one turn in 25 calibers	
Ammunition	Separated	
Primer	Percussion or percussion-electric	

Weight, Complete Round	AP-T M358 Shot(APBC-T) *	107.31 pounds(48.8 kg)
	HEAT-T M469(T153E15) Shell(HEAT-T) #	52.55 pounds(23.9 kg)
	HE-T M356(T15E3) Shell(HE-T) **	89.15 pounds(40.5 kg)
	WP-T M357(T16E4) Shell(Smoke) **	89.15 pounds(40.5 kg)
	TP-T M359E2(T147E7) Shot(TPBC-T) *	107.31 pounds(48.8 kg)
Weight, Projectile	AP-T M358 Shot(APBC-T)	50.85 pounds(23.1 kg)
	HEAT-T M469(T153E15) Shell(HEAT-T)	31.11 pounds(14.1 kg)
	HE-T M356(T15E3) Shell(HE-T)	50.41 pounds(22.9 kg)
	WP-T M357(T16E4) Shell(Smoke)	50.41 pounds(22.9 kg)
	TP-T M359E2(T147E7) Shot(TPBC-T)	50.85 pounds(23.1 kg)
Maximum Powder Pressure	48,000 psi	
Maximum Rate of Fire	5 rounds/minute, manual loading, two loaders (M58)	
	4 rounds/minute, manual loading, one loader (T123E6)	
Muzzle Velocity	AP-T M358 Shot(APBC-T)	3500 ft/sec(1067 m/sec)
	HEAT-T M469(T153E15) Shell(HEAT-T)	3750 ft/sec(1143 m/sec)
	HE-T M356(T15E3) Shell(HE-T)	2500 ft/sec(762 m/sec)
	WP-T M357(T16E4) Shell(Smoke)	2500 ft/sec(762 m/sec)
	TP-T M359E2(T147E7) Shot(TPBC-T)	3500 ft/sec(1067 m/sec)
Muzzle Energy of Projectile, $KE=1/2MV^2$ Rotational energy is neglected and values are based on long tons (2240 pounds)	AP-T M358 Shot(APBC-T)	4318 ft-tons
	HEAT-T M469(T153E15) Shell(HEAT-T)	3033 ft-tons
	HE-T M356(T15E3) Shell(HE-T)	2184 ft-tons
	WP-T M357(T16E4) Shell(Smoke)	2184 ft-tons
	TP-T M359E2(T147E7) Shot(TPBC-T)	4318 ft-tons
Maximum Range (independent of mount)	AP-T M358 Shot(APBC-T)	25,290 yards(23,125 m)
	HEAT-T M469(T153E15) Shell(HEAT-T)	25,290 yards(23,125 m)
	HE-T M356(T15E3) Shell(HE-T)	19,910 yards(18,206 m)
	WP-T M357(T16E4) Shell(Smoke)	19,910 yards(18,206 m)
	TP-T M359E2(T147E7) Shot(TPBC-T)	25,290 yards(23,125 m)

* With propelling charge assembly M46(T38E1) in cartridge case M109(T25)
** With propelling charge assembly M45(T21E1) in cartridge case M109(T25)
# With propelling charge assembly M99(T42E1) in cartridge case M111

The 120mm Gun T123E6 was manufactured from 160,000 psi yield strength steel as a lightweight version of the M58 gun.

120mm Gun M256 (XM256)

Carriage and Mount	120mm Gun Tank M1E1 and M1A1 in a combination mount	
Length of Chamber	23.5 inches (597mm)	
Travel of Projectile in Bore	185.7 inches (4716 mm)	
Length of Bore	208.7 inches (5300mm), 44.2 calibers	
Depth of Breech Recess	11.5 inches (293mm)	
Length, Muzzle to Rear Face of Breech	220.2 inches (5593mm), 46.6 calibers	
Diameter of Bore	4.72 inches (120mm)	
Chamber Capacity	670 cubic inches (10.98 liters)	
Weight, Tube	2590 pounds (1175 kg)	
Total Weight	4200 pounds (1905kg)	
Type of Breechblock	Semiautomatic, vertical sliding wedge	
Rifling	None, smooth bore	
Ammunition	Fixed, partially combustible case	
Primer	Electric	
Weight, Complete Round	APFSDS-T M829 Shot(APFSDS-T)	41.2 pounds(18.7 kg)
	APFSDS-T XM827 Shot(APFSDS-T)	42.1 pounds(19.1 kg)
	HEAT-T-MP XM830 Shell(HEAT-T-MP)	53.4 pounds(24.2 kg)
	TPCSDS-T XM865 Shot(TPCSDS-T)	41.9 pounds(19 kg)
	TP-T XM831 Shell(TP-T)	53.4 pounds(24.2 kg)
Weight, Projectile	APFSDS-T M829 Shot(APFSDS-T)	15.5 pounds(7.0 kg)
	APFSDS-T XM827 Shot(APFSDS-T)	9.9 pounds(4.5 kg)
	HEAT-T-MP XM830 Shell(HEAT-T-MP)	29.8 pounds(13.5 kg)
	TPCSDS-T XM865 Shot(TPCSDS-T)	7.1 pounds(3.2 kg)
	TP-T XM831 Shell(TP-T)	29.8 pounds(13.5 kg)
Maximum Powder Pressure	92,610 psi (6300 bar)	
Maximum Rate of Fire	6/rounds/minute	
Muzzle Velocity	APFSDS-T M829 Shot(APFSDS-T)	5578 ft/sec(1700 m/sec)
	APFSDS-T XM827 Shot(APFSDS-T)	5414 ft/sec(1650 m/sec)
	HEAT-T-MP XM830 Shell(HEAT-T-MP)	3740 ft/sec(1140 m/sec)
	TPCSDS-T XM865 Shot(TPCSDS-T)	5578 ft/sec(1700 m/sec)
	TP-T XM831 Shell(TP-T)	3740 ft/sec(1140 m/sec)
Muzzle Energy of Projectile, KE=1/2MV²	APFSDS-T M829 Shot(APFSDS-T)	3343 ft-tons
Rotational energy is neglected and	APFSDS-T XM827 Shot(APFSDS-T)	2012 ft-tons
values are based on long tons	HEAT-T-MP XM830 Shell(HEAT-T-MP)	2890 ft-tons
(2240 pounds)	TPCSDS-T XM865 Shot(TPCSDS-T)	1531 ft-tons
	TP-T XM831 Shell(TP-T)	2890 ft-tons
Maximum Range (independent of mount)	APFSDS-T XM827 Shot(APFSDS-T)	32,000 yards(29,300 m)
	TPCSDS-T XM865 Shot(TPCSDS-T)	7700 yards(7000 m)

152mm Gun-Launchers XM150E5 and XM150E6

Carriage and Mount	152mm Gun-Launcher Tanks MBT70 (XM150E5 in combination mount) and XM803 (XM150E6 in Mount XM160)	
Length of Chamber (to rifling)	10.5 inches	
Length of Rifling	160.85 inches	
Length of Chamber (to projectile base)	9 inches	
Travel of Projectile in Bore	162 inches	
Length of Tube and Chamber	171.35 inches, 28.5 calibers	
Overall Length	183.25 inches, 30.5 calibers	
Diameter of Bore	6.000 inches (152.4mm)	
Chamber Capacity	285 cubic inches (600 cubic inches w/XM578)	
Total Weight	2715 pounds (XM150E5)	
Type of Breechblock	Semiautomatic, separable chamber, electrically operated	
Rifling	48 grooves, uniform right-hand twist, one turn in 43.5 calibers	
Ammunition	Fixed with combustible case or Shillelagh missile	
Primer	Electric	
Weight, Complete Round	MGM-51C missile (as fired)	61.5 pounds(28.0 kg)
	MTM-51C missile (as fired)	61.5 pounds(28.0 kg)
	APFSDS-T XM578E1 Shot(APFSDS-T)	40.1 pounds(18.2 kg)
	HEAT-T-MP M409 Shell(HEAT-T-MP)	49.8 pounds(22.6 kg)
	WP XM410E1 Shell (Smoke)	49.8 pounds(22.6 kg)
	Canister M625 (10,000 flechettes)	48.0 pounds(21.8 kg)
	TP-T M411A1 Shell(TP-T)	49.8 pounds(22.6 kg)
Weight, Projectile	APFSDS-T XM578E1 Shot(APFSDS-T)	20.2 pounds(9.1 kg)
	HEAT-T-MP M409 Shell(HEAT-T-MP)	42.8 pounds(19.5 kg)
	WP XM410E1 Shell (Smoke)	42.8 pounds(19.5 kg)
	Canister M625 (10,000 flechettes)	41.8 pounds(19.0 kg)
	TP-T M411A1 Shell(TP-T)	41.8 pounds(19.0 kg)
Maximum Powder Pressure	72,000 psi (design pressure)	
Maximum Rate of Fire	4 rounds/minute (manual loading, conventional ammunition)	
Muzzle Velocity	APFSDS-T XM578E1 Shot(APFSDS-T)	4850 ft/sec(1478 m/sec)
	HEAT-T-MP M409 Shell(HEAT-T-MP)	2475 ft/sec(754 m/sec)
	WP XM410E1 Shell (Smoke)	2475 ft/sec(754 m/sec)
	Canister M625 (10,000 flechettes)	2475 ft/sec(754 m/sec)
	TP-T M411A1 Shell(TP-T)	2475 ft/sec(754 m/sec)
Muzzle Energy of Projectile, KE=1/2MV²	APFSDS-T XM578E1 Shot(APFSDS-T)	3286 ft-tons
Rotational energy is neglected and	HEAT-T-MP M409 Shell(HEAT-T-MP)	1781 ft-tons
values are based on long tons	WP XM410E1 Shell (Smoke)	1781 ft-tons
(2240 pounds)	Canister M625 (10,000 flechettes)	1739 ft-tons
	TP-T M411A1 Shell(TP-T)	1739 ft-tons
Maximum Range (independent of mount)	Not available	

The bore evacuator on the XM150E5 was eliminated on the XM150E6 and it was fitted with a closed breech scavenging system.

REFERENCES AND SELECTED BIBLIOGRAPHY

Books and Published Articles

Bell, James P., "Competition as a Requisition Strategy: Impact of Competitive Research and Development on Procurement Costs", Institute for Defense Analysis, Alexandria, Virginia, November 1983

Burba, Major General Edwin H., "The MBT70 Today", Armor, September-October 1968

Hochmuth, Milton S., "Organizing the Transitional", A. W. Sijthoff, Leiden, 1974

Kelly, Orr, "King of the Killing Zone", W. W. Norton & Company, New York, New York, 1989

Spielberger, Walter J., "From Half-Track to Leopard 2", Bernard & Graefe Verlag, Munich, Germany, 1975

Tuttle, George A., Trapp, Eugene W., and Trestrail, Calvin D., "The United States Federal Republic of Germany Main Battle Tank Program", Society of Automotive Engineers Paper 680534, New York, New York, August 1968

Wagner, William, "Continental its Motors and its People" Armed Forces Journal International with Aero Publishers, Inc., Fallbrook, California, 1983

Reports and Official Documents

"Astron Concept Studies", Ordnance Tank Automotive Command, Center Line, Michigan, 17 May 1955

"Chrysler TV8 Tank Concept", Defense Engineering Division Chrysler Corp., undated

"Familiarization Manual for Tank, Combat, Full Tracked, 152mm Gun-Launcher XM803", Detroit Diesel Allison Division GMC, December 1971

"Final Report of Independent Concept and Cost Study of a New Prototype Tank System", Detroit Diesel Allison Division GMC, 16 October 1972

"Fire Control System MBT70", AC Electronics Division GMC, Milwaukee, Wisconsin, January 1968

"Gun Systems for Tanks", Ordnance Tank Automotive Command, Center Line, Michigan, 29 January 1957

"Improved M1E1 Characteristics and Description Book", prepared by Chrysler Corporation for U. S. Army TACOM, Warren, Michigan, October 1972

"M1 Tank Combat Load Plan", U. S. Army Armor School, Fort Knox, Kentucky, May 1985

"M1A1 Characteristics and Description Book", General Dynamics Land Systems Division, September 1987

"Main Battle Tank Components and Concept Review", Ordnance Tank Automotive Command, Center line, Michigan, 20-21 March 1962

"Main Battle Tank MBT70 Driver's Familiarization and Training Data", Allison Division GMC, Cleveland, Ohio, July 1969

"Main Battle Tank Task Force Report" U. S. Army Combat Developments Command, Fort Knox, Kentucky, 1 August 1972

"Medium Gun Tank Studies", Ordnance Tank Automotive Command, Center Line, Michigan, 7 April 1955

"Medium Tank Concept", Engineering Division Chrysler Corp., December 1953

"Modified Rex II Tank", Engineering Staff Ford Motor Company, 1956

"New Evolutionary Tank Study (Final Report)", prepared by Chrysler Corporation for U. S. Army TACOM, Warren, Michigan, October 1972

"New Main Battle Tank Alternative Tank Configuration Data" Detroit Diesel Allison Division GMC, 1 June 1972

"New Main Battle Tank (MR)", presentation at Ordnance Tank Automotive Command, 12-13 October 1960

"Notes on Development Type Materiel, 90mm Gun T208", Watervliet Arsenal, Watervliet, New York, January 1957

"Notes on Development Type Materiel, 90mm Gun T208E9", Watervliet Arsenal, Watervliet, New York, October 1957

"Notes on Development Type Materiel, 105mm Gun T140E3", Watervliet Arsenal, Watervliet, New York, October 1953

"Notes on Development Type Materiel, 105mm Gun T254E2", Watervliet Arsenal, Watervliet, New York, March 1959

"Notes on Development Type Materiel, 120mm Gun T123E6", Watervliet Arsenal, Watervliet, New York, August 1958

"Notes on Development Type Materiel, 152mm Gun-Launcher XM81", Watervliet Arsenal, Watervliet, New York, March 1961

"Notes on Development Type Materiel, Tank T95", Detroit Arsenal, Center Line, Michigan, 1 June 1957 and 1 August 1958

"Notes on Development Type Materiel, Tank T95E1", Detroit Arsenal, Center Line, Michigan, 1 February 1959

"Notes on Materiel for Driver-in-Turret Test Rig", Army Tank Automotive Command, Warren, Michigan, September 1963

"Preliminary Concept Studies, 105mm Gun Four Track Tank", Associated Engineers, Inc., Springfield, Massachusetts, October 1953

"Preliminary Concept Studies of a New Medium Tank", H. L. Yoh Co. Inc., Philadelphia, Pennsylvania, 1 June 1953

"Preliminary Design Notes on Two 25-ton Medium Tanks", Cadillac Motor Car Division GMC, Cleveland, Ohio, 21 December 1956

"Preliminary Development Studies, 105mm Gun Tank T96", Engineering Staff Ford Motor Company, August 1955

"Preliminary Operating and Maintenance Manual for Main Battle Tank, Armored, Full Tracked, 152mm MBT70", prepared by the Allison Division GMC for Dept. of the Army MBT Engineering Agency, Warren, Michigan, December 1969

"Project Astron Final Report", New Products Division, Continental Motors Corp., Detroit, Michigan, 9 May 1955

"Project Astron Phase 1 Preliminary Concept Study", General Motors Engineering Staff, Detroit, Michigan, 18 May 1955

"Questionmark", presentation at Ordnance Tank Automotive Command, Center Line, Michigan, April 1952

"Questionmark III", presentation at Ordnance Tank Automotive Command, Center Line, Michigan, June 1954

"Questionmark presentation at Ordnance Tank Automotive Command, Center Line, Michigan, August 1955

"Questionmark V", presentation at Ordnance Tank Automotive Command, Center Line, Michigan, 1956

"Re-orientation Study, Medium and Heavy Gun Tank Program", Ordnance Tank Automotive Command, Center Line, Michigan, 17 September 1956

"Rex VI Tank, 90mm Gun", Ordnance Tank Automotive Command, Center Line, Michigan, October 1956

"Report of the Ad Hoc Group on Armament for Future Tanks or Similar Combat Vehicles" (ARCOVE), 20 January 1958

"Report of the Fourth Tripartite Conference on Armor and Bridging", Quebec, Canada, November 1957

"Research and Engineering Program Review", Ordnance Tank Automotive Command, Center Line, Michigan, July 1960

"Tank Development", presentation for the Research and Development Board, Detroit Arsenal, Center Line, Michigan, 16 May 1951

"T95E4 . T95E6 Tank Program", Engineering Staff Ford Motor Company, 13 March 1958

"Technical Manual TM9-2350-255-10-1 Tank, Combat, Full Tracked, 105mm Gun M1, General Abrams", Dept. of the Army, November 1981

"Turret Concept Study", Ordnance Tank Automotive Command, Center Line, Michigan, 8 October 1956

"XM1 Tank Program Characteristics and Description Book", prepared by Chrysler Corp. for Program Manager XM1 Tank System DARCOM, 29 February 1980